# MICROSIMULATION AND PUBLIC POLICY

# CONTRIBUTIONS
# TO
# ECONOMIC ANALYSIS

232

# MICROSIMULATION AND PUBLIC POLICY

Selected papers from the IARIW Special Conference on
Microsimulation and Public Policy
Canberra, 5–9 December, 1993

Edited by

Ann HARDING
*National Centre for Social and Economic Modelling*
*University of Canberra*
*Canberra, Australia*

Emerald

United Kingdom – North America – Japan
India – Malaysia – China

Emerald Group Publishing Limited
Howard House, Wagon Lane, Bingley BD16 1WA, UK





**British Library Cataloguing in Publication Data**
A catalogue record for this book is available from the British Library

ISBN: 978-0-444-81894-2
ISSN: 0573-8555 (Series)

Printed and bound by CPI Group (UK) Ltd, Croydon, CR0 4YY

Certificate number 1985

Awarded in recognition of Emerald's production department's adherence to quality systems and processes when preparing scholarly journals for print

INVESTOR IN PEOPLE

# INTRODUCTION TO THE SERIES

This series consists of a number of hitherto unpublished studies, which are introduced by the editors in the belief that they represent fresh contributions to economic science.

The term "economic analysis" as used in the title of the series has been adopted because it covers both the activities of the theoretical economist and the research worker.

Although the analytical methods used by the various contributors are not the same, they are nevertheless conditioned by the common origin of their studies, namely theoretical problems encountered in practical research. Since for this reason, business cycle research and national accounting, research work on behalf of economic policy, and problems of planning are the main sources of the subjects dealt with, they necessarily determine the manner of approach adopted by the authors. Their methods tend to be "practical" in the sense of not being too far remote from application to actual economic conditions. In additon they are quantitative.

It is the hope of the editors that the publication of these studies will help to stimulate the exchange of scientific information and to reinforce international cooperation in the field of economics.

*The Editors*

To my husband, John Sekoranja, with love

# PREFACE

In December 1993, the Australian Bureau of Statistics and the National Centre for Social and Economic Modelling (NATSEM) at the University of Canberra jointly organised a *Special Conference on Microsimulation and Public Policy*. This international conference was held in Canberra, under the auspices of the International Association for Research in Income and Wealth.

The holding of the conference reflected growing realisation of the importance of the exchange of technical and other information among the small - but rapidly growing - number of microsimulation modelling practitioners. The conference papers aimed to both illuminate exciting new directions in microsimulation modelling and to stress the links between such modelling and public policy. This volume contains a selection of the papers presented at the conference.

Many thanks are due to Guy Orcutt, the founding father of microsimulation, for his opening address, and to the five session organisers at the original conference - Peter Saunders (University of NSW), Michael Wolfson (Statistics Canada), Joachim Merz (University of Luneburg), Allan Maslove (Carleton University) and Harold Beebout (Mathematica). Thanks also to the discussants, whose incisive comments are reflected in the final manuscripts contained in this volume - Gordon Keenay (Inland Revenue UK), Tim Smeeding (Syracuse University), Steve Gribble (Statistics Canada), Steven Caldwell (Cornell University), Heinz Galler (University of Eichstaett), Holly Sutherland (Cambridge University), Ingemar Eriksson (Swedish Ministry of Finance), Anders Klevmarken (Gothenburg University) and Michael Wolfson. All of the above contributed greatly to the conference and also helped in the final selection of papers for this volume.

The conference was resoundingly successful in bringing together academics, statistical agencies and policy analysts and modellers from within government. This is a tribute to the many months of work spent organising the conference by Sue Sheridan of NATSEM and Harry Kroon of the Australian Bureau of Statistics.

Finally, very special thanks are due to Louise Carney, who spent many months helping me to edit and finalise the manuscript.

Ann Harding
NATSEM, University of Canberra

# CONTENTS

CHAPTER 4: MICROSIMULATION IN COMPARATIVE SOCIAL POLICY ANALYSIS: A CASE STUDY OF UNEMPLOYMENT SCHEMES FOR BELGIUM AND THE NETHERLANDS

Lieve De Lathouwer



CHAPTER 5: INCLUDING CONSUMPTION EXPENDITURE AND WELFARE SERVICES IN A MICROSIMULATION MODEL

Aino Salomaki



CHAPTER 6: THE USE OF MICROSIMULATION TO UPDATE POVERTY STATISTICS BASED ON HOUSEHOLD BUDGET SURVEYS: A PILOT STUDY FOR THE UNITED KINGDOM

Klaas de Vos and M. Asghar Zaidi

# PART 2: INCORPORATING BEHAVIOURAL RESPONSE

## CHAPTER 7: A MICROSIMULATION APPROACH TO THE DEMAND FOR DAY CARE FOR CHILDREN IN DENMARK

Hans Baekgaard



## CHAPTER 8: MODELLING CONSUMER BEHAVIOURAL RESPONSE TO COMMODITY TAX REFORMS

Elizabeth Symons and Neil Warren



## CHAPTER 9: MARKET AND NON-MARKET LABOUR SUPPLY AND THE IMPACT OF THE RECENT GERMAN TAX REFORM - INCORPORATING BEHAVIOURAL RESPONSE

Joachim Merz

CHAPTER 10: DIRECT AND BEHAVIOURAL EFFECTS OF INCOME TAX CHANGES - SIMULATIONS WITH THE SWEDISH MODEL MICROHUS

Anders Klevmarken and Paul Olovsson



## PART 3: LIFETIME AND RETIREMENT INCOMES

CHAPTER 11: POVERTY ALLEVIATION VS SOCIAL INSURANCE SYSTEMS: A COMPARISON OF LIFETIME REDISTRIBUTION

Jane Falkingham and Ann Harding



CHAPTER 12: SOCIAL SECURITY AND LIFETIME INCOME REDISTRIBUTION: A MICROSIMULATION APPROACH

J.H.M. Nelissen

CHAPTER 13: MICROSIMULATION OF PENSION REFORM PROPOSALS: MODELLING THE EARNINGS OF COUPLES

Heinz P. Galler



CHAPTER 14: PENSION ANALYSIS IN A STATIC MODEL WITH LIFETIME INCOME DISTRIBUTION: INITIAL RESULTS

Bengt Eklind, Ingemar Eriksson, Joakim Hussénius and Mikael Müller



CHAPTER 15: THE FUTURE BURDEN OF PUBLIC PENSION BENEFITS: A MICROSIMULATION STUDY

Leif Andreassen, Dennis Fredriksen and Olav Ljones

## PART 4: MICRO/MACRO LINKS AND FIRM BEHAVIOUR

### CHAPTER 16: FORECASTING CHANGES IN THE DISTRIBUTION OF INCOME: AN APPLIED GENERAL EQUILIBRIUM APPROACH

G.A. Meagher



### CHAPTER 17: XECON: AN EXPERIMENTAL/EVOLUTIONARY MODEL OF ECONOMIC GROWTH

Michael Wolfson



### CHAPTER 18: ENDOGENOUS ECONOMIC GROWTH THROUGH SELECTION

Gunnar Eliasson



### CHAPTER 19: FIRM SPONSORED TRAINING, TECHNICAL PROGRESS AND AGGREGATE PERFORMANCE IN A MICRO-MACRO MODEL

Gérard Ballot and Erol Taymaz

## PART 5: MODELS AND DATA

Ann Harding

NATSEM, University of Canberra
Australia

# Chapter 1: Introduction And Overview

## Abstract

*After a general introduction, the second section of this chapter summarises the different types and general structure of microsimulation models. Section 3 provides an overview of each of the chapters in this volume.*

## 1. Introduction

This volume contains selected papers from the Special Conference on Microsimulation and Public Policy, held in Canberra in December 1993 under the auspices of the International Association for Research in Income and Wealth. The idea of analysing the impact of social and economic policies by simulating the behaviour and characteristics of individual decision-making units was pioneered by Guy Orcutt in the United States in the 1950s (Orcutt, 1957; Orcutt et al, 1961). Most such models to date have been based upon the individual or the family, but models based upon the firm have also been constructed.

Microsimulation models have become very powerful tools in many countries, being used routinely within government to analyse the distributional impact of policy changes to tax and cash transfer programs (such as unemployment benefits). Such models have frequently played a decisive role in determining whether or not particular policies are implemented.

Yet, despite having made a major contribution to the development of tax/transfer policies during the past decade or two, in many ways microsimulation modelling is still a new field. Only slow progress has been made in moving beyond simulating the immediate impact of tax/transfer policies to include, for example, the use of services such as health, the behavioural responses of consumers to policy changes, and the distributional impact of such economic changes as variations in protection or interest rates. Similarly, while the use of models estimating the current immediate distributional impact of tax/transfer policy change has become routine, microsimulation models simulating the future impact of policy changes or the future structure of the population have not yet become widely used by policy makers. Similarly, very few microsimulation models of firms exist.

In part, this is due to the exceptionally demanding computing requirements of some microsimulation models. Dramatic falls in the price of computers and sharp increases in computer hardware capacity are rapidly changing the feasibility of routinely using microsimulation models for policy analysis. To give just one example, when the STINMOD static microsimulation model developed at NATSEM in Australia was released for beta testing in early 1994, it took more than 10 minutes to run on the latest PCs; now, two years later, it takes only 1.5 minutes to run on the new Pentium PCs (Lambert et al, 1994).

While computing obstacles seem likely to continue to diminish rapidly, a more fundamental problem retarding the development of microsimulation models is the lack of good data - both about the current characteristics of the population and about the forces driving people's behaviour. Reliable data are absolutely critical to the development of sound models and, in many countries, the lack of longitudinal data has severely hindered the development of microsimulation models for exploring the future.

## 2. Types and Structure of Microsimulation Models

Models simulating the household sector typically begin with a microdata file. Such a file may be based upon administrative data (such as tax or social security records) or upon a sample survey of the population. In both cases, the microdata usually contain thousands of individual or family records, with a host of variables describing the demographic, labour force, income and other characteristics of each individual or family.

The data frequently require amendment by central statistical agencies or microsimulation modellers before they can be used. For example, this may include adjustment for under-reporting or misreporting of income or expenditure, the imputation of missing values, and the adjustment of weights for non-response. Variables needed by the policy analyst are frequently not present in the major microdata set used as the basis of a microsimulation model. In such cases, new variables have to be created, using a variety of techniques ranging from exact matching to some form of statistical matching (Paass, 1986).

There is frequently a lag of several years between the collection of microdata and its public release by an agency. As a result, the data have to be 'aged' to simulate the impact of current (or future) government policy. Whether data are aged 'statically' or 'dynamically' is a major difference between the various types of microsimulation models.

The ageing in *static* microsimulation models involves two basic steps. *Reweighting* involves changing the weight attached to each individual record in the microdata, to reflect economic and social change since the data were collected. (This step is sometimes also called 'grossing up'.) For example, when a survey was originally conducted, the central statistical agency might have decided that there were 400 people in a particular country with similar characteristics to the first person on a particular microdata file. Four years later, due to an increase in, say, unemployment or the incidence of sole parenthood, more recent data might suggest that there are now 450 people with similar characteristics to the first person in the microdata file. The weight of the first person would thus be increased to reflect this (while the weights of other records in the file might be decreased). Reweighting has typically been used to age sample surveys by a few years, either to bring them up to date or to provide budget estimates for two or three years into the future. It has not generally been used for exploring some decades into the future.

The other key technique, *uprating*, attempts to adjust monetary values within the original microdata to account for estimated movements since the time of the survey or anticipated future movements. For example, earnings or rent paid are typically increased to account for growth since the survey - although the degree of complexity of the uprating procedures can vary greatly.

After these two steps, static modellers typically impute the receipt of social security and other benefits and/or income tax or other liabilities, by applying the rules for eligibility or liability to each of the microunits. Steps may also be taken to make the number of program participants or taxpayers match to external benchmark data by, for example, simulating less than 100 per cent take-up of a particular payment. At this point, a baseline data file has been generated, which usually shows the current incomes and characteristics of each person or family in the microdata file, plus the imputed current incidence of taxes paid to and cash transfers received from government.

Most static models also allow the analyst to vary the rules of eligibility or liability, and produce output showing the gains or losses - and the cost to revenue and the government budget - from the policy change. Importantly, the impact can be assessed for the entire population or for relatively large or small population sub-groups (such as sole parent pensioners with two children or the home-owning elderly living in a particular region). It was these unique features which led a recent exhaustive review in the United States to conclude 'that no other type of model can match microsimulation in its potential for flexible, fine-grained analysis of proposed policy changes (Citro and Hanushek, 1991, p. 115).

The key purpose of static models has usually been to show the *immediate* or *morning after* impact of a policy change, before any of the individuals within the model change their behaviour in response to the policy shock. Yet, in some cases, a policy change may be expected to have a major impact upon behaviour - and may even be designed to do so. Consequently, a growing number of modellers have attempted to simulate the changes *in the behaviour of the individuals directly affected by a policy shock* - for example, by allowing labour supply or consumption patterns to vary in response to a tax change (see Part 2 of this volume). Such static models hold certain characteristics fixed (such as family composition) but allow other characteristics to vary (such as labour force participation or expenditure upon taxed goods).

As explained by Citro and Hanushek, the behavioural responses in the short or long-term by those directly affected by a policy change will subsequently lead to second-round effects - that is, 'to effects that alter the nature of factor or product markets or the level and distribution of consumption, production and employment in the economy or in a sector of it affected by the policy change' (1991, p.178). For example, a cut in marginal income tax rates might initially increase labour supply and subsequently lead to changes in wage rates. Consequently, efforts are now being made to link microsimulation models to macroeconomic models, to capture these second-round or feedback effects of a policy change.

In summary, static models are most frequently used to provide estimates of the immediate distributional impact of policy changes. Static ageing techniques, principally reweighting and uprating, are typically used to either age a microdata file so that it more accurately represents the current world or to provide forward estimates of the impact of policy change during the next few years. Dynamically ageing microsimulation models, on the other hand, involves 'updating each attribute for each micro-unit for each time interval' (Caldwell, 1990, p. 5).

Dynamic models often start from exactly the same cross-section sample surveys as static models. However, the individuals within the original microdata are then progressively moved forward through time; this is achieved by making major life events - such as death, marriage, divorce, fertility, education, labour force participation etc - happen to each individual, in accord with the probabilities of such events happening to real people within a particular country. Thus, within a dynamic microsimulation model, the characteristics of each individual are recalculated for each time period.

There are two major types of dynamic microsimulation models. *Dynamic population* models involve ageing a sample of an entire population, and typically begin with a comprehensive cross-section sample survey for a particular point in time (such as a census). To date, such dynamic models have been used for such purposes as the analysis of retirement incomes, future health status, the long-term impact of social security amendments, future demand for nurses, and the lifetime redistributive impact of the social security system.

*Dynamic cohort* models use exactly the same type of ageing procedures, but usually age only one cohort rather than the many cohorts represented in an entire population. Typically, one cohort is aged from birth to death, so that the entire lifecycle is simulated. For some applications, such models are more cost-efficient than ageing an entire population. Such models have been used to analyse lifetime income distribution and redistribution, lifetime rates of return to education, and repayment patterns for student income-contingent loans.[1]

Like static models, dynamic models face problems when attempting to incorporate either behavioural change in response to government policy changes or second-round effects. This sounds strange, as economists would normally assume that any 'dynamic' model would incorporate behavioural change. However, while dynamic models typically do capture some types of behavioural change (eg. a woman within a model might leave the labour force with the birth of her first child), they do not necessarily allow changes in behaviour *initiated by government policy change*.

Dynamic models are also far more 'data hungry' than static models. Cross-section data are generally not very useful for setting the parameters of dynamic models, as it is the probabilities of *transition* between states which are critical. Yet, unfortunately, many countries do not have comprehensive longitudinal or panel surveys which would allow one to set such probabilities with some confidence. Even where such data do exist, the relevant data embody age, period and cohort effects and there is no

[1] For references to a wide range of recent microsimulation studies see Orcutt et al (1986), Lewis and Michel (1990) and Harding (1993).

guarantee, for example, that the behaviour captured during the time period covered by a survey will continue in the future.

On top of these daunting data requirements, dynamic models have to date required many years to construct, so that they are both expensive and time-consuming. In theory, such models provide a superb technique for answering questions about the future or lifetime distributional impact of policy change - questions which it seems difficult to answer as comprehensively using any other method. On the other hand, the prohibitive computing, data and resource requirements of such models mean that their usefulness to the policy making process has not yet been conclusively established. During the next decade, more progress should be made on evaluating whether the benefits of dynamic microsimulation models are worth their costs.

While most microsimulation models developed to date have focussed on the household sector, a number have been created which simulate the behaviour of business firms, rather than individuals or households. In such models, individual firms decide on prices, production, hiring of labour, wages, investment and so on.

# 3. Overview of This Book

## Part 1: Static models and applications

Part 1 of this book deals with static microsimulation models and some recent pathbreaking applications. One of the key purposes of the IARIW conference was to highlight the links between microsimulation and public policy. Government economic departments across the world have now been using microsimulation models for a decade or two, and their experience can assist those contemplating the development of such models. One of the lessons which the microsimulation modeller learns about very early is the trade-off between accuracy and flexibility.

Typically, for example, administrative income tax databases consist of very accurate data and are excellent for evaluating policy reforms for individuals where all of the variables required for simulating the policy reform are already present within the data. Policy makers, however, frequently require analysis of the impact of new policies, which are based on variables or populations not present within administrative data - or analysis of the combined impact of a range of social and economic policies. In such cases, modellers turn to sample survey data to gain the increased flexibility - and then face such issues as the scope of the survey, missing values, sample size and sampling error.

*Richard Eason's* chapter describes the four microsimulation models used by the UK Inland Revenue, which has been utilising microsimulation for analysis of the distributional and revenue effects of tax policy for the past 15 years. The four models comprise a personal tax model (based on administrative data), a tax/benefit model based on the family expenditure survey, a corporation tax model and a North Sea Oil model. Eason discusses many of the issues which immediately confront the public policy modeller, including desirable characteristics of the data source, imputation and statistical matching, reconciliation of model results with information from such other sources as the National Accounts, and the projection of revenue estimates.

From the opposite side of the world, *Anil Gupta and Vishnu Kapur* describe experience at the Canadian Department of Finance, where a number of microsimulation models have become an integral part of the policy formulation process. The TI model is based on a stratified sample of individual tax returns and thus, like the comparable UK personal tax model, provides exceptionally good estimates for tax policy reforms dealing only with individuals. Given the growing interest in tax reforms which affect families, the Department has also developed the T1-Family model, which consists of the exactly matched personal tax file records of husbands and wives. The TTSIM model is based on the Survey of Consumer Finances and is used for the simulation of both welfare and taxation reforms, including comprehensive analyses of both income and sales tax issues. Finally, a longitudinal file of individual tax records has been constructed for the simulation of inter-temporal reforms.

Gupta and Kapur describe the databases underlying each of these models and the validation processes used to check their accuracy. They also provide a number of case studies of recent successful contributions by the Department's microsimulation models to the policy reform process. For example, the TTSIM model was used to simulate major reforms to child benefit and Federal sales tax, and the authors note that the microsimulation models provided far superior estimates to projections based on aggregate analyses.

As the above two chapters make clear, static microsimulation models have been successfully used for many years to test the impact of alternative policies within particular countries. However, their potential for shedding light on issues of comparative social policy has not yet been comprehensively exploited. The chapter by *Lieve de Lathouwer* of Belgium provides one interesting illustration of the possibilities. De Lathouwer starts from the observation that the Dutch system of income protection to the unemployed is more efficient and produces lower post-transfer poverty rates for the unemployed than the Belgian system. She then explores the extent to which this is due to the actual provisions of the Dutch system (rather than to differences in earnings, social structure etc. in the Netherlands). This is done by simulating the rules of the Dutch system against the Belgian population, as captured in the Belgian static microsimulation model, MISIM.

The results provide an extremely interesting illustration of the challenges and trade-offs facing social policy makers. For the same revenue cost as the Belgian system, the Dutch rules produce better outcomes on a number of measures - that is, lower poverty gaps, a greater proportion of outlays being directed towards those in pre-transfer poverty, and very substantial reductions in the poverty rates of unemployed households, all using an austere poverty line. On the other hand, the simulation suggests that the Dutch system produces higher poverty rates than the Belgian when a more generous poverty line is used - and a one-fifth cut in the proportion of the unemployed receiving assistance. Essentially, these results reflect two key areas where the Dutch rules differ greatly from the Belgian system. First, application of the Dutch rules results in the removal of assistance (via a means test) from couples where one spouse is long-term unemployed and the other earns a reasonable income. Second, however, the Dutch system provides more generous assistance to those still receiving benefits after this greater targeting. As de Lathouwer notes, there is great debate within Belgium about whether such greater selectivity is desirable - and about the merits of more generous levels of assistance to the unemployed.

Another of the interesting new directions in microsimulation is the expansion of the scope of the traditional static models (which usually simulate cash transfers and income taxes) to include usage of services and consumption of goods. Many countries have comprehensive income surveys with sufficient detail about the characteristics of families to allow the accurate imputation of the rules of the direct tax and cash transfer system. However, additionally including the usage of government services and consumption of goods in such static models frequently requires the imputation of completely new variables onto the original microdata base. *Aino Salomaki* was fortunate enough to have an income survey and a household expenditure survey for Finland which had overlapping samples. It was thus possible to test different methods of simulating consumption for households in the income survey and to check the accuracy of the results by examining the small matching sub-sample.

Two methods of imputation are tested - average statistical matching (where average consumption for a specified population sub-group is calculated from the expenditure data and the same sub-group in the income survey is assigned the relevant average) and hot ranking (where households in both surveys within particular sub-groups are ranked by income and the highest ranking household in one survey is matched to the highest ranking in the other survey). Salomaki concludes that hot ranking appears superior on some counts, but that there does not appear to be much difference between the two methods. It would, however, be desirable to conduct more comprehensive testing than was possible in this chapter, where the evaluation is conducted at a fairly aggregated level. (For example, it would be illuminating to assess the level of consumption by a variable not used to define population sub-groups in the average statistical matching method - thereby showing, for example, whether hot ranking or statistical matching more accurately predicted average consumption or indirect tax by, say, family size). The chapter also looks at the impact upon income inequality of widening the definition of income to include services, and demonstrates that broadening the definition of income makes a difference to perceptions about the redistributive impact of welfare programs.

One of the most common uses to which static microsimulation models have been put is to satisfy the demand for up-to-date poverty and income distribution estimates. Because the microdata surveys upon which microsimulation models are based are often run infrequently or are released some years after collection, microsimulation modellers routinely attempt to statically age the original microdata surveys. As noted earlier, such attempts usually involve three key steps: reweighting of the original records to account for socio-demographic change since the original survey; the uprating of all or selected income components to account for income changes; and the imputation of important new tax and social security rules. As Citro and Hanushek have noted, there has been far too little validation of microsimulation models and techniques such as static ageing (1991). The chapter by *Klaas de Vos and Asghar Zaidi* is, therefore, a very welcome addition to the scant literature evaluating the effectiveness of microsimulation techniques.

De Vos and Zaidi start with two UK Household Budget Surveys for 1985 and 1988 and calculate poverty rates for those two years, using both income and expenditure poverty lines set at 40, 50 and 60 per cent of average income and expenditure. They then statically age the 1985 Household Budget Survey to the 1988 world, using a variety of techniques of increasing sophistication. The resulting simulated 1988 poverty rates are then compared with the actual 1988 poverty rates. The results suggest that reweighting alone produces simulated poverty rates which are far below the actual 1988 rates. However, reweighting combined with income uprating produces results which are reasonably close to the actual 1988 poverty rates. While the authors point out that the results must be treated with caution, the results nonetheless suggest the reassuring conclusion that static microsimulation models produce much more accurate poverty estimates for policy makers than the alternative of simply assuming that poverty now is the same as some years ago.

## Part 2: Incorporating behavioural response

As discussed above, traditional static microsimulation models typically replicate the impact of changed rules for government programs, without allowing for any behavioural responses by individuals to the changes - ie. they simulate only the first-round effects of changes. This can be very frustrating for policy makers and economists, who may expect substantial behavioural changes in response to a reform. Even in cases where behavioural responses are not expected and/or decision makers mistrust estimates which take them into account, the actual behavioural responses may be so great that the ultimate cost of the policy option is drastically under or over-stated by microsimulation models. Part 2 of this book contains four chapters where attempts have been made to incorporate behavioural responses in both static and dynamic models.

*Hans Baekgaard's* paper provides a very interesting case study for Denmark, where the LAW static microsimulation model has been frequently used to predict the morning-after distributional impact of a wide range of policy reforms. In late 1991 the Danish government sharply reduced the cost of public day care to parents with two or more children in care. This led to unanticipated behavioural responses, as demand for day care increased greatly and waiting lists grew. As a result, the impact of the policy changes was not accurately estimated by the standard static microsimulation results. A new model was developed to simulate changes in demand for day care as the price of day care changed. This model estimated that the late 1991 reforms would lead to an additional 8.8 per cent increase in demand for day care by January 1993, resulting in a possible total budgetary impact two to three times greater than predicted by the standard static microsimulation model analysis.

Later survey data suggest that this second model overestimated the price responsiveness of the demand for public day places in 1993. This was because neither the original static microsimulation model nor the new behavioural model took any account of the long-term trend towards increased usage of public child care by Danish parents. Baekgaard also points to other pitfalls awaiting microsimulation modellers, including the failure to predict in the behavioural model the responses of actors apart from families (such as the Danish municipalities, who did not behave in the way expected by the central government).

The chapter by *Elizabeth Symons and Neil Warren* examines the importance of incorporating consumer behavioural response when analysing the impact of indirect tax reforms, using a static simulation model with a model of consumer demand that is based on an Almost Ideal Demand System and the 1988/89 Australian Household Expenditure Survey microdata. Symons and Warren examine the impact and revenue estimates of a range of possible indirect tax policy reforms, with and without allowing for behavioural response.

The authors draw a number of relevant conclusions for both policy makers and modellers. Revenue estimates of indirect tax policy changes appear to be lower after taking account of the behavioural responses of consumers. For example, the total tax revenue estimate from replacing the existing Australian wholesale sales tax with a 5.2 per cent general consumption tax is $207 million lower after taking account of behavioural response - although it is notable that this amounts to only 0.2 per cent of total revenue . Once behavioural response is allowed, changes in one tax affect the amount of revenue collected from a wide range of other indirect taxes, as consumers increase or decrease their purchases of various goods and services in response to the tax change. Interestingly, not allowing for behavioural response appears to overstate the regressivity of indirect tax changes and have the greatest impact upon lower income groups.

Moving from indirect to direct taxes, the chapter by *Joachim Merz* also attempts to include behavioural responses when assessing the impact of the 1990 German income tax reforms. However, Merz is not concerned with income inequality, but with the effects of the tax changes on both primary and secondary labour supply and, unusually, on unpaid non-market activities within the household as well. While economists have frequently concentrated upon the impact of taxes upon paid labour supply, relatively little attention has been devoted to unpaid activities. Merz argues that because tax reforms may affect both paid and unpaid labour supply via changed opportunity costs etc., both should be analysed in any comprehensive assessment of the impact of tax changes.

The major thrust of the 1990 German reforms was to decrease marginal tax rates, which resulted in a higher after-tax wage rate. The analysis by Merz suggests that, as a result, time spent in household production activities (such as do-it-yourself activities) would fall by about 2.4 per cent, while hours of work in secondary occupations would increase slightly by 0.3 per cent. However, little behavioural change was predicted for those engaged in a primary occupation, with only an estimated 0.03 per cent decrease in weekly hours of work . Interestingly, Merz also found that the labour supply of higher income families seemed likely to be stimulated by the tax changes, while there was no apparent improvement in the labour supply incentives facing low and middle income families.

While the overall average changes appeared relatively small, the impact differed for sub-groups of the population. It could be argued, however, that these estimated labour supply responses do not appear particularly large, suggesting that the first-round distributional impact estimates produced by traditional static microsimulation models might prove reasonably accurate.

The chapter by *Anders Klevmarken and Paul Olovsson* differs from the other contributions to Part 2, as they attempted to simulate behavioural response in a dynamic rather than a static microsimulation model. Using the Swedish dynamic microsimulation model MICROHUS to examine the impact of the 1991 Swedish tax reforms, they calculate that the direct or first-round effects of the tax/transfer changes would result in a 17 per cent increase in inequality in 1991. They then attempt to simulate both the direct and indirect effects of the policy shock, with the results suggesting that taking account of behavioural response would increase the growth in inequality by a further five per cent, to give a total increase in inequality of 23 per cent in 1991.

However, Klevmarken and Olovsson face the problem that the world which they are seeking to model does not remain the same in all other respects apart from the tax changes between 1989 and 1992. In particular, there was rising inequality in wage rates in Sweden over this period, and this was reflected in their dynamic model. Thus, part of the 23 per cent increase in inequality bears no relation to the tax changes, but is a result of labour force, demographic and other social changes. They thus try to isolate the impact of the tax changes in another way, by simulating the 1989 tax rules against the 1991 Swedish population within their model. This produces a seven per cent rise in inequality in 1991, with this rise obviously being due to factors other than the 1991 tax reforms. This suggests that the 1991 tax reforms produced about a 15 per cent increase in inequality in 1991, after taking full account of both direct and indirect effects: this estimate of change is very similar to that produced using the simpler standard first-round direct effects estimate of 17 per cent.

## Part 3: Lifetime and retirement incomes

During recent years there has been a growing realisation that the ubiquitous studies of the annual redistributive impact of the social security and taxation systems provide only a partial picture of redistribution. Such systems are not only designed to redistribute income from rich to poor, but also to reallocate income across the lifecycles of individuals (eg. from the peak earning years to retirement). There has thus been growing interest in the lifetime redistributive impact of social security and taxation systems. Economists have, of course, always been interested in lifetime redistribution - but have faced the major challenge of inadequate data. Dynamic microsimulation models represent one exciting new way to examine these issues.

One fascinating but to date unanswered question has been whether a social insurance based system would produce a very different picture of lifetime intra and inter-personal redistribution to a poverty alleviation system. The chapter by *Jane Falkingham and Ann Harding* examines the redistributive impact of the tax/transfer systems of the UK and Australia, using two comparable dynamic cohort microsimulation models. The results suggest that the lifetime distribution of cash transfers is more progressive in Australia than in the UK (as would be expected, given the tightly income-tested nature of Australian cash transfers). However, the UK cash transfer system has a greater redistributive effect, as average lifetime cash transfers are double the Australian average. The lifetime impact of the Australian direct taxation system appears substantially more progressive than the British system. As a result, the authors conclude that the net impact of the Australian tax/transfer system is more redistributive over the lifetime than the UK tax/transfer system.

The paper also examines the extent to which the tax-transfer systems of the two countries redistribute income from the lifetime rich to the lifetime poor and, conversely, the extent to which they transfer income between different periods of each individual's lifecycle. Making the assumption that all cash transfers are financed from direct taxes, Falkingham and Harding conclude that about two-fifths of all lifetime benefits received by Australians are paid for by the same individual at another stage during their lifecycle, with the remaining three-fifths representing transfers from one individual to another. Coincidentally, the proportions are exactly reversed in the UK, suggesting that social insurance systems produce greater income smoothing' over the lifecycle and less redistribution from rich to poor than poverty alleviation systems.

*Jan Nelissen's* chapter describes the NEDYMAS dynamic population microsimulation model which he constructed for the Netherlands and examines the lifetime redistributive impact of the various Dutch social security schemes. The Dutch schemes provide a particularly relevant test case, as they range from means-tested social assistance programs, general flat-rate social insurance schemes (such as the state old age pension) and earnings-related insurance schemes for state and private sector employees (such as unemployment and sickness benefits).

Nelissen concludes that the flat-rate schemes may not always be significantly more progressive than the earnings-related schemes, although this depends critically upon whether there are maximum benefit or contribution limits and the structure of any exemptions. Equally interestingly, it appears that there are major differences between generations, with funds being transferred from more recent cohorts to those born before the 1950s. He suspects that the generations born after 1965 may contribute more to the system during their lifetime than they receive as benefits - in contrast to the cohort born in 1930, who receive an estimated 1.9 times as much in benefits as they pay in contributions.

Apart from the above focus on lifetime questions, dynamic microsimulation models also offer great potential for the analysis of issues where long-term patterns of accumulation are important. There has been growing concern in many countries during the past decade about the ageing of their populations and the cost and sustainability of existing retirement income provisions. Microsimulation offers a particularly useful tool for the analysis of possible futures and the impact of different policy changes. Indeed, this area seems likely to emerge as one of the key applications of microsimulation as we head towards the 21st century.

During recent years there has been concern in Germany about female old age poverty, as women's lower earnings and their interrupted work histories as a result of family obligations have led to lower pensions after retirement. One proposal for pension reform has been to split the pension contributions and claims of a couple in a given year, regardless of who actually earned the income. Dynamic microsimulation models offer one of the most appropriate ways of analysing both the cross-sectional and longitudinal impacts of such reforms. However, the extent to which couples make joint decisions on labour supply and the degree of assortative mating between partners are critical factors in determining the impact of such reform proposals.

In the Frankfurt dynamic microsimulation model there is already interdependence between the labour force and earnings behaviour of married women and their husbands. *Heinz Galler* describes an attempt to estimate to what extent these dependencies between the earnings of husband and wife are not captured by the standard earnings functions for the two individuals, but are reflected in additional correlation between the error terms. The results suggest links between the labour force behaviour of husbands and wives, with wives tending to decrease their labour force participation as the earnings of their husbands increase. On the other hand, Galler also found positive assortive matching of partners with respect to their earnings capacities, which increases the dispersion of family income in the population. Overall, however, Galler reports only limited success in estimating such models and suggests that further improvements are required in econometric techniques.

While dynamic microsimulation models are frequently regarded as the most appropriate tool for providing forecasts, the chapter by *Bengt Eklind, Ingemar Eriksson, Joakim Hussénius and Mikael Müller* describes a exciting Swedish application using static modelling techniques. To the envy of most microsimulation modellers, who grapple with small samples and confidentiality restrictions, many of the Nordic countries possess both longitudinal and cross-sectional data which can be exactly matched, due to unique identifying personal numbers within their populations. The Swedish model exploits this capability, exactly matching longitudinal data about earned income during the preceding 30 years to a 1991 cross-section. Future earnings are projected using a statistical matching procedure, by gender and age. Varying assumptions can be made about economic growth, interest rates and so on.

Previous Swedish projections have suggested that payroll taxes will need to increase from about 14 to 21 per cent during the next 20 years to finance the public supplementary pension program. However, simulations with the new model indicate that the future net costs of public pensions as a percentage of income from work may remain stable at about 15 per cent in both 1994 and 2025. The authors speculate that this may be because the forthcoming generation of retirees will receive higher private pensions than the current generation - and will therefore pay more income tax and require less basic pension from the state to top up their incomes.

*Leif Andreassen, Dennis Fredriksen and Olav Ljones* use the Norwegian dynamic population model MOSART to project future labour force and pensioner numbers in the first 50 years of the next century. Their simulation suggests that there will be a large increase in the number of pensioners relative to the number of workers during this period, due mainly to the ageing of the population. The authors examine the contribution rate which will be required to maintain the current old age and disability pensions and suggest a dramatic increase will be needed, from about 18 per cent now to just below 30 per cent in 2035. They note, however, that this conclusion is very sensitive to the assumptions made about both future labour force participation and disability rates and about future pension levels.

## Part 4: Micro/macro links and firm behaviour

Static microsimulation models have been extensively used for estimating the immediate impact of changes in tax and transfer policy. However, many other government policy changes also have a significant effect upon income distribution (eg. changes in the terms of trade, protection and monetary policy), even though their impact is mediated via changes in factor income. Attempts to assess the detailed distributional consequences of such changes are in their infancy, and involve linking macroeconomic and microsimulation models. *Tony Meagher's* chapter describes linkages established between the well-known MONASH applied general equilibrium model of the Australian economy and the 1990 income survey microdata.

The MONASH model is used to produce economic forecasts from 1990-91 to 1996-97 for 112 industries in six regions. These changes are then mapped to the income survey, generating changes in employment by occupation, region and income decile - and hence changes in wages. Changes in business, property and government benefit income are also simulated. The results suggest that those in the lowest decile will experience the strongest growth in incomes during the forecast period, while the percentage increases in income are forecast to decline smoothly as the family income level increases.

The study highlights the interaction between ownership of the factors of production and the business cycle in determining the distribution of income. In a recession, when rates of return are low or even negative, owners of large amounts of capital often have relatively low incomes. However, when the economy recovers and rates of return improve, their incomes increase relatively quickly. It is this mechanism that lies behind Meagher's result of strong income growth for the lowest decile.

While the majority of microsimulation models constructed to date have focussed upon the household sector, a number of models have been established which use individual business firms as the decision making units. The chapter by *Michael Wolfson* sketches the architecture of a new Canadian microsimulation model, XEcon, which will attempt to more accurately capture economic growth processes. Rather than making the simplifying assumptions common to recent theoretical growth models - including omniscient agents and steady state growth equilibrium - the model will allow more realistic assumptions and will 'evolve' rather than being 'solved'. Wolfson envisages a model with a user-friendly Windows-type interface, where alternative simulations can be easily specified and run.

The views of the economy which could be generated from such a model are potentially very wide ranging. They are seen as including the size distribution of firms at a given point in time, changes in measures summarising the degree of inequality or polarisation of income or wages, and factor incomes and other basic national accounts identities over time. Wolfson envisages using the model to shed light on many hotly debated issues in economics, such as firm dynamics, the impact of interest rates upon economic growth and the role of venture financing upon the overall pace of technical progress.

*Gunnar Eliasson's* MOSES micro to macro model is a large empirical model which has been under development in Sweden for the past two decades. Firms in the MOSES model exist in an experimentally organised economy and face such a varied menu of economic choices as to make even barely informed decisions at the micro level impossible. In this chapter, Eliasson examines the impact upon economic growth of firm entry and exit processes. He concludes that in the long-term encouraging new firm entry enhances future economic growth and manufacturing output - and that the decline in the number of new manufacturing firms being established in Sweden each year thus has worrying implications for future economic growth rates.

*Gérard Ballot and Erol Taymaz* report on an interesting extension of the MOSES model in which, for the first time, the role of training and human capital accumulation as a source of innovation and growth is examined within this new 'evolutionary' microsimulation model of firms. MOSES is one of the few micro to macro models to allow for learning by firms. Variables such as the number of firms, the rate of growth or of investment in education and training, and the rate of technical progress are not set by the user, but instead are endogenous. Firms within the model are allowed to learn about global technology, by experimentation, innovation and imitation. They are also allowed to choose different levels of general and firm-specific knowledge and training.

Three experiments run within the model lead the authors to three tentative conclusions. First, that investment in training results in strong innovation and growth. Second, that firms are likely to under-invest in general skill training, leading to a strong case for government intervention. Third, that the type of training is important and that a balance between general skills training and firm-specific training is required.

## Part 5: Models and data

As all of those working in microsimulation soon discover, the quality of the microdata upon which models are based is a critical limiting factor. Those seeking to reweight one data source to another, or to statistically match surveys, immediately realise the importance of consistent variable definitions across surveys. Similarly, the often glaring inconsistencies between macroeconomic and micro data haunt modellers - such as when the micro expenditure estimates of households do not sum to a total which comes anywhere near a national accounts aggregate, or when the national accounts and income surveys suggest divergent conclusions about income trends.

As *Keith Blackburn and Warren Richter* point out in their chapter, enormous international effort has been devoted to ensuring that macroeconomic data collections display a high level of consistency, both across time, across collections and between countries. Yet comparable effort has not been devoted by the international statistical community, by national statistical agencies and by the major users of social data to ensuring that social data are similarly integrated. One consequence is that no internationally agreed and overarching framework for social data has yet been endorsed. Given the importance of suitable microdata for social policy making, the authors describe how one agency, the Australian Bureau of Statistics, is addressing the problem. They also outline possible future steps towards improved social data.

One of the major criticisms frequently levelled at results produced by microsimulation models is that no estimates are usually given of the degree of reliability which can be attached to such results. Thus, policy makers might be keenly interested in knowing exactly how confident they can be about an estimate of a $500 million cost to revenue or of 500,000 losers! The paper by *Stephen Pudney and*

*Holly Sutherland* is thus a very important contribution to attempts to assess the accuracy of microsimulation model results. The authors attempt to provide estimates of the uncertainty attached to tax and benefit policy changes which include behavioural labour supply responses by females. They find that sampling error is the dominant source of uncertainty for most summary statistics, but that measures describing the impact of policy on female participation are very uncertain, principally because of parameter estimation error. Providing an illuminating perspective on the results presented in Part 2 of this volume, Pudney and Sutherland conclude that results containing behavioural responses may be of no practical use to economists, because the uncertainty attached to the results is so great.

The final chapter in Part 5 by *Steven Caldwell* describes the first of a new generation of PC-based dynamic population microsimulation models, the Cornell University based CORSIM model. Caldwell's paper describes the simulation of demographic and family processes, schooling, earnings, assets, transfer income, taxes and consumption within CORSIM. He provides a checklist, which will be useful to those constructing dynamic models, of the different types of output validation which can be undertaken. Finally, he describes how CORSIM can be applied to the analysis of such health issues as dental services and expenditures.

## References

Caldwell, S.B. (1990), 'Static, dynamic and mixed microsimulation', mimeo, Department of Sociology, Cornell University, Ithaca, New York, July.

Citro, C.F. and E.A. Hanushek (1991), *The Uses of Microsimulation Modelling, Vol. 1: Review and Recommendations*, National Academy Press, Washington.

Harding, A. (1993), *Lifetime Income Distribution and Redistribution: Applications of a Microsimulation Model*, North Holland, Amsterdam.

Lambert, S., R. Percival, D. Schofield and S. Paul (1994), 'An introduction to STINMOD: A static microsimulation model', STINMOD Technical Paper No. 1, NATSEM, University of Canberra, October.

Lewis, G.H. and R.C. Michel (1990), *Microsimulation Techniques for Tax and Transfer Analysis*, Urban Institute Press, Washington.

Orcutt, G. (1957), 'A new type of socio-economic system', *Review of Economics and Statistics*, Vol. 58, No. 2.

Orcutt, G., M. Greenberger, J. Korbel and A. Rivlin (1961), *Microanalysis of Socioeconomic Systems: A Simulation Study*, Harper and Row, New York.

Orcutt, G., J. Merz and H. Quinke (1986), *Microanalytic Simulation Models to Support Social and Financial Policy*, North Holland, New York.

Paass, G. (1986), 'Statistical match: Evaluation of existing procedures and improvements by using additional information', in: G. Orcutt, J. Merz and H. Quinke (eds.).

PART ONE

# Static Models and Applications

Richard J Eason

Inland Revenue
United Kingdom

# Chapter 2: Microsimulation of Direct Taxes and Fiscal Policy in the United Kingdom


## Abstract

*This paper compares four microsimulation models used by the Inland Revenue to forecast changes in tax liabilities and yields. The personal tax model and corporation tax model use tax administration data, while the North Sea oil model uses data based on planning returns from companies. The tax and benefit model uses the Family Expenditure Survey, a household interview survey. Choice of data source, sampling and data validation are discussed, as well as imputation methods for estimating missing values and statistical matching for adding new variables. Static and dynamic methods of projecting the data are described. Some aspects of computing are considered.*


## 1. Introduction

For over 15 years, statisticians in the Inland Revenue have been using microsimulation computer models to provide analyses of the United Kingdom's direct tax system. This paper examines that experience to distil some key features of successful microsimulation. It focuses on data issues and projection methods. It therefore concerns subjects of primary interest for statisticians - those of acquiring and using a micro-set of data that enables computer models to produce reliable statistical analyses for policy making. The paper covers data sources, sampling, collection, enhancement and construction of projected databases for analysis of the short and medium term.

* The author is grateful for the advice and assistance of colleagues in preparing this paper. More detailed information about the models and data sources are contained in the papers referenced.

The Inland Revenue's Statistics and Economics Office aims to provide high quality analysis of the direct tax system to Ministers in Her Majesty's Treasury and their supporting advisers. For the statistical service, this involves three main functions:

- statistical analysis of existing direct taxes and their impact (for example, distributional analysis of the tax burden by type of taxpayer and estimation of the impact on revenue of particular sources of income or types of allowance or relief);

- estimating the direct effects on revenue of specified changes to the tax system, including those resulting from taxpayer behaviour (each annual Budget gives rise to a wide range of possible tax changes, for which costs and distributional analyses are needed); and

- forecasting tax yields for each tax over the short and medium term (these forecasts are based on economic forecasts produced by the Treasury, and are regularly revised as economic forecasts are updated and more information becomes available on amounts of tax collected).

For each of these functions, the Office provides a fast and reliable service with appropriate interpretative advice. The Office also has the important task of providing the Central Statistical Office with aggregate statistics for the compilation of the national accounts and making statistics of direct taxation widely available. Each year it publishes over 100 tables of statistics in *Inland Revenue Statistics*, it contributes statistics and analysis to appropriate Government publications, and it responds to large numbers of Parliamentary Questions and other enquiries.

The Office uses several computer simulation models in its work, ranging from quite small to very large. In this paper, four of the larger microsimulation models are considered. The models are:

- the personal tax model (PTM);
- the corporation tax model (CTM);
- the North Sea oil model (NSOM); and
- the interdepartmental group on tax/benefit modelling (IGOTM) model.

## 2. Macrosimulation

Before these models were developed, all statistical work was undertaken using aggregated statistics or sub-aggregates of the totals but, over time, the manipulation of this 'macro-data' was found to have serious weaknesses. Experience varied with the complexity of each tax, the diversity of taxpayers, and the variation in the tax base but, in general, macro-models were insufficiently detailed to produce reliable costings and forecasts. Also, many tasks could not be attempted because the detailed information required was not collected or was lost within the aggregated statistics. Microdata and microsimulation offered a solution to the difficulties.

## 3. The Origins of the Models

The PTM was the first microsimulation model to be introduced by the Inland Revenue. It was clear to statisticians in the early 1970s that providing accurate distributional analyses of income tax liabilities and costings of proposed changes to the rates and allowances needed more than just a large representative sample of tax records. Such a survey was first held for 1937-38, and sample surveys have been regularly held since 1962-63. The statisticians needed the ability to manipulate that data quickly and flexibly but, at that time, they were limited to interpreting distributional analyses and extrapolating patterns in the aggregate data. As computing power increased, it became practical to set up a computer model to examine each of the sampled cases and recalculate tax liabilities on specified bases. The microsimulation model has been continuously used and developed since then. There is also a specimen income model, which calculates the income tax liabilities of individuals with specified levels of income. It can use the tax regime of the day, historical ones, or a variety of alternatives but, unlike the main PTM, it uses hypothetical cases rather than a representative sample of real ones.

The CTM was introduced rather later than the PTM. It was started when forecasts of corporation tax one year ahead were found to be in error by over 25 per cent in the mid-1970s. The error appeared to be caused by the failure of forecasting methods using macro-data to estimate correctly the proportion of allowances and losses that could be set against profits. The macro-approach measured the proportion used historically, but there was no reliable means of assessing how it would change when profits or allowances were forecast to rise or fall. The absence of the ability to simulate this proportion, known as the 'tax bite', led to the creation of a microsimulation model for corporation tax. A new survey, called the corporation tax enquiry (CTE), was set up to provide the detailed basic data. Both the model and the enquiry have been developed significantly since their introduction.

At about the same time, the extraction of oil and gas from the North Sea was starting, and a new tax regime, including petroleum revenue tax (PRT), was introduced. PRT was expected to yield very large amounts as the new oil fields became fully productive. The new tax was charged on the value of each company's share of production in each field. All development and operating expenditure was deductible and, from 1983, expenditure on exploration and appraisal in potential fields anywhere in the North Sea obtained relief. Thus, for a typical oil field there would normally be a considerable time lapse, perhaps several years, before the cumulative value of production exceeded the cumulative expenditure and tax became due. Therefore, forecasting the tax yield needed a microsimulation model (NSOM) which could develop a detailed profile over several years for each company's interest in each field. In addition, each company would be liable to corporation tax on the profits earned in its share of each oil field. As the corporation tax liability was expected to be significant, the new NSOM was developed to simulate both petroleum revenue and corporate tax, and it used as basic data statistical information based on returns by each company to the government department (now the Department of Trade and Industry- DTI) that regulated the oil production.

The tax and benefit model, known as the IGOTM model, was introduced in 1986. Its general purpose was to provide several government departments with a common simulation model operating on a very detailed database of individual and household financial information. Its main use was expected to be the analysis of packages of changes, such as those in the Budget, when the combined impact of changes in income tax, national insurance contributions, indirect taxes and social security benefits would need to be assessed. The model was developed in the Central Statistical Office using the Family Expenditure Survey (FES) as the database. The Inland Revenue was one of the customer departments involved extensively in the model's development and use. The model's very broad database covered all types of income and expenditure as well as demographic, household and employment conditions. It thus provided statisticians and economists in many parts of government with the facility to undertake a wide range of analyses for the first time. For the Inland Revenue, its first main application was a study of the relationship between entitlement to social security benefits and liability to income tax, at a time when both were subject to major reform. Since its introduction, the model has been substantially revised and improved.

These four models were developed to meet the needs of tax and fiscal policy advisers for more accurate analysis and to improve forecasts of tax receipts. The use of microsimulation led to advantages well beyond the possible benefits from analysis of aggregate or grouped data, and it was soon accepted that analyses should not be based on aggregate data only. The use of microsimulation enabled statisticians not only to improve and widen the range of their numerical estimates, but also to improve their interpretation of the statistics by using the insight gained from studying disaggregated results. The policy advisers and Treasury officials on public finance expect good quality interpretation as well as good quality statistics.

# 4. Key Features of Microsimulation

Effective microsimulation requires sufficient accuracy and breadth of coverage in the basic data in order to simulate the workings of the tax system and its impact on taxpayers. For the statistician involved in giving advice about the future, obtaining available historical information is only the first stage. The information collected may need to be supplemented to make it sufficiently complete and it must be consistent with other relevant information. Then, after being satisfied that the information is an accurate statistical representation of the past, the statistician must project it forward, so that reliable advice about the future can be given.

In this paper, these processes are considered in four sections:

(1) the historical data collected;

(2) imputation and statistical matching;

(3) reconciliation with other information; and

(4) projecting forward.

## 4.1 The historical data collected

### Sources of data

For the four models, the type of basic data used varies considerably. For two, the PTM and the CTM, the data are obtained by stratified statistical samples of administrative tax records. For the NSOM, the main data source is the statistical and planning return made by each company to the government department responsible for energy policy; these data are from all taxpayers, not a sample, and they are not from administrative tax records. Last, the IGOTM model relies on the Family Expenditure Survey (FES), a large household survey conducted by the Office of Population Censuses and Surveys (OPCS) on behalf of the Central Statistical Office. Its original purpose was to measure patterns of expenditure for use in calculating the Retail Prices Index, but it now has many other uses inside and outside government.

Why are these three different types of basic data used? The choice of data source is critical to the success of microsimulation. In the Inland Revenue's experience, administrative tax records contain one major advantage over other data sources. If the task is to forecast or cost tax changes, then data taken from tax records is the most accurate starting point for subsequent estimating and forecasting. Only that data will reflect precisely the amounts used in real tax calculations and giving rise to tax liabilities. Any other source of data will almost certainly suffer from definitional and reporting differences, which will introduce error. For example, the tax unit for corporation tax is each separate company, but many non-administrative statistics and published accounts are consolidated for company groups. A further advantage is that a very high response rate can be obtained from surveys of administrative records, so details of the most important cases will not be missed. In the UK, almost one-third of income tax is paid by the top five per cent of taxpayers, so very good coverage of that group is necessary. In addition, data are reported according to the appropriate parts of tax law, and this helps provide consistency and reliability. The first preference, therefore, is to build databases from administrative tax information.

Unfortunately, administrative data may have limitations. The range of variables used in the administration of the tax system may be too limited for some purposes. First, it will be limited to reduce costs for both the taxpayer and the Inland Revenue. Second, it will almost certainly be restricted to information needed for the existing tax system. Tax inspectors may be able to extract additional variables from accounts for companies, but generally there may be nothing relevant in the tax records on which to assess a new tax relief. For example, a proposal for an educational fees allowance cannot be costed from tax administration data when information on such fees is not available from tax returns. In the United Kingdom, taxpayers will not supply information irrelevant to the current tax system just because statisticians might find it helpful for costing potential tax changes. Also, for some complex cases, it can take several months for tax liabilities to be settled or 'agreed' between the taxpayer and the tax inspector. Consequently, tax administration data can be somewhat out of date before they are confirmed as accurate. Furthermore, the statistician needs to be fully aware of the costs that extracting data imposes on those administering the tax system; often information about one taxpayer may be in different files, different locations, and on different computer systems, depending on the individual's sources of income and reliefs.

Another aspect of administrative data is that, when either the tax law or the administrative system changes, the data may also change. This may be beneficial or not. In recent years, some tax reforms have simplified taxes and data have been lost. For example, data on children and company stocks are no longer available following the abolition of child tax allowances and stock relief against corporation tax. On the other hand, the increasing use of large operational computer systems to automate routine tax work has made some data more accessible and more accurate.

Overall, the administration of direct taxes is likely to be the most accurate source of data, but the range of information may be narrow, final 'agreed' figures in some instances may not be timely, and the data available may change as the tax law and administrative systems change.

Data for the North Sea oil model come mainly from the returns submitted by oil companies to the Department of Trade and Industry. These give current estimates of activity and forecasts of key variables based on the plans of those companies. It is the presence of the forecasts that makes these returns so valuable for microsimulation modelling. The need to base forecasts solely on national aggregates of production, operating costs, etc. is avoided. Since much of the activity is on major projects which take several years to plan and bring into operation, these returns are a relatively reliable source of up-to-date and forward-looking information. Their disadvantage is that they do not have the full range of information necessary to simulate all aspects of the PRT and corporation tax systems, so various assumptions and off-model adjustments are required. The choice of data source is therefore determined by the availability of microdata on forecasts, rather than by the accuracy or breadth of historical data.

For analysis of taxation of individuals, the PTM and the survey of personal incomes are supplemented by the use of the IGOTM model, based on the Family Expenditure Survey (FES). The FES is very different from the survey of personal incomes. The latter covers about 75,000 individuals (a 95 per cent valid response rate from a sample of 80,000) and provides relatively accurate data on sources of income and other details needed to calculate income tax liability. By contrast, the FES covers a much wider range of information for 7,000 households containing approximately 13,000 adults (representing a 70 per cent response on a sample of 10,000 households). Despite the best efforts of the interviewers, the FES information on income is generally accepted to be less than totally accurate, and the sample is deficient of some types of taxpayer. There is a fall in response with age, and the highest one per cent of earners are known to be under-represented. People living in hotels, boarding houses and institutions are outside the sampling frame.

The main attraction of the FES is the wide range of information available. The questionnaire covers many different aspects of income, expenditure, employment, financial investments, and demographic and social factors that might be relevant to potential tax reliefs. Using the IGOTM model with FES data makes it possible to model the household's entitlement to various social security benefits and the extent to which members benefit from public expenditure of various kinds. A number of government departments use it to undertake detailed analyses of the distribution of income and the impact of public expenditure and taxation.

There are also important practical aspects which influence microsimulation modellers in their choice of database. The Inland Revenue statisticians are fortunate to have access to tax administration data. The statisticians can discuss the nature, availability, accuracy, and range of data with those responsible. They can negotiate the details of data to be reported on statistical returns or extracted from operational computing systems. They can also argue for their needs when the administrative and operational systems change.

By contrast, the typical researcher outside government would see considerable practical advantages in using a survey like the FES. It offers a much wider range of variables, which thereby increases the range of analyses that can be undertaken. The sample is much smaller, which means that computer processing is relatively inexpensive and may be easily undertaken on a micro-computer. The survey is funded by the Central Statistical Office for the Government Statistical Service. The FES data are supplied to the Economic and Social Research Council's Data Archive at the University of Essex, which allows researchers to have access to the data at virtually no cost. The survey of personal incomes is also available from the Data Archive, which receives it from the Inland Revenue for only the marginal cost of preparing an 'anonymised' database.

For corporation tax simulation, one other possible data source is published accounts information, rather than data drawn directly from tax administration. Like the FES, published accounts have the advantage of breadth of information, but they have proved unsuitable for accurate simulation of corporation tax. Published accounts are often consolidated for a group of companies and, since the tax unit in the United

Kingdom is each individual company, the accounts are at the wrong level of aggregation. Hence, no detail of the tax calculation is disclosed in published accounts (for example, there are no data on group relief surrendered by one company and received by another). The levels of profit and, for example, depreciation, are defined differently from those used in tax assessments. Another disadvantage is that coverage of the published accounts often does not distinguish adequately between activity in the United Kingdom and in the rest of the world. This is necessary for accurate corporation tax simulation because the assessment of tax is based on income and expenditure in the United Kingdom and income remitted from abroad. Another feature of published accounts is that companies have some flexibility in the method of preparation and the detail they publish - and thus consistent reporting cannot be guaranteed.

Nevertheless, the Inland Revenue does have a published accounts computerised database, which is used extensively. The database has only recently been acquired and its full potential has yet to be exploited, but it has already proved useful in monitoring the latest dividend announcements, which are needed for short term forecasting of Advance Corporation Tax. It has allowed investigation of specific aspects of company finance (for example, the issue of scrip dividends and employee share schemes). The statisticians on corporation tax can therefore use published accounts to supplement the data from tax administration in a similar way to the use of FES to supplement the survey of personal incomes.

### How much data to collect

Once the choice of data source is made, the next issue is how much data should be collected. The NSOM uses the whole population of oil fields because they number relatively few (about 50 originally, but now over 200) and variation among them is large. The PTM, FES and CTM use statistical samples because their respective populations are about 30 million, 25 million and one million. Moreover, the models use stratified samples to ensure adequate representation of each type of taxpayer, relative to their importance as taxpayers. In the case of the FES, the sample is structured by region and type of area. In the case of the CTM, the sample is 15,000 companies, but the largest 3,000 taxpayers of the population are all included in the sample to ensure that all the very largest taxpayers are covered. Random sampling stratified by size is the most efficient data collection method when the variation in income for individuals can range from a small pension to a million pounds or more - and, for companies, from a very small profit to one of a billion pounds (or a large loss).

However, stratified sampling can involve practical difficulties. One is the maintenance of a population register with details of size and type of taxpayer for each member. In the Inland Revenue, the local tax offices and the operational computer systems implement central instructions for picking the sample of individual taxpayers, while the Accounts Offices track changes in the population of companies. Nevertheless, the statisticians do need a pragmatic approach when it is clear that a taxpaying company is in the wrong stratum because its size has changed rapidly. If historical information puts a company in a stratum with a sampling fraction of one in ten in which, say, trading income ranges from £10,000 to £100,000, but the latest survey, after checking, reports income as £5 million, then its impact on the sample results after grossing may be distortionary rather than a good representation of nine other cases. So, after sampling, consideration would be given to whether the case should be treated as an 'outlier' or not. On the survey of personal incomes, there are similar procedures in which exceptional levels of income are referred back to tax offices for checking and are then sometimes treated as outliers. The statisticians also have to try to keep their samples up-to-date, and this can impose additional costs on the data suppliers who have to collate information for the new cases.

Great care is therefore necessary to ensure that samples are statistically representative. They need to be stratified if the dispersion of income is wide, in order to reduce sampling errors on measures of total tax liability or total income. However, for use as a microsimulation model database, the samples need to be representative for any aspect of the tax system on which the policy spotlight might fall. For example, surplus advance corporation tax is a topic of considerable interest in the United Kingdom, even though relatively few companies are affected. Nonetheless, the microsimulation model and its database are relied upon to measure this phenomenon and to forecast its future levels. For income tax, the position is similar for non-working wives. Before the introduction of independent taxation in 1990, there was a large demand for analyses of the impact for wives, both those with their own earnings and those without. The latter group was not important under the previous joint taxation system because their (generally) low amounts of investment income were aggregated with that of their husbands. However, they were among the beneficiaries of the new scheme. The sample used for costings was not particularly good in this respect, and alterations were made to the sample design when independent taxation began.

It is important to recognise the limits of each database derived from a sample. The CTE is skewed towards large companies, so its coverage of very small companies is not very good. Thus, for statistics and costings of the special tax rates for small companies (that is, those with low profit levels), a much larger but less detailed sample is used. In effect, a separate microsimulation model has been developed for small company relief.

## Data validation

The statisticians responsible for the data used in the microsimulation models have developed various approaches to checking that the data for each taxpayer are as accurate as possible. For data collected from the tax offices or departmental computer systems, immediate data quality checks are automated to detect fairly obvious errors (for example, data unreadable, data out of plausible ranges, or data failing to satisfy specified relationships). Queries are then referred back to the data supplier. Queries may also be raised later, as the data are subject to scrutiny by the statisticians.

For the major surveys, this process is reinforced by having the statistical and computing staff visit the data suppliers to ensure that issues affecting the data are fully discussed. These visits also serve other purposes. First, they can help increase the data suppliers' appreciation of the value of the surveys undertaken, which typically they consider as low priority compared with their main activity of collecting taxes. Second, the visits also give the statistical staff valuable insight into the mechanics of administration and the effects on the provision of data. Third, they can also be used for piloting new questions and for helping to estimate the costs of collecting the data.

With the increasing reliance on data extracted from operational computing systems, discussions with the systems development teams are very important. The statisticians need to organise the extraction and transfer of the appropriate data, but they also need to understand precisely the process by which the data are collected, stored, and updated. Otherwise, misinterpretation of the data is probable.

The extent of the changes to data arising from the implementation of a new administrative system can be seen from an example. A new system for assessing corporation tax, called 'Pay & File', is being established. It replaces the traditional approach in which the inspector normally issued an estimated tax assessment. This could be based on very little information, because the company had not yet submitted its published accounts and tax calculations. In due course, some tax would be paid and accounts submitted. Then, after often lengthy negotiations, a final assessment to tax would be agreed and any further tax paid. Under Pay & File, the company must pay its own estimate of its tax liability on the due date (nine months after the end of its accounting period) and file its tax accounts on a standard return three months later. When the accounts are finally agreed, any under- or over-payment of tax will have accrued interest. The data from the Pay & File system will therefore be quite different in nature from the previous data; they will be estimates made by the company rather than by the inspector and, although the data will be on a standard return, quality will be unknown. Exhaustive validation of the data will be necessary to understand their strengths and weaknesses and discern changes from data supplied under the previous system.

Pay & File will impact directly on the CTE used for microsimulation modelling, because some of the data will be taken directly from the return through the computer system supporting 'Pay & File' and some will have to be added using special 'screens' which the inspectors will complete. Any discontinuities in data arising from this, or from the transition from the traditional system to the new system, will need to be detected and allowed for.

Validation of the data for the NSOM is rather different. The main validation is undertaken by the Department of Trade and Industry, which receives the returns and makes some modifications, but the statisticians and economists in the Inland Revenue do raise queries about particular returns when figures look implausible. Validation of the FES is undertaken by both the OPCS and the Central Statistical Office.

## 4.2 Imputation and statistical matching

Despite the statisticians' efforts, the historical data collected will almost certainly be incomplete. When no information is returned for a particular case even after following up its absence, the case will generally be treated as a non-respondent, and the grossing factors will be amended accordingly. The more interesting position is when some information for a particular case is available but the record is not complete. For example, if 15 items of information are requested but only 10 are supplied, the analyst will be reluctant to exclude the case from microsimulation because that would be rejecting valuable real data. A method of 'imputation' is needed to substitute estimates for the missing items without distorting the microsimulation.

Similar methods are applied to another case where information is missing. This case occurs when the survey excludes certain important variables, but information from a different source is available on these variables and the functionality of the microsimulation model would be enhanced if the extra variables could, by some means, be added to the micro-database. In recent years, the Inland Revenue has developed statistical matching techniques to handle these situations.

These methods make processing the raw data more complex and therefore add possible delays in time-critical processes. They also involve mixing original real data with estimated data in the databases. On the other hand, the methods bring benefits of greater accuracy and the potential to undertake a wider range of analyses. It is clear that these advantages outweigh the difficulties. Statisticians have to work to very precise and demanding timetables to incorporate imputation and statistical matching in the process of preparing the raw data for microsimulation, and they also have to be alert to problems of drawing invalid conclusions from such databases. Some of these issues are illustrated below.

The main use of imputation for incomplete records is in the processing of corporation tax assessments to estimate the aggregate levels of profits and various tax reliefs. Tax inspectors will often submit partial records because the final details have yet to be agreed with the taxpayer. In this situation, the partial information normally specifies either the amount of tax paid or the amount of taxable profits. This is used, in conjunction with records for the same company in previous years or for similar companies in the same year, to estimate values for the missing information. The process also has to cater for companies making a trading loss rather than a profit. A similar process is undertaken on the corporation tax enquiry when inspectors report only the total capital allowances due to a company and omit the composition. The imputation processes tend to be fairly simple in algebraic terms, but choosing the best method to use in various circumstances requires research.

A different form of imputation is used on the IGOTM model because the FES is not designed specifically as a data source for tax/benefit modelling. For example, there is no question on whether people have opted out of the State Earnings Related Pension Scheme, so this has to be imputed from the recorded national insurance contributions. Also, as annual income is required for income tax modelling, the annual salaries reported in the FES are abated for periods of unemployment and sickness. Checks are also made to ensure that the patterns of employment, unemployment, and entitlement to social security benefits are plausible.

Statistical matching is used to add new variables to the database. For example, the number of children a taxpayer has is not available in tax office data, but it is in the FES. So a suitable variable is added to the database prepared from the survey of personal incomes by taking, from the FES, the distribution of children (within various age groups) by parents' income, marital status, and employment status of spouse, and allocating the children as appropriate to the cases on the database. Similar arrangements are made for mortgage interest payments (using FES data), personal and occupational pension contributions (using data from the General Household Survey), and various forms of investment income (using, *inter alia*, the OPCS Omnibus Survey). For each variable, the total results from the process must reconcile with aggregate estimates, and the distribution must be realistic. Experience has shown that successful statistical matching needs a reliable external source of data, good judgement on the degree of detail in the allocation process, and the ability to hit the control targets after the sample data have been grossed.

This can be difficult to achieve. It can be very difficult when some valid data on a particular variable are available but are limited to certain cases or circumstances. The process must then mix the valid original data for the variable with estimates based on external information and still produce acceptable aggregate and distributional estimates. This sometimes requires careful analysis and some iteration towards acceptable estimates.

The main benefit of statistical matching on the PTM is that, by supplementing the investment income reported in the survey by amounts not known to the individual's tax office (because he or she is not liable to tax or is taxed by deduction at source), better statistics on the total income of each case are constructed. This then improves various costing and other analyses. The difficult area for the statistician is analysis of investment income itself. Once the matching has been undertaken, some analyses do no more than reflect the original assumptions of the imputation. For example, if the external data sources used suggest that one in three basic rate taxpayers have no investment income above £5 a year, when the PTM is used for analyses it will produce the same result. Its reliability will depend on the external data source, not the PTM. Similarly, if the matching took no account of regional variation, analyses by region would produce random results. The statistician must therefore retain a very clear understanding of the value of analyses undertaken using imputed and matched data and they should also be constantly on the alert for new sources of information that may be used to verify or improve the methods.

In summary, imputation and statistical matching are used to improve the completeness of the basic data and to add new variables. They enhance the accuracy and functionality of the models, but the processes can be complicated, and the analyst must distinguish between those analyses which are genuinely improved and those which are not.

## 4.3 Reconciliation with other information

In some subject areas it may be good practice to collect data from a statistical sample, impute for missing information, add new variables by statistical matching, and simply gross the results of microsimulation for sampling fractions and non-response. For direct taxes, that approach would rarely be sufficient because the analysis would ignore other relevant information. The most obvious example is that the estimate of total tax collected derived from the sample cases may not be consistent with the actual amount of total tax collected. In other words, grossing the sample data produce many results that are not consistent with some other aggregate information.

Over recent years, considerable thought has been devoted to solving problems of sample data failing to reconcile with other more accurate information. The database for the PTM is not reconciled directly with aggregate tax receipts because income tax is actually paid by some institutions and entities as well as individuals. It would not be easy to isolate the total amount appropriate to the coverage of the PTM. The sample should also provide a good basis for the key purpose of costing the changes to allowances and to the basic rate of tax. Reconciliation would be needed if the model was used directly for forecasting receipts, but it is not. Instead, it is used to estimate marginal tax rates for specific groups of taxpayers, and these are then applied to the appropriate marginal income changes in the Treasury forecasts.

The largest reconciliation exercise is undertaken for corporation tax, because the model is used directly for forecasting. The grossed sample estimates from the CTE are limited to information about the assessment of corporation tax liability from profit, other income, capital allowances, charges, losses, etc. Reconciliation is in several stages. The profit levels are adjusted to be consistent with the best estimates of profits obtained from a larger sample used in compiling the national accounts estimates for the Central Statistical Office. The adjustments applied to profits are applied to the whole tax assessment in order to preserve the relativities in the original data for each case. For the main sectors (home, industrial and commercial companies, the financial sector), the modified sample assessment data are then reconciled with sector receipts information and with estimates of the time lag between accrual and payment of tax. This is achieved by the analyst overwriting some of the CTE's estimates with alternative values where the CTE data (or the model's simulation) are known to be subject to error.

This process is largely a matter of judgment and relies on the analyst having an extremely good understanding of the strengths and weaknesses of the wide range of data available. For corporation tax, and probably in many other areas, the most important element of constructing a good forecast is establishing a good understanding of the present position. The reconciliation process is the means by which that understanding is developed and utilised.

For the NSO model on oil and gas production, there is no formal reconciliation, because the basic information is not subject to sampling error. Nonetheless, the microsimulation does calculate tax liability in each field for past years, and these estimates are routinely checked against actual figures.

For the IGOTM model based on the FES, the potential for error in the grossed sample values is considerable because of the nature of the Survey, sampling error and biases in the achieved sample, non-response, and inadequate responses to particular questions. As indicated by its name, the FES was primarily a survey of expenditure. Income details were originally collected so that expenditure could be analysed by households grouped by their level of income. The aim was to obtain a reasonable estimate of total income, not accurate estimates of every item of income. Hence, checking that microsimulation results for analysis of income items are consistent with other information is very important - but it can also be virtually impossible. For example, a distributional analysis over households at different income levels of a Budget package of, say, ten different measures covering income tax, indirect taxes, and social security benefits, would be subject to error from many factors.

The approach to reconciliation adopted by the Inland Revenue is to employ detailed grossing factors and check results wherever possible against good national aggregates, (for example, to check that the cost of a change in personal income tax allowances using the FES reconciles with the costs from the more accurate PTM). Twenty one strata based on demographic characteristics - such as sex, age, marital status, and number of children - are used for grossing. However, aggregate results have been shown to be inconsistent with administrative totals because the numbers of unemployed in the survey year differ from current estimates. Another approach is to limit the use of the FES to measuring proportions or distributions. For example, when estimating the mortgage tax relief deducted at source for non-taxpayers, the proportion attributed to non-taxpayers is measured from the FES and then applied to an accurate national aggregate. This is believed to be more accurate than estimating the amount directly from the FES.

Development of methods to reconcile the sample data with other aggregate data has led to a major improvement in estimates based on microsimulations in recent years. Reconciliation for corporation tax is now an integral part of the routine statistical work, and the need to check microsimulations for income tax against other data sources is well recognised.

## 4.4 Projection forward

In today's rapidly changing economies, producing analyses or costings on data two or three years out-of-date would be almost valueless. Even producing analyses on data reflecting today's position is not adequate. One of the main roles of the statisticians in the Inland Revenue is to advise about next year's tax yields and the costs of making tax changes over the next three years or more. The historical data collected for use in microsimulation models must therefore be projected forward, so that estimates about the future can be made.

Such projections are made in the context of forecasts prepared regularly by the Treasury. These include aggregate forecasts for all the major economic variables - for example, GDP, earnings, company profits and dividends. The statisticians have developed methods of applying these aggregate forecasts to the micro-data used in simulations.

In some areas the methods are quite straightforward. For North Sea oil and gas, the microdata contain each company's forecasts of its activity for several years ahead. These are subject to error, of course, and DTI incorporates some adjustments in the field level forecasts. Exceptional events can also have a major effect on the tax take. In 1988, the Piper Alpha oil platform was destroyed by fire with many fatalities. This disaster led to an immediate loss of production and, over the next five years, there was heavy expenditure to reconstruct the platform and introduce additional safety measures in all fields. Forecasting the level of exploration and appraisal is very uncertain, and assessing when new fields will be discovered, appraised and brought into production is also liable to error. The Treasury provides projections of critical entities such as the price of oil (in dollars) and the exchange rate, both of which are subject to considerable volatility. Overall, the companies' own forecasts are very beneficial compared with using national aggregated forecasts, but forecasting error is always likely.

For the PTM and the IGOTM model, straightforward static projection methods are used, but the practice can be difficult. Major demographic trends are taken into account by adjusting grossing factors. For both models, the data for each case are generally uprated by the national aggregate forecast growth (or decrease) in the most relevant economic variable. For the PTM, changes in employment levels and projections of the number of occupational pensions are also incorporated. For the IGOTM model, each year's FES data are divided into four quarters and three regions of the UK, giving up to twelve separate types of case for projecting. Each income variable is uprated by the most appropriate economic index, such as the average earnings index, retail prices index, interest rates and Building Society stocks, and regional increases in rents paid to local authorities. Social security benefits are uprated in line with known changes in benefit rates and, beyond that, by the forecast growth in the Retail Prices Index. The feasibility of projecting by grossing in two dimensions - demographic characteristics and unemployment levels - is currently being tested.

Each case is projected to move in line with the national or regional changes expected, and there is no allowance for any variation from those levels, except in one area. In the second half of the 1980s, when salaries for the highly paid were increasing rapidly, the straightforward methods described above were modified to allow for differential earnings growth. This prevented the forecasts from consistently underestimating the tax liability of the more highly paid. For the IGOTM model, differential uprating of earnings from employment is based on earnings split into seven bands and with different factors for males and females. These factors are derived from the New Earnings Survey, but are rescaled to produce the same figure for total grossed earnings from employment as a set of flat-rate uprating factors derived from the Average Earnings Index. The projections therefore reflect various forecasts by government departments, including the population forecasts made by the Office of Population Censuses and Surveys and the detailed economic forecasts by the Treasury.

The static methods described above do not work for corporation tax, because tax losses by companies can be carried forward or backward to reduce tax liability in other periods. When projecting the data forward, the tax simulation would be unrealistic if all profit-making cases were predicted to remain in profit and all loss-making cases to remain in loss. Projection methods need to allow cases to move from profit to loss and loss to profit, as they do in practice, so that the model can simulate the use of losses carried back and losses brought forward.

Without some such dynamic projecting, tax would be seriously overestimated. For many years, the statisticians tried to use the random variation in specified statistical distributions to simulate these 'crossovers', but recent research has led to much better dynamic methods being introduced. These are based on historical observation of the proportion of companies in various categories (such as large regular profit makers) moving from profit to loss and vice versa. These proportions are now simulated directly by random allocation in the projection methods. To complete the projection, the proportion of total profits in the projected year being made by loss-making companies in the previous year is also measured historically and then simulated in the projections. Similar arrangements are included for simulating the historical proportions of losses being made by companies which made, respectively, losses and profits in the previous year.

Although this dynamic method is essentially a simple one of directly simulating historically observed proportions of certain types of change, in practice the methods have made the computer processing more complex, and the analyst has to specify several inputs. The analyst also has to judge how to adjust observed transition probabilities to allow for changes in the economy (for example, how much should the probability of moving from profit to loss increase during a recession when aggregate profits fall by 10 or 20 per cent). The new methods have been in use for over a year, and they have proved relatively easy to use. Since the economic forecasts only project net profits, the projection methods also have to convert figures for net profit projections into profits and losses. This is done by regression analysis on historical data and by using the results on the forecast levels of net profits.

Projecting the data forward is essential. For some applications, simple static methods can suffice; every variable for every case can be uprated by a relevant national or regional indicator, derived from government economic forecasts. Some structural changes, for example in population, may also need to be incorporated. More complicated methods occur if, in order to simulate important parts of the tax system, dynamic projections are needed to allow for variation from national average changes in the projections. Dynamic methods can be conceptually simple, but in practice they require considerable research and skill to implement well.

## 5. Other Aspects

Although this paper is primarily about collecting and preparing data before they are used in microsimulation models, it would be incomplete without a few further observations on the Inland Revenue's experience of models and computing.

In two of the four models, the PTM and IGOTM, the simulation of the tax system is undertaken separately from the projection processes. In these, the simulation is dependent on only one year's information, and the projection processes can uprate the data from the base year to the projected year without working through any intervening years. In the models for corporation tax and North Sea oil taxation, the tax simulation is dependent on previous years because, for example, losses brought forward or carried back are important in determining tax liability. Therefore, the models have been developed as composite projection and simulation models. As such, they are more complicated logically and in practice.

The Inland Revenue has operated the models through its own mainframe computers or, for IGOTM, through the mainframe at the Central Statistical Office. Statisticians have direct access to the mainframes and can set up and run the models when required. However, the models need extensive support by computer specialists. In liaison with the statisticians, these specialists design the system, prepare the purpose-written programs, liaise with the mainframe operating staff, undertake routine maintenance and updating, and undertake minor and major development work whenever the tax system is changed or better statistical or computing methods are introduced. They also prepare and manage the databases. The task of successfully operating the microsimulation models is therefore a partnership between the statisticians and the computer specialists.

Size and speed are the main advantages of computing on large mainframes. In the months preceding the Chancellor of the Exchequer's Budget, the models are likely to be used repeatedly for costing possible tax changes, examining distributional aspects, and forecasting tax yields. The results of each particular simulation are needed virtually immediately, so that they can be considered by officials and Ministers before further options or refinements are investigated. In this environment, the statistician often has only a few hours in which to clarify and translate the request into a practical simulation, set up the computer runs with the aid if necessary of computing specialists, run the model, examine the results, and report the conclusions. Efficiency of the model, in terms of its capacity to cope with the specific change and its speed of operation, is therefore essential.

With current computing technology, purpose-written programs run on large mainframes can be flexible in terms of additional input, small changes to the rules of processing, and additional output. And the statistician is not seriously disadvantaged if, as a result of the latest changes, the elapsed time for the computing increases by several seconds or even a few minutes. However, substantial changes to models that involve, for example, more than minor programming changes can take considerable time and effort, mainly because each system is large and contains numerous internal dependencies. Although the mainframe models are completely flexible in theory, that flexibility may not be immediately available or cheap.

The Inland Revenue is concerned to increase the efficiency of supporting and developing the mainframe models. This would lead to reductions in costs, but other benefits might also result: more transparent and user-friendly models, quicker development, easier maintenance, and greater flexibility in practice. One present difficulty is that new statistical and computing staff need several weeks or months to become reasonably familiar with the purpose, structure, strengths and weaknesses of a model. A more standardised approach to computing would cut down this learning period. A current project is investigating whether any aspect of modelling can be standardised.

It is also clear that mainframe models will be replaced by more local processing when technology allows this to be done without loss of efficiency. This trend will bring the statisticians important advantages of more autonomy and flexibility. Already, many small systems are run on micro-computers and the mainframe models are supported by micro-computer systems for tasks such as preparing input parameters from the Treasury economic forecasts, making off-model adjustments to improve the simulation, and preparing final output.

In the past decade, there have been improvements to the data collected and used on the models, but the major improvement in the service provided to tax and fiscal policy advisers and to Ministers has probably come from the advances in statistical and computing methods. The main examples are:

- the development of IGOTM to undertake tax simulations on data from an existing household survey;
- improvements to projection methods in the corporation tax model;
- the use of imputation and statistical matching techniques to improve the income data on the personal tax model; and
- the development of the North Sea oil model and off-model adjustments to handle new tax reliefs and other changes to the tax regime.

Finally, it is worth recording how the models handle behavioural changes when the tax system is changed. All the models operate as straightforward tax calculators for specified tax determinants. By running the models on different sets of determinants or different specifications of the tax system, costings are produced. The standard approach to incorporating behavioural effects is to modify the tax determinants. For example, a recent change in the capital allowances rules was expected to increase capital expenditure and bring forward some expenditure over a period of several months. The statisticians and economists liaised to establish estimates of these effects on capital expenditure, and then the input parameters to the corporation tax model were modified appropriately. Thus, the costing of the tax change reflected both the change in allowances on the existing level of expenditure and the new allowances arising from the assumed behavioural change. In this approach the behavioural change is likely to be spread evenly over taxpayers, whereas the response is in practice more likely to be confined to larger changes by only some of the taxpayers. The method may therefore underestimate the effects.

Overall, the models provide some facilities to incorporate behavioural changes, although it is often difficult to find relevant research or previous experience on which to assess the changes likely in the tax determinants. Although some attempts have been made in the past to build microsimulation models which incorporate behavioural responses directly (for example, of labour supply to tax and benefit changes), such approaches are not currently being pursued in Inland Revenue models. Behavioural changes are incorporated into the Treasury's macro-economic model when forecasts are prepared.

## 6. Conclusion

The Inland Revenue's experience of microsimulation in recent years has demonstrated several points which may be of interest to others. Microsimulation has brought considerable advantages over working with statistical aggregates and sub-totals. Greater accuracy has been obtained in distributional analyses, costing tax changes, and forecasting tax yields. Greater functionality has arisen from the ability to perform detailed analyses.

For accurate microsimulation, data from the tax administration are more reliable because they suffer from fewer reporting and definitional problems. Data from external sources can, however, cover a wider range of information - but accuracy is uncertain. Data must normally be collected by a statistical sample. Stratified samples are needed if the variation in the population is large. Valid analysis is limited to those variables for which the sample is representative.

Moving to data adjustments, imputation techniques help make maximum use of the sample response by allowing partial records to be used. Statistical matching can be used to add new variables to the data set or to correct for deficiencies in the reported data. The sample data should also be adjusted to reconcile with other aggregate data to help remove sample error and biases.

Data should be projected to the present (and, normally, the future) when the analysis is to be used for advising decision makers. Projection methods need to reflect the complexity of the subject being simulated. In taxation, projection methods range from the straightforward and static to the complicated and dynamic.

With current computer technology, large microsimulation models are run more successfully on mainframe computers when the models are complicated, or when they need to be run quickly and frequently. Micro-computers are used in support and will, almost certainly, undertake more microsimulation in the future. Adequately supporting the models requires partnership between statistical and computing skills

Finally, behavioural changes are costed within the microsimulation models by estimating the effects on the tax determinants and by using the models to assess the tax impact.

## References

Central Statistical Office (1993), *A Brief Guide to the Family Expenditure Survey*. HMSO, London.

Central Statistical Office (1993), 'The survey of personal incomes', in: *Inland Revenue Statistics, 1993*, HMSO, London.

Church, J. (1987), 'Tax benefit modelling in government: The IGOTM model', *Statistical News*, No. 78, Central Statistical Office, London.

Eason, R. and W. Elmore (1992), 'The United Kingdom's model for forecasting corporation tax', Paper prepared for OECD Fiscal Affairs Committee, Working Party 2. (An abridged version is published in *Inland Revenue Statistics, 1992*, HMSO, London.)

Keenay, G.A. (1995), 'Personal income tax modelling in the UK Inland Revenue', DAE Working Paper No. MU9502, The Microsimulation Unit, Department of Applied Economics, University of Cambridge, London.

Anil Gupta and Vishnu Kapur

Department of Finance
Government of Canada

# Chapter 3: Microsimulation Modelling Experience at the Canadian Department of Finance

## Abstract

*The Canadian Department of Finance uses two large microsimulation models to assess existing policies and analyse proposed policy changes. The first model was developed in the late 1960s and is based on information obtained from individual tax returns. The second model was developed in the later part of the 1980s and is based on the income and expenditure data obtained from an annual survey of the Canadian population. This paper describes how these two different models complement each other for policy formulation and model validation in the Department of Finance. The paper also describes different methodologies used in projections of the data for the purposes of policy analysis. The projections reflect changes in economic forecasts, by linking key parameters of the models with the macro model of the economy maintained for fiscal forecasting by the Department of Finance.*

## 1. Introduction

This paper describes how microsimulation contributes to policy formulation and policy development in the Canadian Department of Finance. It also describes the development and evolution of microsimulation models in the Department.

The Canadian Department of Finance has the mission of supporting the Minister of Finance in carrying out his/her responsibilities, which cover fiscal and monetary policy, taxation and economic policy, and providing advice in social and other policy areas. The Department uses an in-house macroeconomic model to make economic forecasts. The Department has over the years developed, along with other models, two microsimulation models: a model based on tax returns (the T1 model), used

* The views expressed in this paper are those of the authors and do not necessarily represent the views of the Government of Canada.

for tax policy evaluation, and a comprehensive model (the TTSIM model), based on survey data and used both for tax policy and social policy evaluation. These two models play complementary roles. Further, they serve the very useful role of cross-validating each other.

The T1 model was developed in the late 1960s in the context of a major reform of the Canadian tax system (known as the Carter Commission). The model provided analytical support to the Carter Commission. Since that time, the model has seen a number of changes. A projection component was added to the model in 1973, to enable analysis of future impacts of pending tax policy changes (normally planned for implementation two years hence). In the past two decades, the Canadian tax system, though still individual-based, has gone through a number of changes which have made it more comprehensive with regard to the treatment of families (children and spouses). Further, a number of refundable tax credits based on family incomes have been introduced. Many new provisions in the tax code are of an inter-temporal nature (eg. capital gains, cumulative net investment losses, forward averaging). The various changes to the tax system have necessitated tax modelling not only on an individual basis, but also on a family and longitudinal basis. Finance has responded to these needs over time and at present has a number of models dealing with such issues independently and in conjunction with each other. A family model based on administrative (tax returns) data has been created by merging tax data of one spouse with that of the other. Further, typical taxpayer models have been developed for analysing the impact on typical individuals. To deal with issues of an inter-temporal nature, a model based on longitudinal data has been created, derived from individual tax returns. These models have been successfully used over the years in relation to a wide range of issues.

The need for the development of a comprehensive microsimulation model stemmed from a number of tax and social policy initiatives taken by the Department in the late 1980s. These requirements led to the development of the TTSIM model in 1987-88. In the five years of its history, the model has been used extensively for policy-related work. It continues to evolve in response to the ever-increasing demands placed on it.

The policy environment in the Department has a significant influence on the development and use of the microsimulation models. Section 2 of this paper places these models in the policy setting, where high demands are made on the models and on the analysts using them. Section 3 describes how the microsimulation models fit into the general modelling environment at the Department of Finance. Section 4 describes the microsimulation models, while Section 5 provides details on the databases used. In the day-to-day operation of these models, validation plays a key role. Section 6 covers the validation (in particular, cross-validation) experience at the Department. Section 7 presents three case studies relating to the use of microsimulation models in particular policy settings. Section 8 concludes the paper.

# 2. The Policy Environment and Process

## 2.1 The policy environment

The responsibilities of the Department derive from and support the responsibilities of the Minister of Finance. The Minister has a number of statutory responsibilities. The Minister also has a significant role as a central policy adviser to the government. The program (or statutory) responsibilities and core functions of the Minister include fiscal and monetary policy, budget preparation, and taxation policy. The central agency functions include development of the fiscal framework, preparation of economic forecasts, and analysis and assessment of major policy initiatives and options. Sound economic management is the cornerstone of the government's policy framework. The objective of sound economic management is the achievement of sustainable, non-inflationary growth.

Microsimulation models come into play in the Department in two key areas for which the Minister of Finance has direct or indirect responsibility: taxation policy and social policy. In many instances these two areas interact (for example, when a benefit is taxable, or when a benefit is delivered through the tax system). The Department is also required to advise the Minister on social policy issues in his role as a central policy adviser. Such advice takes into account fiscal realities and economic policy considerations.

## 2.2 Policy process

The budget process is by far the most important policy vehicle in the Department of Finance. In the post war era budgets have become increasingly important in the management of the economy. Further, since the early 1980s, there has been an increased tendency for budgets to include wide-ranging policy announcements, including those on taxation policy and social policy. The government has adopted a regular annual budget cycle, with the budget being generally presented in February. The process itself spans a number of weeks, if not months, over which policy proposals are developed. An extensive formal process of budget-related consultations with interest groups and associations is carried out by the Minister of Finance. This has been increasingly implemented through the use of discussion papers and White Papers. For example, the 1992 budget announced a major re-structuring of child benefits and the proposals were contained in a White Paper. The budget proposals are discussed intensively and extensively in the parliament and in parliamentary sub-committees. In addition to the budget, the Department of Finance, in its role as a central agency, reviews policy proposals put forth by various departments.

In the context of microsimulation, the policy proposals, developed during the budget process or otherwise, place heavy demands on policy analysts. Proposals are often required to be developed at short notice for consideration by senior management, the Minister and/or the Cabinet. These deliberations often result in changes to the proposals, or in the development of new proposals, which almost invariably require simulations of these changes or proposals at short notice. In many situations, these require a thorough knowledge of the model and its components, as well as of the database.

Case Study 1 (Section 7) on Child Benefits Reform in 1992, demonstrates how microsimulation modelling is both a feasible and reliable way of bringing basic research to bear on social and taxation policy issues.

## 3. The Modelling Environment at the Department of Finance

Figure 1 provides an overview of the modelling environment at the Department of Finance. The Canadian Economic Forecasting Model (CEFM) is the centrepiece of the modelling effort. This model serves as a forecasting and policy analysis tool. It plays a central role as a sort of clearing house for the wide array of information required to prepare the fiscal and economic forecasts and policy analysis simulations. The model is employed for detailed fiscal budget analysis as well as for the evaluation of alternative policy measures. The forecast horizon of CEFM is medium term, constituting a seven year projection. Over the forecast period, sectoral analysts form a view on the upcoming forecast, based on discussions with the business community and on current economic information. These views are then incorporated into the model simulations. The release of National Accounts historical data signals the update of CEFM historical series and the refinement of short and medium-term forecast profiles. Microsimulation models at the Department of Finance must be consistent with the results produced by the macroeconomic model.

Figure 1: Modelling Environment

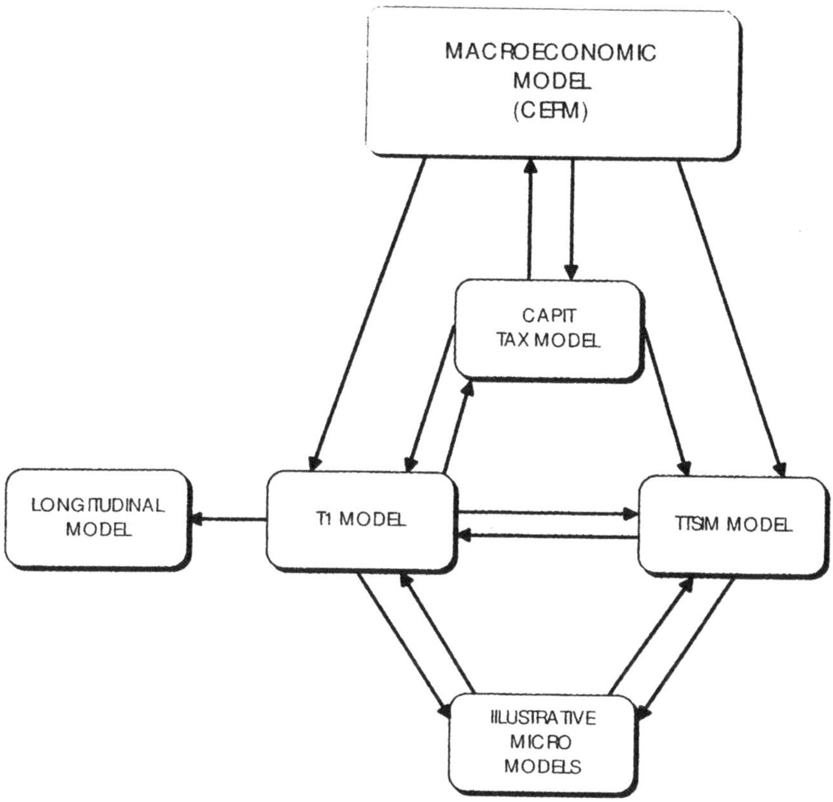


Another model of importance in the Department of Finance is the Canadian Aggregate Personal Income Tax (CAPIT) model, which forecasts aggregate revenues from personal income taxes. It is a macro model making use of time-series data. Besides the TI, TTSIM and CAPIT models, the Department makes use of a number of 'micro models' to simulate the behaviour of typical families. Further, the longitudinal tax data model complements these microsimulation models.

Although these models are not fully integrated, they communicate with each other as depicted in Figure 1. The CAPIT model draws a number of forecasted time series from the CEFM model, lending to it particular time series on personal income tax data. The TTSIM model uses per capita income data from the CEFM model to 'grow' incomes into a future year by source. Lately, it has started to borrow labour force variables to reflect the employment/unemployment situation forecast by the CEFM model.

Similarly, the T1 model uses a number of demographic and income (earnings, investment income, etc.) variables from the CEFM model to 'age' base-year data into a future year. The T1 model lends to the database for the TTSIM model a number of tax-related variables by means of a statistical merge. This merge is normally done on base-year data. The T1 model borrows expenditure-related data from the TTSIM model or from the database used by that model, in order to determine the combined impact of direct and indirect taxation measures.

These are only a few instances of the interchanges between these models. In day-to-day work, and particularly in benchmarking and validation work, it is very common to compare the outputs of one model with those of another.

## 4. Microsimulation Models

This section provides a description of the microsimulation models in use in the Department.

### 4.1 The personal income tax analysis model (T1)

The upkeep and development of the personal income tax analysis model (T1) is the responsibility of the federal Department of Revenue. One of its main uses is to assist the Department of Finance in policy analysis by providing analysis of proposals and approaches to income tax changes, and evaluation of the impact of such changes on government revenues and on various groups of taxpayers. The model can also be used to estimate the revenue effect of an economic forecast.

The first version of the model was completed in 1965. The original intent was to provide a tool to test and compare the effect of alternative fiscal policies on revenues and on taxpayers. The model has undergone extensive technical modification over the years, but the basic design remains unchanged.

The main input to the model is a stratified sample of individual tax returns. The manipulation part of the model consists of modules that perform a very detailed set of tax calculations on the basis of analyst specifications. Each sampled return is read and the tax calculated according to the applicable combination of analyst-specified rules. The new figures are then weighted by the number of taxpayers represented by each record. These results are accumulated. The output of these simulations is either a specific table of impacts or a data set which, generally, acts as input for subsequent analysis via SAS (Statistical Analysis System) at the Department of Finance. Figure 2 provides a flow-chart of a tax model run process.

Figure 2: Tax Model Run Process

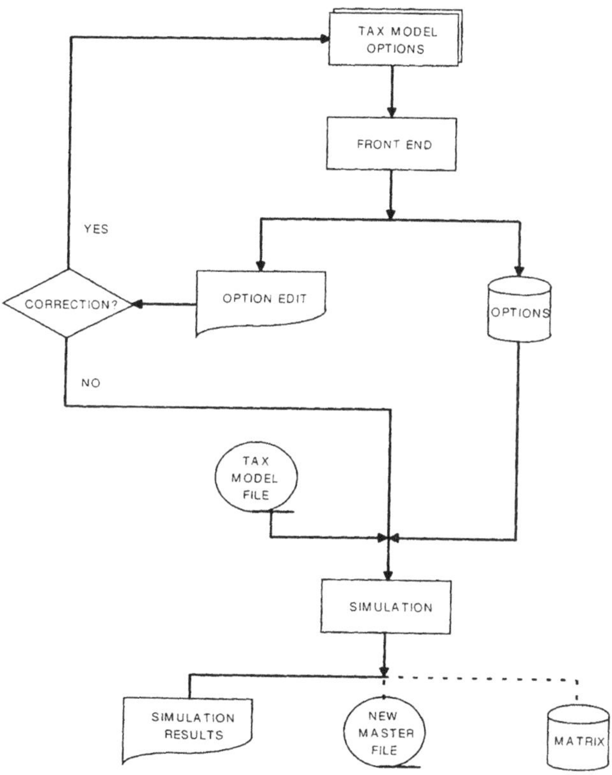

The T1 model also provides, rapidly, a statistically accurate projection of the effects of changes to tax structure. It has the capacity to analyse:

- population growth, growth of incomes and of deductions. (This capacity enables analysts to account for trends in population growth, rates of inflation and productivity, and changes in deductions that are geared to income, etc.);
- modification of existing income, deductions, or tax components and the creation of new components. (This option is especially useful in analysing the effects of changes to the definitions of income, the revenue effect of increasing the size and scope of deduction items, and the results of modifications to the tax calculation rules); and
- specification of the tax calculation. (This option allows the analyst to alter the income levels that form the boundaries of the 'tax brackets' and to specify the rate that will be applied to each bracket).

## 4.2 The T1-family model

In the Canadian taxation system over the past five to ten years, increasing attention has been focused on family income, particularly as a means of income-testing benefits. The unit of taxation, however, remains individual. This has created increasing pressure to develop new databases and models, which can perform both individual and family-based analyses, while calculating taxes on an individual basis.

As a result, in 1985, the Department of Finance, in conjunction with Revenue Canada, created a family model based on administrative data (T1-Family). The database for this model is constructed by using the social insurance numbers (SINs) of the two spouses to exactly match their tax files. The resulting data file is then weighted, according to a weighting scheme derived from the T1 individual file stratification criteria. The tax calculation routines have been modified to calculate taxes for both spouses, including transfer of unused credits and exchange of other relevant information between two tax-filing spouses, to permit family-based analysis. The T1-Family model is validated against the T1 model, both in terms of aggregates and of tax impacts on basic programs. The family model is mostly used for assessing the distributional impacts of policy changes, whereas aggregate impacts and distributions on an individual basis are derived from the T1 model. The database is now created every year, along with the T1 individual file, and simulation routines are modified to accommodate tax changes.

Efforts are continuing to expand the T1-Family model file to include children residing with the family unit. Also, Finance intends to bring the non tax-filing population into the database through statistical match, in order to increase coverage on the T1 and T1-Family models. Further improvements in the merging and weighting scheme under the T1-Family model are also underway.

The T1-family model in its present form has limitations arising from coverage and family definition problems. For example, the T1 model does not cover non tax-filers. Further, not all tax-filers belonging to the same family can be linked. These problems are, to some extent, overcome by the TTSIM model.

## 4.3 The TTSIM model

The TTSIM model is a static microsimulation model which is designed to assess the fiscal and distributional impacts of a number of tax-transfer programs and changes to these programs. Presently, the model simulates a number of programs:

- child benefits (see the case study in Section 7);
- elderly benefits;
- a guaranteed annual income type benefit;
- payroll taxes;
- federal and provincial income taxes;
- federal Goods and Services Tax; and
- refundable sales tax credit, etc.

It is proposed to add to this list the capability of simulating unemployment insurance and social assistance benefits. The model uses survey data (specifically Survey of Consumer Finances data), aged and enriched by statistical merges with individual tax data and family expenditure data. Structurally, the model is designed to simulate a base set of programs and an option set. Simulation of the base set offers certain advantages. First, the base parameters may change. Second, because the model is highly parameterised and these parameters are a part of the extensive output of a typical run of the model, it is very easy to identify what the model exactly simulated in the run. The main model is divided into 'pre-loop', 'loop-through-cases' and 'post-loop' sections. It is highly modularised, invoking some 30 subroutines. Its output consists of an aggregate impact table and a microdata file containing a number of key base and difference variables for subsequent analysis on

SAS. The model is implemented in FORTRAN-77 on a mainframe AMDAHL machine. A typical run on the model takes about 5 minutes of elapsed time on this configuration. The model experiences heavy day-to-day use in the tax and social policy divisions of the Department, particularly as a social policy analysis tool.

## 4.4 A comparison of the models

Table 1 compares the TTSIM and T1 models and databases. While, with the introduction of non-refundable credits, T1 is starting to cover a greater proportion of the Canadian population, the coverage is by no means complete. On the other hand, TTSIM covers the Canadian population and contains comprehensive family/household definitions. While T1 is representative of incomes of individual Canadians beyond low income levels (especially beyond the tax threshold), TTSIM is stronger on low incomes but relatively weak on high incomes. Thus, if an option involves individuals rather than families at high enough incomes, T1 is the obvious choice. However, for options involving family-related variables, TTSIM is preferred. For simulations on specific taxation issues, T1 is the obvious choice. However, there is a fair degree of overlap between the two models with respect to coverage, availability of variables, and simulation capabilities. This overlap allows cross-validation of results between the models. For example, if an issue relates to a particular tax expenditure (eg. age credit) in a family context, an analyst would start by comparing the amount and the incidence of the tax expenditure obtained from the TTSIM model with respective numbers obtained from the T1 model before starting to use the TTSIM model.

Table 1: Comparison Between TTSIM and T1 Models

| Attribute | TTSIM | T1 |
|---|---|---|
| Population coverage | Complete | Incomplete |
| Family definition | Comprehensive | Weak |
| Common-law variable | Present | Absent |
| Incomes (general) | Not very strong | Strong |
| Low-incomes | Acceptable | Weak |
| Transfer incomes | Present | Incomplete |
| Tax variables | Some (borrowed from TI) | Strong |
| Special tax items | Absent | Strong |
| Labour force variables | Present | Absent |
| Flexibility | Strong | Good |
| Responsiveness | Good | Good |

The use of the TTSIM and T1 models is illustrated in Case Study 2 on the 1987 reform of the Canadian personal income tax system (see Section 7).

# 5. Databases for Microsimulation

This section provides a description of the databases used by microsimulation models in the Department.

## 5.1 T1 tax model

The T1 model sample file is a stratified systematic sample of immediately assessed T1 returns from the Taxpayer's Master File (TAPMA) of individual tax returns. Its main stratification criteria are source of income, geographical region, size of income and status of taxpayer as taxable or non-taxable. The subsample file contains about 350,000 records for the 1991 taxation year. Over the years the sample size has increased from one per cent to about three per cent of the total tax-filing population.

### Statistical sample and coverage

Broadly defined, there are two types of strata - 'regular' and 'special' strata. There are 1,050 regular strata and 4 special strata. The special strata are provided for tax-filers with unusual characteristics, including filers with unusually large total incomes, non-resident filers, filers with large negative total incomes or large deductions, and filers who for some reason do not fit into the other strata. Filers who are outside the special strata criteria are designated as falling into one of the remaining 1,050 regular strata.

The three stratification criteria are as follows:

- Primary Source of Income (employment income, investment income, farming or fishing income, professional income, or business income);
- Place of Residence (urban or rural areas); and
- Tax Status and Total Income Range (tax-filers owing at least one dollar of tax are classified into one of four income classes, while those not owing tax are classified into three income classes).

A manual transcription process enables the collection of certain items which are not captured by the computer at the time of assessment. For instance, self-employed gross income and occupation type are not required in the processing of returns and therefore can only be obtained through transcription. These data items further enrich the database.

### Data ageing of T1 database

The data for the T1 model are aged to future years (4 years hence) using regression methods. The data ageing is a two part process. First, the data are grown to reflect population change between the base and the future year. Second, after the population growth has been completed, the data are further adjusted to achieve the income growth over and above the growth due to population change. While TTSIM ages the population as a whole, T1 grows the number of tax-filers to reflect population change.

For the first part, to grow the number of tax-filers, the T1 data are broken into three dimensions according to sex, occupation (largest source of income), and age. For each of these dimensions, time-series regression methods are employed to project aggregate number of filers, using historical tax-filer data together with the time series of total population projections available through the Department's macro model CEFM. Checks are made to ensure that these projections are internally consistent. For example, the number of projected males and females should agree with the number of total filers projected by these models. Raking techniques are then used to fill in the cells in the three dimensional matrix by sex, occupation and age. Having obtained the growth in population for each of the cells in the three dimensional strata, a T1 model run then reweights the T1 data by the growth factors so computed.

The projection of income components is done independently of the population projection. Income is projected similarly using major components. The per capita growth of these income components is derived from the ratio of the projected income component to that of the value from the first step, by each strata. All the calculated fields, such as income deductions, calculation of net income, taxable income and taxes, are computed by the model. Remaining income components are grown using the projected consumer price index growth or the growth in average wage index.

## 5.2 TTSIM model

A major requirement of the TTSIM model is the ability to simulate the whole spectrum of Canadian tax-transfer programs. This in turn requires a comprehensive database meeting the following conditions:

- the database must represent the whole of the Canadian population;
- it should contain sufficient detail on socioeconomic characteristics of individuals - for example, age, sex, income, demographic information, labour force characteristics;
- it should be able to link individuals in the same household/family;
- it should contain tax-related variables, such as tax deductions, in order to make tax calculations;
- it should contain information on household expenditures so that indirect taxes (sales tax) can be computed;
- it should contain work history and other related variables for the simulation of the Unemployment Insurance (UI) program;
- it should preferably contain sub-annual data for simulation of welfare programs; and,
- it should represent the period of time for which simulation results are required, which is normally two to five years ahead of the period for which survey data are currently available.

No database (based on either sample surveys or administrative data) fulfilling these requirements exists in Canada. However, a number of databases meet some of these requirements:

- Canada's national income survey (Survey of Consumer Finances (SCF));
- sample of tax-filer data (T1 model database);
- Family Expenditure Survey (FAMEX);
- Labour Markets Activity Survey (LMAS) (short longitudinal data on work and sub-annual data); and

- UI administrative data.

SCF is the most comprehensive database by virtue of its coverage, availability of income and demographic data, coherence of household and family definition, and availability of labour force related variables. However, the SCF fails to capture sizeable amounts of incomes from certain sources (eg. investment income, public and private pension benefits, UI benefits, and social assistance). The database for the TTSIM model is built around the SCF. First, missing incomes are imputed to the SCF. Second, statistical merges are performed with the T1 database and with FAMEX respectively. For the purpose of the merges, each file is partitioned into several hundred strata on the basis of a number of common variables (eg. region, age, family composition). The final merge is conditioned on income within each such strata. A similar merge with the LMAS database is planned in order to incorporate the capability of simulating the UI and welfare programs. Next, the sample is reweighted, and income and other variables are grown in order to age the SCF data to an appropriate future year or years for simulation purposes. The recent recession created the need for adjustments to the labour force data (in particular, number of weeks worked) to reflect the most recent, or a projected, unemployment/employment situation. An attempt has been made to make changes stochastically to this and related variables. The method makes use of an 'annualised' version of monthly labour force data which, in terms of availability, leads the SCF incomes data by almost a year and a half. The results produced by this database are benchmarked against available external data and against the data generated by the T1 model database for the base and projection years. This benchmark exercise helps identify problems with either of the databases and make corrections for differences in either of them.

## 5.3 Longitudinal database

In the early 1980s, a need arose to analyse tax issues with multi-year impacts. Issues such as capital gains exemption with an inter-temporal time horizon could not be analysed using cross-sectional data alone. Some other examples related to tax assisted savings for retirement contributions (known in Canada as Registered Retirement Savings Plans), charitable donations, farming losses, and the responsiveness of the tax system, in general, to income fluctuations over time.

As a result, the Department of Finance, in conjunction with Revenue Canada, created a longitudinal file of individual tax records using the administrative tax data. A 10 per cent longitudinal file of the individual tax-filers was created from the master file, by selecting individuals with social insurance numbers (SINs) ending in the number

5. Since then, the project has been annualised and a one per cent longitudinal file, with SINs ending in 15, is constructed along with the ten per cent file. The file now represents a seven year moving block of tax information on individuals for the latest seven years. The most recent longitudinal file is for the period 1983-1989.

Because of the added time dimension, the longitudinal data file provides promise as well as a challenge to the analyst. While the file has been useful in long-term projects, one major hurdle in dealing with the file so far has been its responsiveness on short-term projects. Work is under way to create a stratified longitudinal file which, because of the improvement in response time achieved through a much smaller sample size, will permit analysts to use it more frequently.

## 6. Validation

The Department has developed a number of procedures to validate its microsimulation models and the databases serving these models. First, the computer code is checked for accuracy by running a number of diverse test cases through the model and performing a micro-check on the results. The process of database assembly and ageing is subject to thorough validation at each stage. Typically, a report on the procedures and assumptions used in the process is produced. This report incorporates validation (usually external validation) of the procedures used. Further, the new database is fed into the appropriate microsimulation model and the results (usually aggregates but sometimes distributions) are compared with external data. This part of the validation can be divided into two parts - validation before ageing (including re-weighting of data) and validation of the aged database. This two part validation is intended to bring out any problems with the base data (as distinct from problems with data ageing). Tables 2 and 3 contain tabular data showing part of base validation. Cross-validation between the T1 model results and those produced by the TTSIM model at this stage of data validation is an important part of this exercise. Inconsistencies between the databases can be resolved at this stage or noted for future reference.

The day-to-day results of the models are usually validated against external data. At the initial stages of a major study or policy proposal, a thorough cross-check of statistics on variables of interest produced by one model and those of the other is conducted. Further, comparisons with outputs of the CAPIT and CEFM models are commonly made.

Table 2: Benchmarking the 1988 TTSIM Database - Comparison of Distribution of Taxable Filers by Income

| Total Income | TTSIM(1) | | Taxation(2) | |
|---|---|---|---|---|
| $ | 000 | % | 000 | % |
| < 10.000 | 1.224 | 9.3 | 1.268 | 9.9 |
| 10,000-20,000 | 4,022 | 30.6 | 3,986 | 31.0 |
| 20.000-30,000 | 3,219 | 24.5 | 3.178 | 24.7 |
| 30.000-40,000 | 2.180 | 16.6 | 2,053 | 16.0 |
| 40,000-50,000 | 1,202 | 9.1 | 1,135 | 8.8 |
| 50,000-60.000 | 601 | 4.6 | 543 | 4.2 |
| 60,000-80.000 | 434 | 3.3 | 377 | 2.9 |
| 80,000-100,000 | 139 | 1.1 | 122 | 0.9 |
| 100.000+ | 131 | 1.0 | 181 | 1.4 |
| Total | 13,153 | 100.0 | 12.844 | 100.0 |

Notes: (1) TTSIM model (2) TI model

Table 3: Benchmarking the 1988 TTSIM Database: Comparison of Average Taxable Income by Level of Income

| Total Income | TTSIM(1) | Taxation(2) | Ratio |
|---|---|---|---|
| $ | $ | $ | |
| < 10.000 | 8.077 | 7.879 | 101.2 |
| 10.000-20,000 | 14.447 | 14,358 | 100.6 |
| 20.000-30,000 | 23,160 | 23.163 | 100.0 |
| 30.000-40,000 | 32.050 | 31.933 | 100.4 |
| 40,000-50,000 | 40.543 | 40.549 | 100.0 |
| 50,000-60,000 | 49,228 | 49,047 | 100.4 |
| 60.000-80,000 | 59,978 | 59.508 | 100.8 |
| 80.000-100.000 | 79.778 | 74.206 | 107.5 |
| 100,000+ | 127,539 | 173,049 | 73.7 |
| Total | 26.199 | 26,627 | 98.4 |

Notes: (1) TTSIM model (2) T1 model

Case Study 3 (Section 7) on the reform of federal sales tax implemented in 1991 illustrates how cross-validation was helpful in arriving at GST parameters.

# 7. Case Studies

This section presents three case studies demonstrating the role of microsimulation modelling in the policy process.

### Case Study 1: Child benefits reform, 1992

In 1992, the federal government undertook a reform of its child benefits system. This reform was urged by two parliamentary committees, by social policy groups outside government, and by many individual Canadians. The Department of Finance was a key player in the process

Before reform, federal child benefits consisted of three elements - a universal family allowance (demogrant) of $35 per month paid to Canadian families (generally mothers) on behalf of each child under age 18; an annual non-refundable tax credit of $71 for each dependent child age 18 or under; and a refundable annual tax credit of up to $601 a year per child under age 18, reduced at the rate of five per cent of family annual income in excess of $25,921.

The system's main shortcomings were identified as inadequate support for low-income families (including working families), inconvenient patchwork of benefits, inadequate targeting, and delays in payments. The challenge was to come up with a new benefit which would take care of these problems without resulting in many losers - and while also avoiding windfall gains to others. There were three main objectives of the reform - to increase and target benefits to low and moderate-income families; to supplement earnings of low-income parents in the work force; and to consolidate benefits into a single monthly payment.

At the initial stages of the reform, a decision had to be made whether the T1 family model or the TTSIM model would be appropriate for this exercise. A number of factors led to the choice of the TTSIM model as the main tool for providing estimates of the fiscal and distributional impact of the changes involved in various scenarios. First, it offered better coverage on families. Second, it could identify certain demographic groups not available through other models (eg. common-law families). Third, the results of the model had already been validated against actual tax and transfer program data, so that a fair degree of confidence could be placed on the results from the model. However, TTSIM was complemented all along by the T1 family model, especially in validating its results.

The process of reform was quite onerous, especially because of a very tight time frame within which it had to be developed and implemented. At each stage of this policy process, a constant flow of numbers (mainly outputs of the two microsimulation models on the cost and distributional implications of each scenario) guided the policy makers.

By providing a projection of the number of potential recipients of the new credit, and thereby estimating the number of cheques that would be issued to recipients of the credit, these models provided substantial help in the administrative process.

Throughout the process, access to the macroeconomic forecast model enabled the microsimulation models to be adjusted for projected changes in income growth. This was an important factor in view of a forecast slow recovery during the proposed period of implementation. Furthermore, cross-validation of cost estimates from the T1 family model provided confidence on cost estimates and in the winner/loser analysis. The actual data that have since started to come out confirm the predictions made by the model.

### Case Study 2: Personal income tax reform

In 1987 the federal government announced a major reform of the Canadian tax system. The reform proposals were announced in a White Paper, *Tax Reform 1987*. The need for reform of the personal income tax system arose because of the proliferation of special tax preferences introduced during the 1970s and early 1980s, which had put increasing pressure on the equity, efficiency and revenue stability of the income tax system. The objective of the reform was to make the tax system fairer. It was accomplished by lowering tax rates and broadening the tax base. It was also the objective of the tax reform to reduce the taxes of low income Canadians and to take many of them off the tax roll. Specific reform components included reduction of the number of tax brackets from ten to three; reduction of the top tax rate from 34 per cent to 29 per cent; conversion of personal exemption amounts and of some of the tax deductions into non-refundable tax credits calculated at the lowest tax rate in order to provide the same benefit to all taxpayers; elimination of employment expense deduction and of investment income deduction in order to broaden the tax base; and restriction of special preferences such as meals and entertainment expenses, home office expenses and automobile expenses. The total revenue cost of the income tax reform was estimated to be $2 billion.

To simulate these changes, a major modelling effort was required. Simulation of the impact of base broadening measures entailed the imputation of certain income and deduction items. Except for these imputations, the existing tax data had most of the necessary information.

Prior to the reform, the taxation policy branch of the Department had already initiated the development of a family model based on administrative data (see Section 4.2 above). This model proved useful for assessment of distributional impacts of the tax reform on a family basis. Most of the simulations (and all of the simulations for the final option) for income tax reform were done concurrently on a survey-based model (a predecessor of TTSIM) to ensure the integrity of the estimates. The establishment of cross-checks between the models also helped to provide distributional impacts on a more comprehensive family basis than was available from the T1 family model.

Based on the analysis provided by these microsimulation models, the government was able to determine in advance that the tax reform would reduce the number of taxpayers by 850,000 - 250,000 of whom would be seniors. These numbers along with the total revenue cost of tax reform, compare well with the actual data that have since become available.

### Case Study 3: Federal sales tax reform

During the period 1987-89, the federal government embarked upon a major reform of an antiquated federal sales tax system which had been in place since 1924. Before reform, the tax was applied to a manufacturer's selling price. This was particularly damaging to the Canadian economy. Reform of the system had been recommended by a number of commissions and committees over the last 50 years. The reform was designed to improve the overall fairness of the tax system, ensuring that lower and moderate-income Canadians would be better off once the new sales tax, called the Goods and Services Tax (GST), replaced the existing Federal Sales Tax (FST).

Sales taxes have been viewed as inherently regressive. In order to enhance the progressivity of the federal tax system (direct as well as indirect), two steps were taken. First, a refundable credit for low and moderate-income Canadians was introduced. Second, several integrated income tax changes that would target tax assistance to lower and moderate income Canadians were introduced.

It was immediately apparent that there was no existing model with the capacity to handle all the different simulations necessary to estimate the cost and distributional impacts of such sales and income tax measures. There was, therefore, a need for an enhanced database and for models that would permit cost and distributional impact analysis on various sales tax changes. Further, to meet the government objective of making sure that families with incomes under $30,000 benefited from this change, a comprehensive database incorporating both personal and sales tax information was required. The existing administrative data and models lacked expenditure information, while survey data did not have enough tax information to simulate the two impacts simultaneously. This led to the need and subsequent creation of a comprehensive database for TTSIM (see Sections 4.3 and 5.2 for details).

A special family-based model was created to perform both the tax and transfer simulations of the personal tax system, and to simulate the GST/FST impact of the existing and proposed sales tax systems. This model provided expenditure distributions for 20 income groups and ten family categories. Using input/output analysis, both GST and FST impacts for these groups were computed. These distributions were then brought into the model to provide the impact at a household level. Within each cell the impacts varied with the level of income. Once both the sales tax and personal income tax impacts, including the refundable credit, were available on a family basis, a comprehensive analysis was completed of the overall impact using the microdata.

During the year-long work on the design of the GST, winner/loser analysis was crucial at each stage of the process. The government was committed to ensuring that families with low to moderate incomes would be better off, and that the tax system would be more progressive after the GST was implemented along with other income tax changes.

Estimates of the number of recipients of the new refundable credit were critical for administrative purposes. As the number of cheques to be issued (about 36 million per year) was large, an accurate estimate of the number of recipients was important. Also, the projected cost of the proposed credit had to be very close to successfully change from FST to GST. Another key government commitment was that the GST would be implemented in a revenue neutral manner. The models were not only able to project the cost and distributional impacts, but also provided substantial help in the administrative process.

The estimates provided by the microsimulation models were far superior to any projections based on aggregate analysis. The final estimates of both the cost and the number of recipients of the new refundable credit came out to be exceptionally close to the actual numbers that have since become available.

# 8. Conclusion

This paper demonstrates how microsimulation has been contributing to taxation and social policy development in the very demanding environment of the federal Department of Finance in Canada. Microsimulation models are required to operate within the constraints of the modelling environment in the Department, where the Department's macroeconomic model plays a central role. This means that macro-micro linkages have an important bearing on microsimulation.

The paper also brings out the mutual dependence of the administrative-based model (T1) and the survey-based model (TTSIM) in database creation, database validation, and the validation of model results. The authors believe that further linkages between microsimulation and macroeconomic models could be developed, so that microsimulation models could start contributing to macro forecasts.

## Reference

Minister, Department of Finance Canada (1987), *White Paper - Tax Reform 1987*, Department of Finance, Ottawa.

Lieve De Lathouwer

Centre for Social Policy
University of Antwerp (UFSIA)
Belgium

# Chapter 4: Microsimulation in Comparative Social Policy Analysis: A Case-Study of Unemployment Schemes for Belgium and The Netherlands

## Abstract

*This paper investigates the degree to which social security schemes as such provide an explanation for a better coverage of poverty risk. Microsimulation offers the possibility of assessing, on a representative sample, the redistributive potential of different social security schemes while controlling for socioeconomic context. Institutional factors are thus separated from other factors which determine welfare outcomes. The application of microsimulation in comparative research is illustrated by means of a case study of two very distinct systems, the Belgian and Dutch unemployment schemes. The empirical findings result from research using the static microsimulation model for Social Security and Taxation (MISIM), developed by the Centre for Social Policy in Belgium.*

## 1. Introduction

In international comparative research, one important field is the study of the effectiveness of social security systems and their impact on welfare outcomes. So far, comparative empirical studies are primarily of a descriptive nature (Deleeck et al., 1992a; De Lathouwer, 1993; Smeeding et al., 1990; Mitchell, 1991; Jäntti and Danziger, 1992; Ramprakash, 1994). However, less is known about the factors which determine the differences in welfare outcomes. Why are social security schemes in some countries more effective, in terms of guaranteeing a minimum income, than those in others?

Welfare outcomes are influenced on the one hand by redistribution policies (social security, social and fiscal policy) and, on the other hand, by the socioeconomic context (and consequently by the profile of needs) within which these policies operate. To find the determining factors methods must be devised which allow researchers to analyse and distinguish the various spheres of influence.

Given that some social security schemes generate better results than others, this paper investigates the degree to which social security schemes as such provide an explanation for a better coverage of poverty risk, and which distribution mechanisms contribute to this. It is difficult to derive an answer to this question from comparative poverty research because welfare outcomes are the result of the social security system and of the economic and social context, both of which are unique to each country.

This paper emphasises the contribution of institutional features to the redistributive impact of social security. Microsimulation is presented as an instrument for undertaking comparative system analysis. Microsimulation models offer the possibility of assessing the redistributive potential of different social security schemes, while controlling for socioeconomic context variables and for welfare effort. Institutional factors are thus separated from other factors which determine welfare outcomes.

These issues are approached by means of a case study for two very distinct systems, the Belgian and Dutch unemployment schemes. The empirical findings result from research using the Belgian microsimulation model for Social Security and Taxation (MISIM), developed by the Centre for Social Policy. MISIM is a static microsimulation model, which enables simulation of policy alternatives for social security and taxation on Belgian samples (Cantillon et al., 1993). The simulation technique makes it possible to simulate the Dutch unemployment scheme for the Belgian benefit-entitled unemployed, to calculate the first-order results for cost and social effectiveness, and to compare them with the outcomes of the Belgian unemployment scheme. The case study presented here must be seen in the broader context of the research activities undertaken by the Centre of Social Policy. In 1986 the Centre enlarged its field of the study of Belgian income distribution and poverty to cover comparative analysis of welfare outcomes (Deleeck et al., 1992a). Starting from descriptive empirical welfare outcomes the impact of distinct social security systems, or parts thereof, is studied. Other case studies include simulations of various family benefit models, basic pension schemes, and various taxation policies.

## 2. Advantages and Limits of Microsimulation in Comparative Analysis

The study of the determinants of welfare outcomes must deal with various spheres of influence, namely legislation and regulations in the social, fiscal and occupational spheres, the implementation of the regulations, and the socioeconomic and demographic profile of the population and the related pattern of needs. (These spheres of influence determine what Hill and Bramley called the 'production of welfare' - Hill and Bramley, 1986; Mitchell, 1991.) The investigation of welfare outcomes, thus, requires techniques for distinguishing and controlling these spheres of influence (Haveman and Hollenbeck, 1980).

The methodological complexity of studying determinants of welfare outcomes is manifested in studies of welfare outcomes over time and over regions in one and the same country (for regional analyses see, for example, Deleeck et al., 1989). In comparative research the complexity is compounded by distinct institutions operating in distinct national contexts. For example, comparatively generous systems in one country do not automatically result in better welfare outcomes when they do not benefit those in need. On the other hand, good outcomes may still be produced by a less generous system if other arrangements exist which provide security of subsistence (such as work, family structures and private arrangements).

Microsimulation offers an analytic tool for assessing the redistributive potential of foreign social security systems, or parts thereof, while controlling for socioeconomic variables (and thus the profile of needs) and for differences in spending. The redistributive potential is itself the result of allocative principles embedded in the system (who gets what) and of the total amount of resources available (spending). By attaining budget neutrality in the simulation model, the distributional effects of various institutional arrangements can be assessed.

Traditional research methods have specific limitations. For example, comparative poverty research does not allow assessment of the distributional potential of social security as such because welfare outcomes are the result of allocative principles in the legal (and extra-legal) arrangements, of welfare efforts and of the economic and social context, all of which are unique to each country.

The application of microsimulation in comparative research adds a new instrument to the range of more traditional methods (cross-national comparisons on empirical data-sets, standard simulations, evidence from administrative data, and comparative analysis of law and national social policies) (Hauser, 1993).

Compared with standard simulations, for example, microsimulation models are based on representative samples of households. It is therefore possible to examine the entire distributional effect of change and not just the outcome on a case by case basis. In contrast, one of the main problems of standard simulations in comparative analysis is that it is necessary to start from hypothetical 'model cases' (see for some applications OECD, 1991; Einerhand et al., 1993). Because of the diversity of the characteristics of the population in the various countries, the question of how to choose such 'model cases' remains.

At the same time, microsimulation has its limits. The results of microsimulation models can never be taken as predicting the real consequences of alternative policy models, although micromodels are based on representative empirical databases. The reason for this is that, in judging the impact of changes in policy, microsimulation models ideally have to take into account very diverse effects (Haveman and Hollenbeck, 1980; Mot, 1991; Atkinson and Sutherland, 1988; Hancock and Sutherland, 1992).

First, for example, policy changes influence the economic efficiency of the social security system. Changes in the social security system affect macroeconomic parameters and generate behavioural responses at the micro level. These responses are usually not included in the simulation model. Behavioural responses with respect to the labour market, saving attitudes or household formation are often not taken into account because there is an almost complete lack of empirically validated theories about the behaviour of micro-units. Similarly, macroeconomic effects are usually not built in because micro- and macro-models do not link up adequately in this respect (Sociaal en Cultureel Planbureau, 1990). Second, factors relating to administrative feasibility (for instance, when a shift occurs from insurance to social assistance or vice versa), have consequences for the relationship between the potential beneficiary and the administration - and consequently also for the process of assignment and (non) take-up of benefits. Finally, policy changes have implications for the political legitimacy and, consequently, for the political robustness of the system. The existing literature indicates that there is, as yet, no micro-model in which the issues distinguished above are taken into account in a reliable way.

Because microsimulation models are extracting an existing social security system from its particular social, institutional and political context, simulation models remain 'scientific constructions'. As Atkinson, Bourguignon and Chiappori put it, a simulation model is: "... arithmetic rather than economics, but in our view it provides a valuable means of summarising the quantitative significance of the first-round effects of variations in taxes and benefits" (1988, p. 345).

On a more operational level, microsimulation models are often limited by the availability of data and the size of the samples which they use. Especially when modelling existing social protection systems, one is soon confronted with the discrepancy between the details of legal rules and the availability of the required data. The existing data sets are often not constructed in accord with the needs of simulation models. Some variables might be lacking in the data set because of inherent institutional differences between the countries. A variable which is essential in one social security system might not be relevant in another system and thus will be lacking in the database. Therefore, modelling of various policy models can only be done broadly, rather than in detail.

## 3. Evaluation of the Impact of Social Security and of its Objectives

In this paper welfare outcomes of social security transfers are measured in terms of poverty risks and of the impact on the lower end of the income distribution. The evaluation of social security is thus centred on poverty alleviation. In this respect it is important to bear in mind that the social security system, besides aiming to guarantee a minimum income also aims, for example, to maintain the living standard (up to a certain point) when social risks occur or to 'smooth' income over the life cycle (see Barr, 1992). In any assessment of the overall impact of the social security system these objectives must also be taken into account. The approach commonly taken to the measurement of the effectiveness and efficiency of the social security system is the one conceptually developed by, amongst others, Weisbrod (1970) and implemented by Beckerman (1979). The effectiveness (or horizontal efficiency) of the social security system measures the proportionate reduction in the extent of (pre-transfer) poverty. The efficiency (or vertical efficiency) of the social security system measures the proportion of total transfers received by families in (pre-transfer) poverty rather than by other families. For a given level of expenditures on transfers both effectiveness and efficiency will be greater, the more the benefits are concentrated on those families who would otherwise be in poverty. Pre-transfer and post-transfer incomes are compared to the poverty line in net terms.

Such a numerical exercise can serve as a first indicator of the adequacy of social security, but the method obviously has its shortcomings (eg. behavioural effects are not taken into account).

In order to assess the effectiveness and the efficiency of transfers, a poverty line is needed. The choice of the level and the structure of the poverty line is one of the most debated issues in economic literature. Because the outcomes in terms of poverty reduction and efficiency are sensitive to the applied poverty standard, the best one can do is to consider a range of possible poverty lines. Poverty estimates can be presented in the form of the head-count measure (percentage of poor population) or the (aggregated) poverty gap (difference in the total amount between the income of the unit and the poverty line) (see Mitchell and Harding, 1993; Mitchell, 1991; Atkinson, 1993). For the simulation output below, two poverty lines are presented.

The first is the 'ec-50 per cent standard', which defines the poverty line as 50 per cent of the average equivalent household income based on samples (O'Higgins and Jenkins, 1990). It must be emphasised that this norm cannot (as yet) be regarded as the official EC poverty standard. This standard yields a relatively stringent poverty line, with a rigid equivalence scale (equivalence factors of 1.00 for the first adult, 0.70 for each extra adult and 0.50 for each child). This poverty standard is recommended for cross-national comparative research (Van den Bosch, 1993) and therefore is not used when comparing poverty risks within some European countries. Second, the 'csp-standard' poverty line, developed by the Centre for Social Policy, is used. This poverty standard is based on the subjective income method, which constructs a poverty line on the basis of minimum estimates provided by the population itself. The csp-standard represents a much more generous minimum level, with different poverty standards for adults (active age) and elderly and with a flat equivalence scale (equivalence factors of 1.00 for the first adult, 0.40 for the second adult; 0.30 for the first child; 0.20 for the second child and 0.14 for the third child) (for a more technical outline see Deleeck et al., 1992a; Van den Bosch, 1993).

Standardised deciles are calculated based on disposable household income and using the OECD-equivalence scale (0.66 for a single; 1.0 for a couple; 1.25 for 3 persons; 1.45 for 4 persons; 1.60 for 5 persons and 0.15 for each supplementary person in the household - OECD,1976).

# 4. A Case-Study of Belgian and Dutch Unemployment Schemes

Comparative poverty research studies have consistently found that the poverty risk is considerably higher for the unemployed than for the average citizen, is higher for the unemployed than for other risk categories and is considerably higher than for the employed.

However, there are considerable differences between countries, which leads one to wonder to what extent various unemployment schemes as such provide an explanation for better coverage of the unemployment risk in terms of minimum guaranteed income. For example, the poverty risk for unemployed heads of households is lower in the Netherlands than in Belgium (19 per cent compared with 26 per cent). The number of entitled pre-transfer poor unemployed households that are made non-poor due to unemployment benefits is higher in the Netherlands than in Belgium (80 per cent compared with 74 per cent). The reduction of the aggregate poverty gap is virtually equal, however, in the two countries, at 92 per cent for Belgium and 91 per cent for the Netherlands. (This is explained by the larger average poverty gaps for pre-transfer and post-transfer poor unemployed households in the Netherlands.) The Dutch scheme is more efficient than the Belgian scheme. Among benefit-entitled households, more means are targeted to the reduction of the poverty gap in the Netherlands than in Belgium (44 per cent versus 39 per cent) (all figures from Deleeck et al., 1992a).

Since the Dutch unemployment scheme produces lower poverty risks for unemployed heads of households in the Dutch context than does the Belgian scheme in Belgium, the questions arises as to whether the Dutch unemployment scheme as such is better able to ensure a guaranteed minimum income, and why.

## 4.1 The major characteristics of the Dutch and the Belgian unemployment schemes

In the description of unemployment schemes, a distinction is traditionally made between the principles of unemployment insurance (benefits linked to previous wage, limited duration and financed by earmarked contributions) and those of unemployment assistance (means-tested minimum benefits financed out of general revenues). In practice, compensation schemes can involve both principles, and the rules governing their operation are often very complex, which makes it difficult to describe them in a simple and synthetic way.

The Dutch unemployment scheme, like most other unemployment schemes in Europe, can be characterised as a dual income protection scheme, consisting of an unemployment insurance scheme and an unemployment assistance scheme. The unemployment insurance scheme, however, provides a two-stage protection. In the first stage, wage-related benefits are provided for a limited benefit period, so long as the unemployed worked at least 26 weeks in the 12 month period immediately preceding the first day of unemployment. The duration of the wage-related period is at least six months and this period can be prolonged depending on the former work history and age of the applicant. (In order to be entitled to prolongation, one must have received a minimum of 52 days of wages in at least three years during the five years preceding unemployment). The maximum wage-related period is five years. The wage-related benefit comprises up to 70 per cent of the lost income, and the maximum ceiling is very high (around 1.5 times the average wage).

In the second stage, after the wage-related period has expired, a flat-rate, non-means-tested benefit is guaranteed (at 70 per cent of the minimum wage) for the duration of one year (for certain older unemployed persons this period may last for 3.5 years). The wage-related benefit and the flat-rate benefit are both part of the *Werkloosheidswet* (WW, Unemployment Act). They are financed by social contributions from employers and employees and are limited to a maximum wage ceiling. Unemployment benefits under the insurance scheme are subject to social insurance contributions and to income tax in almost the same way as income from labour. The minimum protection under the 'insurance pillar' is guaranteed by a separate regulation, the *Toeslagenwet* (Supplementary Benefits Act). This is a means-tested supplement, mainly affecting breadwinners. The rates are the same as those under the assistance scheme (although guaranteed in gross terms), but a more generous means test (without asset test) is applied. These supplements are fully financed out of general taxation.

When the WW benefit has expired, the unemployed person falls back on an assistance benefit, the *Rijksgroepregeling Werkloze Werknemers* (RWW, State Regulation for the Unemployed), which is also payable to new entrants to the labour market or persons who do not satisfy the conditions for wage-related benefits. The scheme is part of the general social assistance legislation, *Algemene Bijstandswet* (ABW, Public Assistance Act) and is means-tested. The minimum benefit levels are set at a very generous level as a result of linking minimum protection to minimum wages. The minimum benefit level for a couple amounts to 100 per cent of the (net) minimum wage for a one-earner family. From this reference income the minima for other household types are calculated: 90 per cent for single parents and 70 per cent for a single person. Persons who share a house (without being a couple) receive only

60 per cent. As a percentage of the minimum wages for employed people these amount to 100 per cent, 88 per cent, 77 per cent and 65 per cent respectively (see Table 1). Special lower rates apply for younger claimants and for very young persons living at home. Since 1993 lower benefits also apply for school leavers between 21 and 27 years of age for the first six months of unemployment. Older unemployed people enjoy more generous assistance programs which do not include an assets test nor a deduction for home-sharing (for persons who share a house without being a couple). Social assistance is financed by general taxation (90 per cent nationally, 10 per cent locally). All benefits are indexed for inflation and, up to 1981, were automatically linked to actual wages (wage adjustments). This principle of linking benefits to wages has since been replaced by ad hoc adjustments. As a result of socioeconomic developments and of policy changes, the proportion of unemployed (as a percentage of all benefit entitlements, corrected for duration) covered by the social assistance scheme increased from 33 per cent in 1980 to 64 per cent in 1995.

The Belgian unemployment scheme is exceptional compared with other West-European unemployment schemes as it consists only of an unemployment insurance scheme with unlimited duration of benefits. Unemployed who satisfy the qualifying conditions (employment record according to age) are entitled to unemployment benefits. School leavers are also insured under the insurance scheme but, for them, lower flat-rate benefits are provided. Until the austerity measures of autumn 1993, the entry conditions for the Belgium scheme were rather generous for younger age groups; qualifying conditions for workers and school leavers have become more stringent since 1994. Despite the generous coverage of the Belgian unemployment scheme, groups who are not or are insufficiently covered can claim general social assistance when they have insufficient means for subsistence.

The general social assistance scheme really functions as a 'last resort' and the number entitled to it is relatively low (although growing) compared with the number entitled to unemployment insurance. (As at 1 January 1995 the latter covered about one million persons (including early retirement and other benefit categories) compared with 69,300 persons covered under the general social assistance scheme.) The 'insurance' character of the Belgian unemployment scheme is, however, very weak because:

- benefits are strongly related to family circumstances (assumed need according to three categories - heads of households, singles and co-habiters);

- benefits are very weakly related to previous wages (due to a low maximum ceiling, which is 78 per cent of the average wage level); and
- duration is not related to employment record.

For singles and, in particular, co-habiters the replacement rate declines over time (from 60 per cent to 40 per cent after one year for singles and, for co-habiters, from 55 per cent in the first year to 35 per cent in the next 3 months followed by a low flat-rate non-means-tested benefit). Unemployment insurance guarantees minimum benefit levels which are higher than the general social assistance level but lower than minimum wages. As a percentage of the net minimum wage the minimum benefits are up to 78 per cent for a couple, 84 per cent for a one-parent family, 62 per cent for a single and only 38 per cent for a co-habiter (when receiving the flat rate benefit; for shorter term unemployed this is higher) (see Table 1). Older unemployed persons receive a 'seniority supplement' over and above the unemployment allowance. The unemployment allowances are indexed for inflation (cost-of-living adjustments), but not for the evolution of wages (wage-adjustments).

Although the duration of benefits is not time-limited, unemployed co-habiters (a category which, in practice, mainly applies to married women) can have their benefit entitlements suspended in cases of very prolonged unemployment and under specific conditions (they remain, nevertheless, insured). Grounds for over-turning the suspension are being able to provide proof of exceptional and uninterrupted efforts to find work; a very restricted aptitude for work; or a net taxable family income that does not exceed 600,000 Belgian francs a year (increased by 24,000 francs for each dependent). This income ceiling can be considered as a 'de facto' means test, but at a generous level (for a family without children 54,536 francs net a month, in 1992). Furthermore, the average duration before suspension is implemented is long. The incidence of suspension based on long-term unemployment has, however, doubled between 1992 and 1993 (from 17,245 to 35,886 persons).

In addition to the actual unemployment benefits unemployment insurance has, in recent years, also included different allowance schemes which enable or even encourage people to leave the labour market partially, for a while or even for good. There is a separate early retirement scheme, *brugpensioenen*, which involves supplements from the employer over and above the legal unemployment benefit.

Table 1: Some Quantitative Indicators of the Belgian and Dutch Unemployment Schemes (as at 1 January 1994)

| | Net Maximum benefit (1) | | Net Minimum benefit (1) | | Duration of benefit |
|---|---|---|---|---|---|
| | amount BF/PPP(2) | % of net minimum wage (3) | amount BF/PPP(2) | % of net minimum wage | |
| **Belgium** | | | | | |
| 1. Unemployment insurance | | | | | |
| - head | 32,864 | 90 | 28,522 | 78 | unlimited |
| - one parent | 32,864 | 97(4) | 28,522 | 84(4) | unlimited |
| - single | 32,864 | 100 | 20,410 | 62 | 1 year |
| | 23,010 | 70 | 20,410 | 62 | after, unlimited |
| - cohabiting (5) | 27,120(6) | 94 | 14,836(6) | 52 | 1 year |
| | 17,246(6) | 60 | 14,836(6) | 52 | 13-15 months |
| | | | 12.298 | 38 | after, unlimited |
| 2. Social assistance | | | | | |
| - head/one parent | | | 25,980 | 72 | unlimited |
| - single | | | 19,485 | 60 | unlimited |
| **The Netherlands** | | | | | |
| 1. Unemployment insurance/wage-related (7) | | | | | |
| - family/head household | 57,445 | 170 | 35,011 | 100 | 6 mths to max. 5 years |
| - one-parent | 56,527 | 166 | 31,510 | 88 | |
| - single | 52,819 | 175 | 24,508 | 77 | |
| - cohabiter (8) | 52,819 | 175 | 24,508 | 77 | |
| 2. Unemployment insurance | | | | | |
| - flat-rate | | | 35,011 | 100 | 1 year, |
| - family/head household | | | 31,510 | 88 | extension to |
| - one-parent | | | 24,508 | 77 | 3.5 years when |
| - single | | | 24,508 | 77 | 57.5 years of age |
| - cohabiter | | | | | older |
| 3. Unemployment assistance | | | | | |
| - family/head household | | | 35,011 | 100 | unlimited |
| - one-parent | | | 31,510 | 88 | |
| - single | | | 24,508 | 77 | |
| - cohabiter | | | 21,007 | 65 | |

(1) Net benefits after yearly taxation. (2) Corrected for PPP (purchasing power parities), 1 Dutch guilder=18.73 Belgian francs/PPP, at 1 February 1994 (source: Eurostat). (3) Net minimum wages for employed, inclusive of holiday allowances. For the Netherlands net minimum wages are calculated exclusive of contributions for early retirement and occupational pension funds, as these schemes are not legislated nationally and differ between sectors. (4) Net minimum wages are calculated for a divorced parent with one child (for unmarried single parents an extra tax allowance exists). (5) Co-habiting persons are all persons not living solely on unemployment benefit as well as married (two-income) partners. For the present benefit levels a married person, living with a partner earning an average wage (which is relevant for the impact of taxation), was chosen. (6) The legal guaranteed minimum and maximum amounts are higher, but married co-habiters who do not receive the flat-rate benefit must pay 10 per cent taxation on a monthly base. For low family incomes there is some compensation through the system of tax allowances. (7) Wage-related benefits are individual, but net benefits are family related through the tax system. (8) Co-habiting in the Dutch system has a different definition than under the Belgian, which includes sharing a house without being a couple.

The unemployment insurance scheme is financed by three parties - employees and employers (who pay contributions on their entire wages - that is, there is no maximum wage ceiling) and government. The latter contribution has been reduced since the 1980s and been replaced by diverse 'solidarity' contributions of various socioeconomic groups. There exists a scheme of tax allowances for the unemployed, under which unemployment benefits are not taxable for those unemployed, singles and breadwinners who live entirely on the allowance and for co-habiters with low incomes. For the latter these tax allowances result in a reduction of taxes, depending on the level of total household income and the proportion of the benefit in this income.

In summary the Belgian scheme can be characterised by its more universal scope and, in particular, unlimited duration of entitlement. On the other hand, there is strong categorical targeting and benefit levels are relatively low. In contrast, the Dutch scheme is a dual income protection scheme, with a strong equivalence principle for short-term unemployment, followed by needs-related (means-tested) minimum income protection through social assistance. Despite the assistance character Dutch minimum benefits are nevertheless at a generous level. Although benefit levels are more generous in the Netherlands than in Belgium, total outlays on unemployment compensation are at a lower level in the Netherlands (2.17 per cent of GDP in the Netherlands against 2.82 per cent in Belgium in 1992 - OECD, 1994).

## 4.2 Simulation model and simulated benefit levels

In this paper the Dutch system is simulated onto the Belgian base population using two scenarios. In the first scenario, the Dutch unemployment scheme is simulated in its entirety (ie. wage-related benefits under insurance for limited time and minimum benefits under assistance). As Table 2 indicates, three variants are distinguished by changing the *level of minimum protection*:

- Variant A has minimum benefits derived from Belgian minimum wages (with the minimum benefit for a family head being 100 per cent of Belgium minimum wage (except holiday allowances);
- Variant B has minimum benefit levels derived from the Dutch minimum wages (with the minimum benefit for a family head being 100 per cent of Dutch minimum wage, inclusive of holiday allowance); and

- Variant C has further increased minimum benefits in order to attain budget neutrality (that is, the total amount of spending equals the spending under the current Belgian scheme).[1]

In the second scenario (particularly relevant for the actual Belgian debate) the present Belgian insurance system is maintained for those who are unemployed for only two years. After these benefits are exhausted, a social assistance scheme comes into operation, based on the actual Belgian social assistance level. At this point only one aspect of the logic of the Dutch system is simulated, namely income protection for a limited duration under unemployment insurance, without altering wage-related or minimum benefits. The present (low) level in the Belgian social assistance scheme is retained. This scenario thus reduces outlays on the Belgian unemployed.

In interpreting the results it is important to take into account the following options and abstractions made when simulating the Dutch scheme.

First, in the simulations of wage-related benefits in Scenario 1 a net replacement rate of 70 per cent was assumed. Thus, social contributions and taxation were not simulated directly, but indirectly. Standard simulations of the Dutch unemployment benefits suggest that the imputed net replacement rate of 70 per cent is somewhat underestimated (for an average production worker the net replacement rate is up to 75 per cent) (De Lathouwer, 1993; Einerhand, Eriksson and Hansen, 1993).

Second, because of the lack of relevant data in the sample, a number of hypotheses concerning the duration of the employment record were used. These were especially significant for the prolongation of benefits under unemployment insurance. Under Scenario 1 it was assumed that for everyone with past employment, the employment record required for entitlement of the basic benefits and the prolonged benefit was satisfied.

---

[1] In order to analyse the proper distributional potential of the Dutch system the total cost of the alternative system needs to be kept at the same level as that of the current Belgian system. In theory this can be done according to various options. Because in this paper the emphasis is on poverty reduction, the difference in spending was assigned to a further raising of the level of minimum income guarantee. This assignment is debatable because it disrupts the balance between maximum and minimum benefits.

Table 2: Level of the Minimum Net Benefits for the Unemployed in the Simulations According to Family Situation, 1985

| Household Situation (2) | Current Belgian System | Scenario 1 | | | Scenario 2 |
|---|---|---|---|---|---|
| | | Variant A | Variant B | Variant C | |
| | BF/PPP/ mth (1) | BF/PPP/ mth (1) | BF/PPP/ mth (1) | BF/PPP/ mth (1) | BF/PPP/ mth (1) |
| family | 23,348 | 26,536 | 28,751 | 30,765 | 18,690 |
| one-parent family | 23,348 | 23,882 | 25,876 | 27,687 | 15,752 |
| singles (4) | 16,406 | 28,575 | 20,126 | 21,536 | 14,776 |
| cohabiting partners (3) | 10,504 | 0 | 0 | 0 | 0 |
| cohabiting others (4) | 10,504 | 15,922 | 17,251 | 18,458 | 9,845 |
| school leavers (5) | 10,200 | 15,922 | 17,251 | 18,458 | 9,945 |

Scenario 1, Variant A: Dutch benefit structure, Belgian minimum wages (excluding holiday allowance)

Scenario 1, Variant B: Dutch benefit structure, Dutch minimum wages (including holiday allowance)

Scenario 1, Variant C: Hypothetical model, Dutch benefit structure, budget neutrality

Scenario 2, Hypothetical model, Belgian benefit structure for short-term unemployed, long-term unemployed benefits according to Belgian social assistance scheme.

(1) PPP (purchasing power parities), 1 Dutch guilder=18,73 Belgian francs (BF) at 1 February 1994 (source: Eurostat). Under Variant B net Dutch minimum for couples is 1869.3 guilders/month (gross minimum wage level is 2163.2 guilders a month).

(2) For families with children child allowance is not included in the presented benefit levels. The latter play a significant role in guaranteeing a minimum income. These allowances are higher in Belgium than in the Netherlands and, in addition, the increase with the number of children is steeper. This leads to a diminution of the differences between families with (many) children.

(3) The benefit level under the current Belgian scheme represents the minimum for 'co-habiting partners' after 18 months of unemployment (for shorter term unemployed, see Table 1). Long-term unemployed 'cohabiting partners' lose their benefit under Variants 1A, 1B and 1C due to the means-test for long term unemployed when other (in particular, partner income) household income exceeds the income ceiling.

(4) Amounts for singles and co-habiting persons (without being a couple) presented under Variant B are applicable for those aged 23 years and up (younger age groups and youngsters living at home receive lower benefits). Under Variants 1A and 1C they are applicable from the age of 21 years on.

(5) Amounts for long-term unemployed co-habiting school leavers. In the current Belgian system from the age of 18 years on; in Variants 1A and 1C from 21 years on; in Variant 1B from 23 years (younger age groups and youngsters living at home receive lower benefits).

Third, possible interactions between institutional arrangements were not taken into account. No changes were made in the child allowance system (which are lower in the Netherlands, thus overestimating Dutch outcomes) and no simulation was done of the Dutch housing benefit scheme (thus underestimating Dutch outcomes). No attention was paid to a possible switch to other branches of social security (such as the more generous disability regulations in the Netherlands). No alteration was made to the current system of early retirement benefits (an advantageous system for certain categories of older unemployed persons).

Fourth, no account was taken of possible behavioural changes affecting the labour market (for instance, changes in labour market behaviour, hidden unemployment, etc.) via saving patterns and in household formation. Possible implications for the implementation cost, the potential for non-take-up of (means-tested) benefits, and the consequences for the legitimacy of the system were abstracted from the simulation.

Finally, the simulations are static in nature and do not take into account a potentially changing profile of needs under the influence of rising unemployment.

A comparison of the above minimum benefit levels shows that the current Belgian minimum unemployment benefits are considerably lower than the Dutch minima in absolute terms (Variant 1B). Linking the minima to Belgian minimum wages (Variant 1A) in line with the Dutch benefit logic, raises the minimum benefit level considerably - but the minima still remain lower than the Dutch ones under Variant 1B.

## 4.3 Simulation results

From the simulations of the entire Dutch benefit scheme for the Belgian unemployed, represented by Scenario 1 (Tables 3 and 4), four observations may be made.

First, compared with the current Belgian scheme, the Dutch benefit structure generates higher benefits for fewer unemployed. This means that if the Belgian unemployment scheme were changed in accordance with the Dutch model its coverage would be less generous - 22 to 26 per cent of those now entitled would be excluded from the system. The considerably smaller coverage of the Dutch system is explained by the large proportion of long-term unemployed from multiple income households (chiefly married women) among the Belgian unemployed. About 40 per cent of them would lose their allowance, because of the introduction of a means test for longer term unemployed. Opposed to this, however, there would be higher allowances for the entitled, both for those receiving wage-related benefits (because of a stronger insurance principle) and for those receiving minimum benefits (as the minima would be higher). The former must, however, be put in context. High wage-related benefits seem to be a theoretical (legal) rather than a social reality. Because of the relatively weak former labour incomes of the unemployed, a strong equivalence principle does not result in high wage-related benefit levels. By contrast, the Belgian system is more generous in coverage, but provides lower allowances than the Dutch scheme. The Dutch system implies that over 50 per cent of those receiving an unemployment benefit in the current Belgian system would no longer qualify for a benefit under unemployment insurance because their rights would have been exhausted or because of lack of rights. About 30 per cent would qualify for social assistance. All in all, a quarter to a fifth of the Belgian population who are now entitled would lose their right to an allowance under the Dutch scheme.

Second, given the assumptions of the modelling (eg. non-simulation of the cost of implementation) the coverage and benefit level under Scenario 1 together lead to a limited reduction of the current budget of about 7 per cent (Variant 1A) to 5 per cent (Variant 1B). A selective system, following the logic of the Dutch benefit scheme (in all respects), consequently leads to smaller savings than one might have expected. In addition there is a significant shift in the method of financing the unemployment scheme (from social insurance to social assistance).

Third, the reorganisation of the current budget in accordance with the Dutch model would lead to greater effectiveness and redistributive impact for unemployment benefits. Poverty would be lower and the welfare effort would be more targeted towards poverty reduction. The impact on the effectiveness of the system depends on the simulated variant and the results are very sensitive to the poverty line (and the equivalence scale) adopted for the evaluation of welfare outcomes (see also Van den Bosch, 1993; Mitchell and Harding, 1993). When concentrating on more severe poverty (measured by the more stringent EC-standard) Variant 1A produces lower poverty rates among the unemployed than the current Belgian system (10.3 per cent versus 12.2 per cent for the current system - Table 3), although reduction is rather modest. Variant 1B (minima derived from Dutch minimum wages) generates a further reduction in poverty to only 8 per cent. Finally, Variant 1C (minimum benefits now exceeding the minimum wage level) leads to very substantial differences in poverty rates compared with the current Belgian system. Note that even in the case of a substantial increase in the minima in accordance with the Dutch model, the allowances are still not sufficient to eliminate poverty altogether.

Table 3: Key Simulation Results for Unemployed Households

| | Current system | Scenario 1 | | | Scenario 2 |
|---|---|---|---|---|---|
| | | Variant A | Variant B | Variant C | |
| *Cost price* | | | | | |
| * total (in % of current cost price) | 100.0 | 92.4 | 94.6 | 00.0 | 66.0 |
| - insurance | | 64 | 63 | 65 | 49 |
| - assistance | | 29 | 32 | 35 | 17 |
| *Generosity* | | | | | |
| * Coverage (% unemployed with benefits) | 100.0 | 74.3 | 76.0 | 77.7 | 67.4 |
| * Average benefit in BF/month | 15.373 | 19.129 | 19.147 | 19.808 | 15.040 |
| *Welfare Outcomes* | | | | | |
| * % households in poverty: | | | | | |
| EC-poverty line | 12.2 | 10.3 | 8.0 | 4.6 | 25.6 |
| CSP-poverty line | 34.3 | 42.8 | 42.5 | 38.8 | 52.4 |
| * reduction of poverty gap for unemployed households: | | | | | |
| EC-poverty line | 92.4 | 94.1 | 95.7 | 97.1 | 80.5 |
| CSP-poverty line | 79.3 | 78.7 | 81.4 | 85.8 | 60.2 |
| * % households in: | | | | | |
| lowest standardised decile | 19.2 | 16.6 | 15.2 | 11.3 | 31.5 |
| lowest standardised quintile | 29.9 | 36.5 | 34.1 | 27.0 | 45.9 |
| * Share of transfers to needy households below poverty line (efficiency): | | | | | |
| EC-poverty line | 39.9 | 44.0 | 43.7 | 41.9 | 52.7 |
| CSP-poverty line | 59.6 | 64.0 | 64.6 | 64.5 | 68.6 |

Source: MISIM (Micro Simulation Model for Social Security and Taxation), csp-sample 1985
Scenario 1, Variant A: Dutch benefit structure, Belgian minimum wages.
Scenario 1, Variant B: Dutch benefit structure, Dutch minimum wages.
Scenario 1, Variant C: Dutch benefit structure, constant input of transfers.

Scenario 2: Current Belgian unemployment scheme for short-term unemployed up to 2 years, For long-term unemployed, benefits according Belgian social assistance scheme.

However, measured by more generous income standards (the broader CSP-poverty line) the apparent improved effectiveness under Scenario 1 is less convincing. The poverty rate (head-count ratio) is actually higher under each of the three variants in Scenario 1 (from 39 to 43 per cent poverty rates) than the 34 per cent recorded under the current Belgian System. However, the total poverty problem is still reduced to a greater extent (except in Variant A), when measured by the aggregate poverty gap indicator. This means that more households are now under the poverty line but on average they are not as far below.

The increase in effectiveness is the result of a significant decrease in poverty among single income providers, while multiple-income families, including two-income providers, are more or less severely affected, depending on the poverty line taken. For many two-income providers, unemployment benefits are an important source of keeping the household income up to (higher) standards (see also Cantillon, 1992).

Fourth, the results in Table 3 also suggest that the Dutch model would lead to greater efficiency from the point of view of eliminating poverty, as a larger share of the transfers help to close the poverty gap. The new distribution of income and welfare generated by the Dutch model confirms the effectiveness results. The major improvements in income and welfare for the unemployed are achieved at the bottom of the income and welfare distribution. The proportion of unemployed households in the bottom decile, which in the current system is 19 per cent, decreases to 16.6 per cent under a Dutch scheme where minima are linked to Belgian minimum wages (Variant 1A); to 15 per cent in a scheme where minima are linked to higher Dutch minimum wages; and 11 per cent in a scheme where minima are at a higher level than minimum wages (Variant 1C).

More disaggregated analysis on income distribution effects, which is not presented in Table 3, shows that welfare improvements are the largest for singles and one income earners and also among multiple income families with a very weak income profile. However, a more generous welfare line suggests that the two scenarios generally result in an increase in the proportion of unemployed at lower levels (eg. 29.9 per cent of the unemployed are in the bottom quintile under the current Belgian system, compared to 36.5 per cent under Variant A).

Scenario 2 shows that, if the present Belgian system was maintained for unemployment lasting up to two years, with subsequent transfer of the longer term unemployed to the current low level of Belgian social assistance benefits, there would be considerable decreases in expenditure and the available means would be used in a considerably more efficient way. Opposed to this, however, a very high social cost would be paid, with a doubling of the poverty risk (measured by the stringent EC-norm). The severe social assistance means test would sharply reduce coverage (more than 30 per cent of those entitled at present would drop out of the system). Consequently, a selective application of too small a budget, in view of existing needs, would lead to considerable ineffectiveness and increased poverty.

# 5. Conclusions

The application of microsimulation in comparative research is illustrated in this paper by means of a case-study of the Belgian and Dutch unemployment schemes.

Recent research has shown that the poverty rate for unemployed heads of households is lower in the Netherlands than in Belgium, even though outlays on the unemployed are higher in Belgium. Reflecting both these factors, the Dutch scheme is also more efficient, in the sense that a higher proportion of outlays are directed towards those below the pre-transfer poverty line. This raised the interesting question of whether the Dutch benefit structure would generate better outcomes for the unemployed in Belgium than the existing Belgian system.

This proposition was tested by simulating the Dutch system against the sample of Belgian households captured in the Centre for Social Policy's MISIM static microsimulation model. Three progressively more generous alternatives were modelled under Scenario 1, all of which essentially involved simulating the key aspects of the Dutch system. Variant C, which probably provided the fairest comparison as it involved the same outlays as the current Belgian system, produced outcomes which illustrated very clearly the trade-offs facing social policy makers. For example, it reduced poverty rates using the EC poverty line to less than five per cent (compared to 12 per cent for the current Belgian system). It also reduced the aggregate poverty gap using either the EC or CSP poverty line. On the other hand, the percentage of unemployed with benefits dropped from 100 per cent under the current system to 78 per cent under Variant C. Similarly, using the more generous CSP poverty line, the poverty rate was actually higher under Variant C than under the current Belgian system. In summary, these results reflected the impact of two key factors: the removal of assistance (via a means test) from couples where one spouse was long-term unemployed and the other earned a reasonable income, and more generous assistance to those still receiving benefits after this change.

Table 4: Simulation Results: Percentage of Poor Unemployed Households According to Family Situation (*)

| | Current system | Scenario 1 | | | Scenario 2 |
|---|---|---|---|---|---|
| | | Variant A | Variant B | Variant C | |
| *EC-poverty line* | | | | | |
| - all households | 12.2 | 10.3 | 8.0 | 4.6 | 25.6 |
| - singles(1) | 13.8 | 5.5 | 5.5 | 4.6 | 25.7 |
| - one-income providers(2)** | 55.9 | 32.3 | 24.7 | 9.7 | 81.9 |
| - two-income providers(3)** | 4.0 | 7.8 | 4.8 | 2.2 | 16.1 |
| - others (in particular, inmates) | 6.4 | 7.4 | 7.4 | 6.9 | 17.2 |
| - all one-income households(4) | 33.2 | 17.8 | 14.4 | 6.9 | 51.5 |
| - all multiple-income households(5) | 4.9 | 7.6 | 5.7 | 3.8 | 16.5 |
| - households without benefits | - | 8.0 | 6.9 | 5.6 | 13.4 |
| *CSP-poverty line* | | | | | |
| - all households | 34.3 | 42.8 | 42.5 | 38.8 | 52.4 |
| - singles(1) | 69.7 | 71.6 | 70.6 | 63.3 | 78.9 |
| - one-income providers(2)** | 86.0 | 81.7 | 80.6 | 75.3 | 92.5 |
| - two-income providers (3)** | 15.6 | 36.6 | 36.0 | 33.1 | 40.3 |
| - others (in particular, inmates) | 26.0 | 21.1 | 22.1 | 19.6 | 42.2 |
| - all one-income households(4) | 77.2 | 76.2 | 75.2 | 68.8 | 85.1 |
| - all multiple-income households(5) | 19.3 | 31.1 | 31.1 | 28.3 | 41.0 |
| - households without benefits | - | 46.3 | 43.7 | 40.0 | 52.6 |

Source: MISIM (Micro Simulation Model for Social Security and Taxation), csp-sample 1985

(*) Proportions of the household categories in the total population of entitled were unemployed: singles (8.5%), one-parent (4.6%), one-income provider (12.2%), two-income providers (49.8%), inmates (25%, of which 17% children living with their parents). In 1992 the household formation of the unemployed remained roughly the same.
(**) The categorisation into one-, two- or more income households concerns initial categories. Alterations in the relationship between these categories (owing to greater selectivity in scope) are not represented in this Table.
(1) Singles are with and without children
(2) One income provider - one partner unemployment benefit, other partner no income;
(3) Two-income provider - one partner unemployment benefit (under Belgian scheme), other partner other income (mainly non labour), under simulations including unemployed without allowances.
(4) All one-income households (all households living solely on one unemployment benefit).
(5) All multiple income households (households where unemployment benefit is cumulated with other income sources), under simulations including unemployed without allowances.

Given the available Belgian budget, increases in the level of unemployment allowances would only be possible if there was such greater selectivity in coverage.

However, debate in Belgium indicates that neither higher minimum benefits (linked to minimum wages) nor the use of a means test as an instrument for greater selectivity are favoured policy options. Such debate reflects the on-going discussion across the world about the relative merits of universal and selective welfare systems (Atkinson, 1993; Mitchell et al, 1994).

The second scenario modelled only involved simulation of a single aspect of the Dutch system: the imposition of a two year time limit for the receipt of unemployment insurance (rather than the current practically unlimited protection provided by the Belgian insurance scheme). The simulation showed that such an option would slash outlays on and reduce coverage of the unemployed by about a third, would double EC poverty rates and lead to greater poverty gaps.

The simulation findings were the result of a static microsimulation model. This means that no account was taken of behavioural changes and second-round effects. The literature on selectivitity versus universality makes it clear, however, that more selectivity can lead to such 'costs' as economic inefficiency (through raising poverty traps, administrative complexity, problems of non-take-up and fraud and political illegitimacy). Although there exists no micro-model to deal with all these issues one must be aware that possible side effects could influence the static simulation results.

The case study presented shows that the use of microsimulation in comparative research is a valuable instrument, providing a better insight on the distributional impact of social security - but also generating new questions in the search for understanding of the effectiveness of social support systems. In this way microsimulation, combined with traditional instruments for comparative research, generates further knowledge about the impact of social security systems. As the simulations showed, family composition - in particular, the tension between one-income families and the growing proportion of two-earner families (resulting from increasing female labour force participation) - has important implications for the distributional impact of transfer schemes (Cantillon, 1991). The results force us to question whether selectivity can/must be achieved through 'assumed' needs (the Belgian model) or through 'actual' needs (the Dutch model for the long-term unemployed). Together with growing budgetary constraints and competing public responsibilities, this leads to new challenges on rethinking instruments for the allocation and redistribution of transfers.

## References

Atkinson, A.B. (1993), 'On targeting social security: Theory and western experience with family benefits', Discussion Paper WSP/99, Welfare State Programme, London School of Economics, London.

Atkinson, A.B., F. Bourguignon and P.A. Chiappori (1988), 'What do we learn about tax reform from international comparison?', France and Britain, *European Economic Review*, Vol. 32, No.

Atkinson, A.B. and H. Sutherland, (eds.) (1988), *Tax-benefit Models*, Suntory-Toyota International Centre for Economics and Related Disciplines, London School of Economics, London.

Barr, N. (1992), 'Economic theory and the welfare state: A survey and interpretation', *Journal of Economic Literature*, Vol. 30, No.

Beckerman, W. (1979), *Poverty and the Impact of Income Maintenance Programmes*, International Labour Office, Geneva.

Cantillon, B. (1991), 'Socio-demographic changes and social security', *International Review of Social Security*, Vol. 4, No.

Cantillon, B., K. Lambert and G. Verbist (1993), 'Het micro-simulatiemodel van de sociale zekerheid (MISISZ), CSB-berichten', Centrum voor Sociaal Beleid, Universiteit Antwerpen, UFSIA, Antwerp, June.

De Lathouwer, L. (1993), 'Universalism versus selectivity in unemployment schemes', in: J. Berghman and B. Cantillon (eds.), *The European Face of Social Security, Essays in Honour of Herman Deleeck*, Avebury, Aldershot.

Deleeck, H. (1992), 'Social Security and '1992', in: A. Pijpers (ed.), *The European Community at the Crossroads*, Martinus Nijhoff Publishers, Dordrecht.

Deleeck, H., L. De Lathouwer and K. Van den Bosch (1989), 'Regional differences in the distribution of social security benefits in Belgium: Facts and causes', *Cahiers Economique de Bruxelles*, Brussels, Vol. 123.

Deleeck, H., L. De Lathouwer and K. Van den Bosch (1992a), *Poverty and the Adequacy of Social Security in the EC: A Comparative Analysis*, Aldershot, Avebury.

Deleeck, H., L. De Lathouwer and K. Van den Bosch (1992b), 'Adequacy of social security in seven EC-countries', *Cahiers Economiques de Bruxelles*, Brussels, Vol. 123, No. .

Einerhand M., I. Eriksson and H. Hansen (1993), 'Unemployment insurance in the Netherlands, Denmark and Sweden', Studies in Social Security and Personal Taxation, DGSZ/FEBO Research Memorandum 93/1, October.

Hancock, R. and H. Sutherland (1992), *Microsimulation Models for Public Policy Analysis: New Frontiers*, Suntory-Toyota International Centre for Economics and Related Disciplines, London School of Economics, London.

Hauser R. (1993) 'Approaches to comparative social policy analysis', in: J. Berghman and B. Cantillon (eds.), *The European Face of Social Security*, Avebury, Aldershot.

Hill, M. and G Bramely (1986), *Analysing Social Policy*, Basil Blackwell, Oxford.

Jäntti, M. and S. Danziger (1992), 'Does the welfare state work?, Evidence on anti-poverty effects from the Luxembourg income study', Working paper **NO WHAT** from the Luxembourg Income Study.

Ministerie Sociale Zaken en Werkgelegenheid (1992), *Sociale Nota*, SDU uitgeverij, 's Gravenhage.

Mitchell, D. (1991), *Income Transfers in Ten Welfare States*, Avebury, Aldershot.

Mitchell D. and A. Harding (1993), 'Changes in poverty among families during the 1980s: Poverty gap versus poverty head-count approaches', Discussion Paper No. 2, October, National Centre for Social and Economic Modelling, University of Canberra, October.

Mitchell, D., A. Harding and F. Gruen (1994), 'Targeting welfare', *The Economic Record*, Vol. 70, No. 210.

Mot, E.S. (1990), 'Verkenning van microsimulatiemodellen, Inventaris en aanbevelingen, Ministerie Sociale Zaken en werkgelegenheid, COSZ publicatie, Vuga-uitgeverij,'s Gravenhage, Vol. 30

OECD (1991), 'Unemployment benefits rules and labor market policy', *Employment Outlook*, OECD, Paris..

OECD (1994), *Employment Outlook*, OECD,Paris.

O'Higgins, M. and S. Jenkins (1990), 'Poverty in the EC 1975, 1980, 1985', in: R. Teekens and B.M.S. Van Praag (eds.), *Analysing Poverty in the European Community*, Policy Issues, Research Options and Data Sources, Eurostat News Special Edition, Luxemburg.

Ramprakash, D. (1994), 'Poverty in the countries of the European Union: A synthesis of Eurostat's statistical research on poverty', *Journal of European Social Policy*, Vol. 4, No.

Smeeding, T., M. O'Higgins and L. Rainwater (1990), *Poverty, Inequality and Income Distribution in Comparative Perspective*, The Luxemburg Income Study, Harvester Wheatsheaf, New York.

Sociaal en Cultureel Planbureau (1990), *Gecumuleerd Beleid*, Cahier nr 76, Sociaal Cultureel Planbureau, Rijswijk.

Van den Bosch, K._(1993), 'A comparison of poverty in seven European countries and regions using subjective and relative measures', *Journal of Population Economics*, Vol. 6, No.

Weisbrod, B.A. (1970), 'Collective action and the distribution of income: A conceptual approach', in: R. Haveman and J. Margolis (eds.), *Public Expenditures and Policy Analysis*, Markham, Chicago.

Aino Salomaki

Government Institute for Economic Research (VATT)
Finland

# Chapter 5: Including Consumption Expenditure and Welfare Services in a Microsimulation Model

## Abstract

*The paper discusses the extension of an income database of a microsimulation model by the inclusion of imputed data on consumption and welfare services. A possible use of such an integrated database in the field of welfare policy, where a broader scope of analysis is demanded, is also discussed. The Finnish microsimulation model 'TUJA' has been expanded to incorporate information on the use of welfare services and the consumption expenditure of households. Problems of statistical matching are discussed, with a specific focus on various merging methods. Two merging methods have been tested and their effects on the merged data and on the estimated distributional impact of the replacement of the turnover tax with a value-added tax have been analysed.*

## 1. Introduction

The Finnish static microsimulation model TUJA has been in use since 1987 and has played a key role in the analysis of the financial and distributional impacts of tax and transfer reforms. The model is of a general nature and can deal with many tax and transfer policy issues. Originally, the model data covered income, cash transfers and income taxes. The scope of welfare policy analysis is, however, wider than this, so a natural development path has been to aim at expanding the scope of the model. A more general model with an integrated database is considered preferable. Accordingly, the incorporation in the database of additional information on individuals and households from other sample surveys has been underway for some time.

The TUJA model is primarily based on the full data file of the household survey for income distribution statistics. The data are very rich, with information about income sources, cash transfers, income taxes, tax deductions, and the characteristics of households and individuals. The total number of variables for each person in the database is about 600. The income and transfer data are mostly derived from several administrative registers (notably tax files) and matched directly to the survey individuals using personal identification numbers. The database is updated annually with the latest income distribution statistics.

The income distribution statistics file does not contain any information on consumption expenditure or the use of welfare services by households. These data are collected through household budget surveys carried out every five years, most recently in 1990.

## 2. Merging the Income Distribution and Household Budget Data

The data in the income distribution survey and the household budget survey of 1990 are unique in the sense that, for the first time, they are based partly on the same sample population. Both surveys consist of two subsamples, one of which is a joint subsample used in both the income distribution statistics and the household budget survey data. For the households in this joint subsample, there are not only income data but also interview and bookkeeping data on consumption and on the use of services. This subsample thus provides an excellent opportunity to test the accuracy of different kinds of merging methods. This is of great value when new annual information on consumption and welfare services is not available. Two methods have been tested in this study - average statistical merging and so-called 'hot rank' merging.

The basic idea of *average statistical merging* is to form a number of similar categories of households in both surveys. The households should be relatively similar in their characteristics - so that the structure of consumption expenditure or welfare services used is adequately homogeneous within categories and that the variables to be imputed are correlated with the classification variables chosen for matching. Thus, the profiles observed in one survey can be transferred to the households of the other survey. The choice of matching variables is a crucial issue in an imputation procedure. This is often addressed by survey statisticians - see, for example, Cohen (1991) and Ford (1983).

A prerequisite of successful merging is that the survey households can be classified by the same characteristics in both surveys. For the purpose of merging the consumption data of households, a total of 160 groups of households were classified - by type of family (four categories), by the population density in the community (four categories), by the ownership of dwelling (two categories) and by the quintile of disposable income per equivalent adult.[1]

Welfare services usage data were merged into the income file by forming categories of individuals (because the data on the use of services were available by individual). The groups were defined according to the characteristics of a typical user of services. A total of 100 groups of individuals were classified - by the population density in the community (four categories), by the age class (five categories) and by the quintile of equivalent disposable income.[2]

The consumption expenditures were grouped into 56 expenditures categories and the welfare services into 43 groups of services. For each type of consumption expenditure in each category of household, a coefficient of consumption propensity was estimated. This expressed consumption expenditure as a share of disposable income. By means of these coefficients, of the similar classification of households and of the similar definition of disposable income, it was possible to impute the desired variables onto the other survey. Similarly, for each variable of services in each category of individuals, the average value of services per user and the frequency of use in each category was calculated.

The *hot rank merging method* operates as follows. A number of similar categories of households or individuals were defined in both surveys. The categories of households in this study were the same as described above. After classifying, the households were ranked by income. The household with the highest income in each category in the donor file was then matched with the household with the highest income in each category in the recipient file and so on. If the two files were of different size, some households in the donor file were duplicated or left out of the categories so that their sizes corresponded to those in the recipient file (British Department of Social Security, 1992).

---

[1] Equivalent adult is an unit of analysis in which the household size and composition are adjusted by an equivalence scale. The scale used in this study is known as the OECD scale, and gives the first adult a value of 1, second and subsequent adults a value of 0.7, and children a value of 0.5.

[2] In summary, the four population density categories were capital areas, other towns, densely populated communities and rarely populated communities; the five age categories were 0-6 years, 7-16, 17-25, 26-64 and 65 + years; and the four family categories were families with children, pensioners, single persons and other.

## 3. Some Characteristics of the Two Surveys

The surveys on income distribution and household expenditure are not identical, and neither are the two subsamples in each survey. There are several features affecting the comparability of the non-identical subsamples in the two surveys. First, their sizes are quite different. The size of panel 1 in the income distribution statistics is 1.9 times that of the corresponding subsample in the household budget survey (Figure 1). Second, a household budget survey involves heavier respondent load. As a consequence, there is a much higher non-response rate in the household budget surveys (about 30 per cent) than in the income distribution survey (about 15 per cent).

A household participating in the household budget survey has to undertake two weeks of book-keeping on non-durable goods. Additionally, there is an interview at the beginning of the book-keeping period and another at the end of the reference year. The interviews cover a large number of questions, including the characteristics of the household and its dwelling, some income issues, the purchase of durable goods, housing costs and the use of welfare services.

There are also a number of differences between the income distribution survey and the household budget survey in their objectives, data gathering procedures and other technical details. Despite this, they are highly comparable with regard to their definitions and classifications of households, as well as their income concepts. Thus, the results of these two surveys are very similar. Table 1 describes some similarities and differences between the household budget survey (HBS), the income distribution statistics (IDS) and the independent estimates of national totals based on the joint subsample (Djerv and Lindquist, 1993).

Figure 1: Sampling Structure of the 1990 Household Budget and Income Distribution Surveys

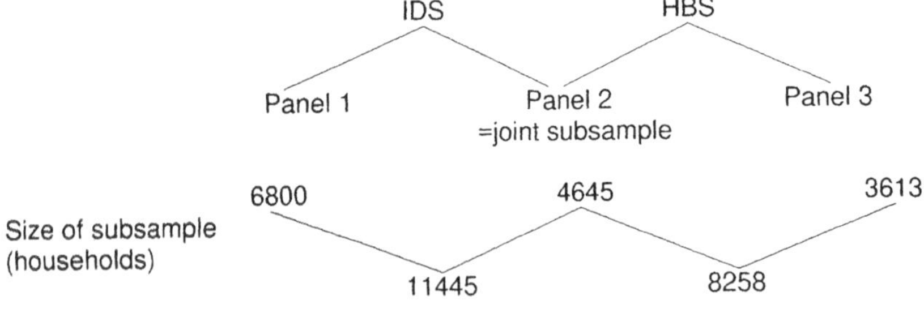

Table 1: Some Characteristics of the Household Budget Survey (HBS) and of the Income Distribution Statistics (IDS) in 1990

| | HBS 1990 | IDS 1990 | IDS / HBS difference in % | Joint subsample |
|---|---|---|---|---|
| *Sample size* | | | | |
| - households | 8 258 | 11 445 | 39.0 | 4 645 |
| - persons | 22 627 | 31 470 | 39.0 | 12 366 |
| *Estimated population* | | | | |
| - households | 2 154 281 | 2 170 574 | 0.8 | 2 164 439 |
| - persons | 4 945 963 | 4 973 852 | 0.6 | 4 962 396 |
| - average size of household | 2.30 | 2.29 | | 2.29 |
| *Number of households by socioeconomic group* | | | | |
| - entrepreneurs | 202 895 | 203 613 | 0.4 | 192 758 |
| - wage and salary earners | 1 233 471 | 1 253 609 | 1.6 | 1 234 476 |
| - inactive households | 717 915 | 713 352 | -0.1 | 737 204 |
| *Number of households by size* | | | | |
| - 1 person | 739 847 | 766 502 | 3.6 | 754 679 |
| - 2 person | 658 733 | 639 777 | -2.9 | 646 880 |
| - 3 person | 327 227 | 321 736 | -1.7 | 328 287 |
| - 4+ person | 428 475 | 442 530 | 3.3 | 434 592 |
| *Some estimates for average income per household (FIM)* | | | | |
| - wages and salaries | 110 200 | 109 300 | -0.8 | 109 400 |
| - entrepreneurial income | 13 400 | 13 000 | -3.0 | 14 200 |
| - property income | 8 600 | 8 800 | 2.3 | 9 300 |
| - transfers received | 39 100 | 36 900 | -5.6 | 37 300 |
| - transfers paid | 42 600 | 41 800 | -1.9 | 42 400 |
| - disposable income | 128 800 | 126 200 | -2.0 | 127 300 |

Note: FIM = Finnish Mark

# 4. Comparison of the Merging Results

The two merging methods described in Section 2 were tested only with respect to consumption expenditure on durable goods and housing costs. The objective of a complete comparison of the two methods covering all consumption items could not be met, because of problems resulting from data-gathering practices. The quality of consumption expenditure data varies according to the data-gathering methods. The data on non-durable goods are gathered during the two week book-keeping periods. The annual expenditure on these goods by a household is estimated to be 26 times that of the book-keeping period. Evidently, this can result in great variations in the amounts of expenditure and does not necessarily represent the real amount of annual consumption. There are many obvious reasons for this. For example, the book-keeping period may be exceptional (eg. coinciding with holidays or due to personal purchase patterns). Therefore, the Central Statistical Office did not recommend the individual matching of the book-keeping data, even in the case of the joint panel (in which the same households are included in both surveys). The consumption expenditure on non-durable goods (roughly 55 per cent of the total consumption of households) was merged in both studies using average statistical matching only - ie, by using the coefficients of consumption propensities that were calculated for the selected categories of households from the whole survey data.

Thus, comparisons of the two merging methods only take into consideration the consumption of goods and services on which data were gathered for the whole reference year. Primarily these are durable goods and housing costs, which together account for 45 per cent of total consumption expenditure.[3] Only these data were regarded as being of sufficient quality to represent the consumption expenditure of individual households. It is still important to note that these items represent that part of consumption where the frequency of purchases is low and the variation in consumption expenditure is wide.

## Average statistical merging

Average statistical merging gives very reasonable estimates of the *totals* of expenditure. The sum of imputed consumption expenditure differs by only -0.8 per cent from that of the household budget survey (Table 2). The estimates on various items of consumption are also good. The deviations of the estimates of durable goods and housing are within a range of a few per cent - on average, -1.1 per cent (Table 4). One exception to this rule is the estimate of child care fees, which exceed the estimate of the household budget survey by 23 per cent. The low frequency of households paying child care fees may explain this.

As far as the *frequency* of expenditure on various items is concerned, average statistical merging results in gross overestimates for durable consumption goods which are not purchased very frequently (Table 3). The method does not maintain

---

[3] In Table 6, the title "Durable goods and housing" refers to this group of commodities which, strictly speaking, does not exactly correspond with the title.

the original frequency within the stratum. If at least one household in the original stratum has an expenditure upon an item it means that, in the merged file, every household in the respective stratum has an average expenditure upon that item. An attempt was made to eliminate this problem. When merging the consumption data, the households had been stratified by taking into account some typical consumption pattern differences - for example, those associated with the ownership of dwelling and family type. However, it turned out that this stratification did not manage to separate the consumption items well enough. One might interpret this as suggesting that the merged consumption items represent average consumption expenditure in the longer run but not in a particular year.

## Hot rank merging

The results of the hot rank merging fit the original information even more closely than those of average statistical merging. The estimate of total consumption differs only by -0.2 per cent from that of the household budget survey (Table 2). The estimate for durable goods and housing differs by 0.5 per cent (Table 4). The estimates for various consumption items are also relatively good, although the variation in the accuracy of totals is greater than that achieved by using average statistical merging. The estimates for durable goods and housing differ within a range of -6 to +6 per cent (Table 4). One advantage of this method, as compared with the average statistical merging method, is the improved accuracy of frequencies. The frequencies of the imputed data in the income distribution survey correspond very well with both those of the joint subsample and with those of the household budget survey (Table 3, Colums 1 to 3).

One problem which affects the results from both merging methods is that the two subsamples which make up the Income Distribution Survey respondents differ in some respects. The key reason for this appears to be that the second panel had higher non-response rates, because the respondents were participating in the survey for the second time. Non-response was especially high in 1990 because it was a joint panel with the household budget survey. In line with higher non-response rates, single person households and households with very low and very high income were under-represented. This seems to hold even though the non-response bias was supposedly corrected in the estimation coefficients of these surveys (that is, the weights of households with a high non-response rate were increased and *vice versa*). Nonetheless, despite the sample differences, the imputed consumption expenditures for the two subsamples in the Income Distribution Survey are very similar.

Table 2: Some Characteristics of the Original Household Budget Survey and of the Imputed Data

| | Original HBS 1990 data | Imputed Data | | |
|---|---|---|---|---|
| | | Entire IDS | IDS, panel 1 | Joint subsample panel 2 |
| | | Hot rank merging | | |
| *Consumption expenditure, FIM mill.* | 232 400 | 232 000 | 232 700 | 231 200 |
| Number of households | 2 541 300 | 2 170 500 | 2 176 700 | 2 163 600 |
| *Durable goods and housing expenditure, FIM mill.* | 101 100 | 101 600 | 102 500 | 100 800 |
| Frequency, number of households | | 2 165 900 | 2 172 600 | 2 159 100 |
| Mean per household, FIM | | 46 900 | 47 200 | 46 700 |
| *Owner-occupied dwelling, FIM mill.* | 36 400 | 36 500 | 36 500 | 36 400 |
| Frequency, number of households | | 1 760 500 | 1 751 800 | 1 769 300 |
| Mean per household, FIM | | 20 700 | 20 900 | 20 600 |
| *Rents, FIM mill.* | 6 100 | 6 300 | 6 500 | 6 100 |
| Frequency, number of households | | 630 600 | 650 700 | 610 600 |
| Mean per household, FIM | | 10 000 | 10 100 | 10 000 |
| | | Average statistical merging | | |
| *Consumption expenditure, FIM mill.* | 232 400 | 230 600 | 230 200 | 231 000 |
| Number of households | 2 541 300 | 2 170 500 | 2 176 700 | 2 164 400 |
| *Durable goods and housing expenditure, FIM mill.* | 101 100 | 100 300 | 100 000 | 100 600 |
| Frequency, number of households | | 2 170 500 | 2 176 700 | 2 164 400 |
| Mean per household, FIM | | 46 200 | 45 900 | 46 500 |
| *Owner-occupied dwelling, FIM mill.* | 36 400 | 35 200 | 34 600 | 35 700 |
| Frequency, number of households | | 2 170 500 | 2 176 700 | 2 164 600 |
| Mean per household, FIM | | 16 200 | 15 900 | 16 500 |
| *Rents, FIM mill.* | 6 100 | 6 500 | 6 800 | 6 100 |
| Frequency, number of households | | 1 830 400 | 1 858 700 | 1 802 100 |
| Mean per household, FIM | | 3 500 | 3 650 | 3 400 |

Note: FIM = Finnish Mark

## 5. Estimates of the Impact of Value-added Tax Reform

Another way to test the two types of imputation of consumption expenditure is to examine a possible tax policy reform. In 1994, the Finnish government submitted a proposal to the Parliament to replace the former turnover tax with a value-added tax (VAT). It was a structural reform with little revenue effect. The old system was

Table 3: Percentage of Households Purchasing Various Commodities Each Year According to the Household Budget Survey and to the Imputed Data

| Commodity | HBS 1990 whole survey data | HBS 1990 joint panel data | IDS imputed data by hot ranking | IDS imputed data by average statistical merging |
|---|---|---|---|---|
| | % | % | % | % |
| Clothing | 82 | 72 | 73 | 100 |
| Footwear | 66 | 61 | 62 | 100 |
| Owner-occupied dwelling | 85 | 82 | 81 | 100 |
| Rents | 23 | 28 | 29 | 84 |
| Fuel and power | 97 | 99 | 99 | 100 |
| Summer house costs | 18 | 19 | 19 | 99 |
| Furniture | 56 | 52 | 53 | 100 |
| Household appliances | 58 | 57 | 58 | 100 |
| Transport equipment | 75 | 75 | 75 | 100 |
| Travelling | 36 | 30 | 30 | 99 |
| Durable recreation goods | 24 | 23 | 23 | 99 |
| Doctor's and hospital fees | 55 | 45 | 46 | 100 |
| Child care fees | 9 | 8 | 8 | 50 |

Table 4: Discrepancy Between Totals of Actual and Imputed Consumption Expenditure on Durable Goods and Housing

| Commodity | HBS 1990 FIM mill. | IDS / HBS hot rank % | IDS / HBS average statistical % |
|---|---|---|---|
| Clothing | 6,800 | -0.8 | 0.2 |
| Footwear | 2,400 | -2.6 | 0.8 |
| Owner-occupied dwelling | 36,400 | 0.1 | -3.5 |
| Rents | 6,100 | 3.0 | 5.3 |
| Fuel and power | 11,100 | 2.6 | -2.3 |
| Summer house costs | 1,100 | -4.0 | -0.3 |
| Furniture | 4,300 | 4.4 | 0.0 |
| Household appliances | 2,800 | 2.3 | 0.0 |
| Transport equipment | 13,100 | 1.8 | 0.2 |
| Travelling | 5,700 | -5.8 | -0.7 |
| Durable recreation goods | 5,000 | 3.8 | 3.1 |
| Doctor's & hospital fees | 5,000 | 0.0 | -2.7 |
| Child care fees | 1,400 | 6.1 | 23.2 |
| Total | 101,100 | 0.5 | -1.1 |

largely based on VAT principles but, strictly speaking, was not a VAT system. The deviations from a general VAT system mainly related to the limited taxability of services and construction and to restrictions on the deductibility of input purchases. The proposed VAT system involved a broadening of the tax base to include service and construction industries. It also changed the incidence of the general consumption tax. As the tax is rolled over to the ultimate consumer, the tax burden on private consumption is increased, even though the reform is revenue neutral.

The impact of the VAT reform on income distribution was calculated as a static estimate with the TUJA model. The tax parameters - that is, how much turnover tax and value-added tax were included in various prices for consumption goods - were estimated by means of an input-output analysis. The distribution of consumption was imputed onto the TUJA model from the household budget survey. Previously, only average statistical merging was applied in imputing the data. In the work reported here, the data matched by the hot ranking method were used as an alternative.

The impact of the VAT reform on income distribution was measured by the change in the value-added tax rate, calculated as a proportion of consumption expenditure. The reform was estimated to increase this tax rate by, on average, one per cent. This result was the same, irrespective of whether the average statistical or hot ranking method was used to impute consumption expenditure. In addition, the estimated changes for various decile groups and types of families were reasonably similar, irrespective of which of the two imputation methods was used (Table 5).

The estimated changes are not exactly identical. Hot rank merging results in a greater variation in tax rate changes. This is to be expected, given the wider variation in the amounts of consumption expenditure. Conversely, the average statistical merging method tends to equalise the impacts. Table 6 shows the effect of the different merging methods more clearly, because it displays only those consumption items that were imputed by hot ranking.

These results have a bearing on the level of disaggregation of households that can be used in the analysis of distributional impacts. It is evident, however, that the level of stratification affects the reliability of the imputed data and has to be taken into account in the disaggregation of the results.

Table 5: Percentage Point Change in the Value-added Tax Rate Due to Tax Reform, All Commodities*

| Decile group, gross income per household | All house-holds | Families with 1 child | Families with 2+ children | Pensioners | One person households under 65 | Childless couples under 65 | Others |
|---|---|---|---|---|---|---|---|
| | | | HOT RANK MERGING | | | | |
| 1 | 1.2 | - | - | 1.3 | 1.1 | - | - |
| 2 | 1.1 | 1.1 | - | 1.1 | 1.2 | 1.1 | 1.1 |
| 3 | 1.0 | 1.3 | 1.4 | 1.0 | 1.1 | 0.8 | 1.0 |
| 4 | 1.0 | 1.1 | 1.1 | 1.0 | 1.0 | 0.8 | 0.9 |
| 5 | 1.0 | 1.2 | 1.2 | 0.9 | 1.1 | 1.0 | 0.9 |
| 6 | 1.0 | 1.2 | 1.1 | 1.0 | 0.8 | 1.0 | 0.9 |
| 7 | 1.0 | 1.1 | 1.2 | 0.9 | 0.9 | 0.9 | 0.8 |
| 8 | 1.0 | 1.0 | 1.0 | 0.9 | 0.8 | 1.0 | 0.8 |
| 9 | 1.0 | 1.1 | 1.0 | 0.9 | 0.8 | 1.0 | 0.8 |
| 10 | 0.9 | 1.1 | 1.0 | 0.7 | 1.2 | 0.9 | 0.8 |
| Total | 1.0 | 1.1 | 1.1 | 1.0 | 1.0 | 0.9 | 0.8 |
| | | | AVERAGE STATISTICAL MERGING | | | | |
| 1 | 1.3 | - | - | 1.4 | 1.2 | - | - |
| 2 | 1.1 | 1.0 | - | 1.0 | 1.2 | 0.7 | 1.0 |
| 3 | 1.0 | 1.0 | 1.2 | 1.0 | 1.1 | 0.9 | 1.0 |
| 4 | 1.0 | 1.1 | 1.1 | 0.9 | 1.0 | 0.9 | 1.0 |
| 5 | 1.0 | 1.1 | 1.3 | 1.0 | 0.9 | 0.9 | 0.9 |
| 6 | 1.0 | 1.1 | 1.2 | 0.9 | 0.9 | 0.9 | 0.9 |
| 7 | 1.1 | 1.1 | 1.3 | 0.9 | 1.0 | 0.9 | 0.9 |
| 8 | 1.0 | 1.0 | 1.1 | 0.9 | 0.9 | 0.9 | 1.0 |
| 9 | 1.0 | 1.0 | 1.1 | 0.9 | 0.9 | 0.9 | 1.0 |
| 10 | 1.0 | 1.1 | 1.0 | 0.9 | 0.9 | 1.0 | 1.0 |
| Total | 1.0 | 1.0 | 1.1 | 1.0 | 1.0 | 0.9 | 1.0 |

* The mark "-" in some cells of Tables 5 and 6 indicates that the number of households in the sample is not representative.

## 6. The Redistributive Impact of Government Programs

The integrated database, including data on income, consumption expenditure, and value of welfare services, has only been in use at VATT for a short time. A previous version merely included consumption data, in addition to income. The old version

Table 6: Percentage Point Changes in the Value-added Tax Rate Due to Tax Reform, Durable Goods and Housing

| Decile group, gross income per household | All households | Families with 1 child | Families with 2+ children | Pensioners | One person households under 65 | Childless couples under 65 | Others |
|---|---|---|---|---|---|---|---|
| | | | HOT RANK MERGING | | | | |
| 1 | 1.1 | - | - | 1.2 | 0.8 | - | - |
| 2 | 1.0 | 1.2 | - | 1.0 | 1.0 | 1.1 | 1.1 |
| 3 | 0.9 | 1.3 | 1.4 | 0.9 | 0.9 | 0.7 | 1.0 |
| 4 | 0.9 | 1.1 | 1.1 | 0.9 | 0.9 | 0.8 | 0.9 |
| 5 | 1.0 | 1.2 | 1.3 | 0.7 | 0.9 | 1.0 | 0.8 |
| 6 | 1.0 | 1.1 | 1.1 | 0.8 | 0.7 | 1.0 | 0.8 |
| 7 | 1.0 | 1.1 | 1.1 | 0.7 | 0.7 | 0.8 | 0.8 |
| 8 | 0.9 | 1.0 | 1.0 | 0.7 | 0.6 | 0.9 | 0.8 |
| 9 | 0.9 | 1.0 | 1.0 | 0.6 | 0.6 | 0.9 | 0.7 |
| 10 | 0.8 | 0.9 | 0.9 | 0.5 | 1.0 | 0.7 | 0.6 |
| Total | 0.9 | 1.0 | 1.0 | 0.9 | 0.9 | 0.9 | 0.8 |
| | | | AVERAGE STATISTICAL MATCH | | | | |
| 1 | 1.2 | - | - | 1.3 | 0.9 | - | - |
| 2 | 0.9 | 1.1 | - | 0.9 | 1.0 | 0.6 | 1.0 |
| 3 | 0.9 | 1.0 | 1.2 | 0.9 | 0.9 | 0.8 | 1.0 |
| 4 | 0.9 | 1.1 | 1.1 | 0.8 | 0.8 | 0.8 | 1.0 |
| 5 | 0.9 | 1.1 | 1.3 | 0.8 | 0.8 | 0.8 | 0.9 |
| 6 | 1.0 | 1.1 | 1.3 | 0.7 | 0.7 | 0.9 | 0.9 |
| 7 | 1.0 | 1.0 | 1.3 | 0.7 | 0.8 | 0.8 | 0.9 |
| 8 | 1.0 | 0.9 | 1.1 | 0.8 | 0.8 | 0.8 | 0.9 |
| 9 | 0.9 | 0.9 | 1.0 | 0.7 | 0.8 | 0.8 | 0.9 |
| 10 | 0.9 | 0.9 | 0.9 | 0.7 | 0.8 | 0.8 | 0.9 |
| Total | 0.9 | 1.0 | 1.1 | 0.9 | 0.8 | 0.8 | 0.9 |

has been used in analysing the distributional impact of the value-added tax reform; the impacts of budget proposals where several measures were proposed concurrently; and the impact of several reforms concerning one single program at a time. The new integrated database makes it possible to estimate the combined impact of amendments to transfer rules, welfare services, and direct and indirect taxation.

Relatively little is known of the total tax rates and progressivity of the total tax system. The same is true of cash transfers and welfare services and their mutual interaction. The following preliminary analysis suggests that indirect taxes and welfare services change the picture of the distribution of welfare, compared with that of the distribution of cash income only.

First, transfers-in-kind and indirect taxes are of major significance. In Finland indirect taxes account for 30 per cent of all taxes paid by households and transfers-in-kind for 40 per cent of all transfers received by households. Second, the progressivity of the total tax system is lower than that of income taxes, because indirect taxes are regressive (Table 7). Third, welfare services have a greater redistributional impact than do cash transfers (Table 8).

Table 7: Tax Paid as a Proportion of Gross Income, by Decile of Equivalent Disposable Income

| Decile of equivalent disposable income | Direct tax as % of gross income | Indirect tax as % of gross income | Total tax as % of gross income |
|---|---|---|---|
| 1 | 8.8 | 16.4 | 25.2 |
| 2 | 12.0 | 14.9 | 27.0 |
| 3 | 16.0 | 12.8 | 28.8 |
| 4 | 19.1 | 12.4 | 31.5 |
| 5 | 21.1 | 11.9 | 33.0 |
| 6 | 23.7 | 11.3 | 35.0 |
| 7 | 25.5 | 10.5 | 36.0 |
| 8 | 27.3 | 10.9 | 38.2 |
| 9 | 29.5 | 8.9 | 38.3 |
| 10 | 33.7 | 7.9 | 41.6 |
| Total | 24.9 | 10.7 | 35.7 |

Table 8: Cash Transfers and Welfare Services as a Proportion of Gross Income, by Decile of Equivalent Disposable Income

| Decile of equivalent disposable income | Cash transfers as % of gross income | Welfare services as % of gross income | Total transfers as % of gross income |
|---|---|---|---|
| 1 | 53.5 | 41.9 | 95.4 |
| 2 | 49.4 | 29.9 | 79.3 |
| 3 | 39.0 | 23.3 | 62.3 |
| 4 | 29.7 | 22.2 | 51.9 |
| 5 | 26.8 | 18.3 | 45.2 |
| 6 | 21.0 | 15.4 | 36.3 |
| 7 | 16.6 | 11.9 | 28.4 |
| 8 | 15.1 | 9.9 | 25.0 |
| 9 | 11.4 | 8.0 | 19.5 |
| 10 | 11.4 | 4.8 | 16.2 |
| Total | 21.4 | 14.2 | 35.6 |

In summary, when households are ranked by disposable income per equivalent adult, the distribution of final income (ie. welfare services added and indirect taxes subtracted from disposable income ) is clearly more even than that of disposable income (ie. gross income minus income taxes). However, a re-ranking of households changes this conclusion. If households are ranked by final income per equivalent adult, which could be regarded as representing their final welfare position, the distribution of final income is slightly more uneven than that of disposable income (Table 9).

Table 9: Share of Disposable Income and of Final Income, by Decile

| Decile of equivalent disposable income | Share of disposable income | Share of final income | Decile of equivalent final income | Share of disposable income | Share of final income |
|---|---|---|---|---|---|
| 1 | 4.2 | 5.2 | 1 | 4.1 | 3.5 |
| 2 | 5.9 | 6.6 | 2 | 5.9 | 5.6 |
| 3 | 7.1 | 7.7 | 3 | 7.0 | 6.9 |
| 4 | 8.5 | 9.2 | 4 | 8.3 | 8.4 |
| 5 | 9.3 | 9.7 | 5 | 9.4 | 9.6 |
| 6 | 10.5 | 10.6 | 6 | 10.4 | 10.5 |
| 7 | 11.5 | 11.2 | 7 | 11.3 | 11.6 |
| 8 | 12.1 | 11.4 | 8 | 12.4 | 12.8 |
| 9 | 13.6 | 12.8 | 9 | 13.9 | 14.1 |
| 10 | 17.2 | 15.7 | 10 | 17.5 | 17.0 |
| Total | 100.0 | 100.0 | Total | 100.0 | 100.0 |

The results described above invite further questions about the interactions between cash transfers and welfare services, on one hand, and between direct and indirect taxes on the other. The combined effects of the entire tax and transfer system are also of interest. Further interest arises because of the economic situation in Finland. Due to the deepest recession ever experienced by Finland in the first half of the 1990s, excluding the war years, the public sector deficit has grown to enormous proportions and the indebtedness of the public sector is rapidly reaching an alarmingly high level. The structures of welfare policy programs have come into focus in a totally new way, with the need to improve the efficiency and effectiveness of the transfer system being obvious.

Behind the difficulties of the public economy lie a long period of economic growth and the rapid development of a welfare state. The average annual growth rates of GDP in the 1970s and 1980s were 3.6 per cent and 3.2 per cent respectively. In the same periods, public expenditure grew by 5.4 per cent and 4.5 per cent and cash transfers by 7.7 per cent and 6.6 per cent

During the long period of rapid economic growth, a great number of welfare programs were implemented. Usually they were planned and implemented one by one, the new ones supplementing the existing ones. In general, the previous programs were not changed when a new program was introduced. This meant that the objectives of various programs were set narrowly, without considering the general aim of welfare policy directed to certain groups of households, and without fully appreciating the contribution of the new programs. Because of this historic development, there is probably an excessively wide range of welfare policy programs, such as cash transfers, welfare services, tax expenditures and indirect financial supports. Furthermore, these programs often have incoherent objectives.

In analysing the complex structures of welfare policy programs one needs to have a broad framework including all forms of redistribution policy. In addition, the objectives of specific welfare policy programs in their entirety, directed to various groups of households, need to be reinvestigated. In such work there is ample room for methodologies like the integrated database outlined above.

## 7. Conclusion

There is an obvious need in Finland for a better understanding of the entire range of welfare policies and their effects. Statistics on cash transfers and direct taxes are widely available, but information on the distribution of indirect taxes and welfare services is relatively scarce. Analyses of the combined effects of tax and transfer systems, as well as of the targeting and effectiveness of cash transfers and welfare services, are very scarce. A new motivation for this kind of analysis has arisen, now that many so-called welfare states are in serious economic crisis and have large budget deficits. Reforms aimed at reducing the total costs of welfare policies and at improving their effectiveness have been demanded.

To reach a better understanding of the entire tax and transfer system, one needs an integrated database on the distribution of income, cash transfers, welfare services and direct and indirect taxes. Often such databases are not readily available. Methods of integrating data from different samples have to be examined.

The above study tests two methods of merging consumption data from the household budget survey onto the income distribution data. One of the methods matches data by individual merging, so-called hot rank merging, and the other by average statistical merging. The results suggest that better accuracy can be reached by hot rank merging than by average statistical merging, although the estimated totals are good in both cases. Hot rank merging is clearly superior in the accuracy of the frequency of expenditure. Average statistical merging loses a great deal of variation and overestimates the frequencies. Nevertheless, if due attention is paid to the purpose of use and the desired disaggregation of results, the results suggest that these two different merging methods can both be used successfully to provide sufficiently reliable overall results.

## References

British Department of Social Security (1992), 'Recent developments in microsimulation in the British Department of Social Security', Paper presented to the Helsinki Seminar on Microsimulation Models, Helsinki, March.

Cohen, M.L. (1991), 'Statistical matching and microsimulation models', in: C.F. Citro and E.A. Hanushek (eds.), *Improving Information for Social Policy Decisions: The Uses of Microsimulation Modelling (Vol. II): Technical Papers*, National Academic Press, Washington DC.

Djerv, K. and M. Lindqvist (1993), 'Kotitaloustiedustelu 1990, Laatuselvitys', (Household Budget Survey 1990. Quality Report), *Working papers, 3/93*, Statistics Finland, Helsinki.

Ford, B.L. (1983), 'An overview of hot-deck procedures', in:. W.G. Madow, I. Olkin and D.B. Rubin, (eds.) *Incomplete Data in Sample Surveys (Vol. 2): Theory and Bibliographies*, Academic Press, New York.

Mot, E.S. (1992), *Survey of Microsimulation Models: Inventory and Recommendations*, Social Security Research Committee, UUGA, The Hague.

Paass, G. (1986), 'Statistical match: Evaluation of existing procedures and improvements by using additional information', in: G.H. Orcutt, J. Merz and H. Quinke (eds.) *Microanalytic Simulation Models to Support Social and Financial Policy*, North Holland, Amsterdam.

Radner, D.B. (1981), 'An example of the use of statistical matching in the estimation and analysis of the size distribution of income', *Review of Income and Wealth*, Vol. 27, No. 3.

Salomäki, A. (1992), 'Estimating the distributional effects of value-added tax reform in Finland', Paper presented to the Helsinki Seminar on Microsimulation Models, Helsinki, March.

Woodbury, M.A. (1983), 'Statistical record matching for files', in: W.G. Madow and I. Olkin (eds.), *Incomplete Data in Sample Surveys, Vol. 2, Proceedings of the Symposium*, Academic Press, New York.

Klaas de Vos
M. Asghar Zaidi

Erasmus University
The Netherlands

# Chapter 6: The Use of Microsimulation to Update Poverty Statistics Based on Household Budget Surveys: A Pilot Study for the United Kingdom

## Abstract

*There is great demand for up-to-date poverty estimates. This chapter evaluates the use of microsimulation to update poverty statistics derived from the Household Budget Surveys of the United Kingdom. It appears that ageing the 1985 survey data from 1985 to 1988 by reweighting produces poverty estimates markedly different to the actual 1988 poverty estimates. However, also applying differential increases in incomes results in estimated poverty statistics which are quite close to the poverty statistics obtained by making use of the actual 1988 data. Although the methodology has a number of limitations, updating the statistics appears to be preferable to using the base year statistics when up-to-date poverty estimates are required.*

## 1. Introduction

In order to be valuable for policy makers, poverty statistics should be as up-to-date as possible. If the figures are derived from Household Budget Survey data (Hagenaars et al., 1994) problems occur in those countries where such surveys are not held every year - or when the figures become accessible to researchers only after a considerable delay (as is usually the case). Hence, it would be useful if poverty statistics could be updated, making use of independent information for the most recent period. De Vos (1991) showed that one possible approach to updating

* Research for this paper was carried out at Erasmus University, Rotterdam, in the framework of the project 'Living Conditions of the Least Privileged in the European Community'. It was commissioned by Eurostat and supervised by the late Professor Aldi Hagenaars. The final revision was completed at the Economics Institute, Tilburg. Responsibility for the statistics in this paper rests solely with the authors. The results do not necessarily reflect the views of the National Governments or the European Commission.

poverty statistics, based on regression of the parameters of the expenditure distribution on a number of macro-economic variables, did not produce updated poverty statistics of a better quality than the non-updated poverty statistics. De Vos also suggested that a possible alternative would be to use microsimulation in order to obtain a (simulated) microdata set for the year for which the poverty statistics are needed.

This chapter presents a pilot study on the feasibility and usefulness of microsimulation as a means of updating poverty statistics. Section 2 provides a short description of the methodology. Section 3 discusses some limitations of this pilot study. Section 4 presents the aggregate results of the simulations and Section 5 presents some disaggregated statistics. Section 6 presents conclusions. Details on the methodology and the figures used in the computations are presented in Appendix A.

The poverty statistics to be updated are the poverty rates obtained by assuming 40, 50 and 60 per cent of average equivalent income and expenditures as poverty cut-offs for the microdata derived from household surveys. The equivalence scale used here is the modified OECD equivalence scale which assumes that for every additional person of 14 years of age and older, and for every person younger than 14, a household would need 0.5 and 0.3 times the resources of a one person household to reach the same welfare level. (For the merits and problems of this poverty concept see Hagenaars et al., 1994).

Based on the pioneering work of researchers such as Guy Orcutt, microsimulation models have now been in use for over 30 years, especially for quantifying the effects of different social security and taxation policy measures (for reviews see, eg. Haveman and Hollenbeck, 1980; Orcutt et al., 1986; and Brunner and Petersen, 1990). One popular example of the microsimulation approach is to use a sample of micro-units (eg. persons, households, tax units or firms) of a base year and apply a model to produce a new, simulated set of micro-units for the target year. The model should account for relevant changes during the simulation period. The simulated sample produced in this way is assumed to be representative of the total population in the target year (in the same way that the actual sample is representative of the base year). The simulated sample can then be used to produce statistics for the target year on the basis of information available for the base year. In addition, one can perform winner-loser analysis, by showing which groups have benefited and which have suffered from a given (or expected) change in the socioeconomic system of a country.

The updating process can range from a relatively simple to a relatively complicated one. For example, a simple option is to simulate new tax or benefit rules and leave the socio-demographic characteristics of the original sample untouched. Another option is static ageing, where the weights attached to each record in the original sample are changed to reflect socio-demographic changes and incomes are uprated to capture income changes since the original survey was conducted. More complicated options include dynamic ageing where the attributes of each micro-unit are changed according to certain behavioural assumptions for the group to which that unit belongs (Hellwig, 1990).

The two methods can also be mixed. For instance, discrete variables, such as demographic characteristics, can be simulated using static ageing and continuous variables, such as income, can be modelled using dynamic ageing. One key distinction between the two methods lies in the treatment of variables which are not included in the weight table of static ageing and not modelled in the dynamic modelling. In static ageing the distribution of these variables is affected in an uncontrollable way through the possible correlation with the variables included in the weight table. In dynamic ageing these variables remain unchanged (for some discussion of this, see Bradbury, 1990a).

## 2. Methodology

The purpose of this study is to carry out a pilot exercise using the Household Budget Survey data for the United Kingdom. Broadly speaking, two different microsimulation approaches are tested. The first approach amounts to a reweighting of the observations based on information from the target year, implicitly assuming that the distribution of resources within and between the subgroups upon which the reweighting is based does not change. The second approach also takes into account the fact that the changes in household resources may be different within - and especially between - subgroups.

Ideally, microsimulation should transfer the microdata for the base year to microdata for the target year. Obviously, a large number of events between the two periods may affect the information on the households involved. Similarly, a large subset of these events may affect poverty status: such events include births, deaths, marriages, divorces, leaving school, (re)entering the labour market, leaving the household, becoming unemployed, getting a better job or a higher salary, (early) retirement, acquiring a new dwelling, changes in the income tax rates, and shifts in tastes and habits. The dynamic approach in which all or most of these events are simulated requires a huge research effort and a large amount of information, or a large number of assumptions. For instance, to simulate a decrease in unemployment, it is necessary to know or to formulate assumptions about which persons are likely to find a job. It is clear that a large number of factors will play a role here, both with respect to the characteristics of the persons involved (education, work experience, wage demands) and with respect to the labour market (local, regional, national). Many factors will not be known and, in that case, the researcher will have to rely on more or less arbitrary assumptions. In particular, when two or more related events have to be simulated (eg. the employment status of different household members), or when the interaction between individuals (marriage, divorce) has to be taken into account, the number of possibilities soon becomes astronomical.

A less sophisticated approach is to simulate certain developments by simple reweighting on one or more variables which are crucial for the phenomena in which one is interested. For instance, while demographic events, such as the birth of a baby, may affect the poverty status of a household (eg. by affecting the income per equivalent adult even if the actual income remains, on average, the same) the demographic situation in a country will not change very fast in a limited period of time. Therefore, this kind of event will be of less relevance if one is interested in the development of poverty statistics.

On the other hand, unemployment at the national level may change much more quickly, and this is likely to considerably affect poverty statistics. One possible way to simulate this is to reweight the unemployed and employed persons in the sample, so that the (reweighted) percentage of unemployed matches that of the target year, as obtained from independent information. The implicit assumption underlying this approach is that the distribution of resources within the groups of unemployed and employed remains the same. The fact that relatively well-off unemployed may be more successful in securing a job, but may end up in the lower part of the distribution of resources of the employed, is not taken into account. This chapter tests a number of possible variables on which to reweight the 1985 survey data and compares the resulting poverty statistics with poverty rates derived from the actual 1988 data.

As mentioned above, reweighting does not affect the distribution of resources within certain subgroups. Nor does it affect the differences *between* different subgroups. It is possible that certain differential developments in the resources of different groups could affect poverty statistics, especially if they resulted in changes in overall inequality. To the extent that information on these differential developments is available, or that plausible assumptions can be formulated, a simple way to simulate them is to proportionally increase the resources of all households in a subgroup. Once again, this would not affect the shape of the distribution within a subgroup - but it would affect the shape of the differences between different subgroups.

This chapter includes several approaches to approximating the *income* increase for different subgroups and the consequences for poverty statistics based on an income cut-off. However, information on differential changes in *expenditures* in different subgroups is not available. Therefore, to obtain simulated poverty statistics in terms of expenditures, certain assumptions about the relationship between income and expenditures must be made (eg. that the estimated relationship between income and expenditures in the base year also holds in the target year) (see Appendix A for a detailed description of the alternative methods tested).

## 3. Limitations

There are a number of limitations to this pilot study. First, the study is limited to one country (the United Kingdom) for which there are survey data for two years (1985 and 1988). Obviously, the results do not allow definite conclusions about the usefulness of the approach in other countries. For instance, because income is sometimes heavily under-reported in household surveys, the quality of the simulated results may be limited.

Second, highly sophisticated modelling of demographic changes and changes in resources of households is beyond the scope of this project. As mentioned above, the study tests the impact of reweighting and the use of fairly simple uprating assumptions about changes in income and expenditures.

Third, the study focusses on those assumptions which are relevant for poverty statistics. There is no point in deriving detailed projections of changes in the shape of the distribution in the higher income brackets, because the results would not influence the poverty rates.

Fourth, in this chapter simulated results for 1988 are compared with actual results from the 1988 Household Budget Survey (HBS). Changes in the actual HBS-figures are not always consistent with the macroeconomic figures (ie. National Accounts) upon which some of the income uprating assumptions are based (see de Vos and

Zaidi, 1993). Thus the exercise should not be evaluated only on the basis of this comparison with the HBS-outcomes. Even if the microsimulation results do not differ much from the HBS-outcomes, it is possible that they might both be biased as a result of sampling errors, measurement errors and/or the application of incorrect assumptions. If they do differ considerably, the HBS results are not necessarily to be preferred, if they suggest patterns inconsistent with independent macro-figures.

Fifth, the quality of the microsimulation results can never be better than the data and assumptions upon which they are based. As the base-year HBS is subject to sampling error as well as possible measurement error, and the assumptions about changes in the distribution of income and expenditures are rather crude, the error margins around the microsimulation results are wide. At most, one could expect to obtain some indications about the trend and the extent to which the overall poverty incidence has changed between the base year and the target year. When results based on different assumptions contradict each other, it would probably be better to use the base year figures only. In that case there would be no solid evidence for assuming that poverty changed considerably between the base year and the target year.

## 4. Results

Table 1 gives the overall poverty percentages resulting from several versions of the approaches sketched in Section 2 and compares them with the actual statistics based on the HBS-data of 1985 and 1988. Simulation 1 uses household weights which ensure that the frequency distributions of records by household type and region of the updated HBS of 1985 matches the distribution shown in the 1989 Labour Force Survey (the 1988 survey being unavailable)[1]. In this simulation, *the incomes and expenditures of individual households are kept unchanged*. (For the purpose of deriving poverty statistics this is equivalent to assuming that all amounts have increased by the same percentage.) The 1985 poverty rate when the poverty line was set at 50 per cent of average expenditure was 14.3 per cent (Table 1). Using the actual 1988 HBS data, the 1988 poverty rate for this poverty line was 17.2 per cent. Reweighting the 1985 HBS data to account for changes by household type and region (Simulation 1) produced a 14.3 per cent poverty rate - suggesting that reweighting alone would not produce reliable poverty estimates in the target year.

---

[1] Simulations 2 to 5 are included only to show the impact upon poverty rates of changes in the socio-demographic variables typically used in reweighting matrices. They are based on reweighting matrices calculated from the actual 1988 HBS, and therefore could not actually be used to update the 1985 HBS to the 1988 world.

Simulations 6 to 9 add increasingly detailed information on differential developments of income between subgroups, to the changes implied by reweighting. Simulation 6 uses the LFS weights computed for the 1988 HBS with respect to region and household type (as in Simulation 1), but also takes into account that income has increased faster for working persons than for non-working persons. As a first approximation it is assumed that the incomes of households with a working head have increased by 18 per cent between 1985 and 1988, but that incomes of households with a non-working head (expressed in 1985 prices) have not increased (see Appendix A for the basis of this approximation).

For the results with respect to expenditures the first set of approaches, (a), are based on the assumption that the relation between income and expenditures estimated for 1985 also holds for 1988. In the second set of approaches, (b), expenditures are computed assuming that the savings rate for each individual household did not change (see Appendix A).

As Table 1 indicates, using the more sophisticated reweighting and uprating procedures of Simulation 6 results in 50 per cent of expenditure estimated poverty rates for 1988 which are fairly close to those derived from the actual 1988 date. Similarly, moving to the 50 per cent of income poverty line, Simulation 6 produces an estimated poverty rate for 1988 which is very close to the actual 1988 poverty rate as shown in the 1988 HBS data (21.6 and 22.4 per cent respectively).

Simulation 7 to 9 essentially represent other reweighting/uprating combinations, with Simulation 8 also capturing the changes in tax rates between 1985 and 1988 (Appendix A). Generally speaking, such changes appear to make only a minor difference to the estimated poverty rates.

The picture that emerges from Table 1 is quite clear. Reweighting (Simulations 1 to 5) hardly affects the simulated poverty rates. This implies that the changes in poverty incidence between 1985 and 1988 cannot, in large part, be attributed to changes in the composition of the population according to the characteristics used in reweighting. Still, these were the most notable changes that occurred with respect to the characteristics available in the HBS (as apparent from the intertemporal comparison presented in De Vos and Zaidi, 1993). Thus, it appears that differential changes in the levels of resources of different households are the underlying cause of the increase in poverty rates during the three years. That conclusion is confirmed by the results of simulations 6 to 9, in which these changes have been implemented in several ways. In those cases the poverty rates based on the income poverty line rise considerably to a level fairly close to the one found from the actual HBS-data for 1988. The poverty rates based on the expenditure also rise, but are rather sensitive to the way in which the savings rate is assumed to change. The rise is quite small when the relationship between income and expenditures estimated for 1985 is also assumed to hold in 1988. It is higher when the savings rate of every individual household is assumed to remain constant.

The overall poverty rates are not very sensitive to the method by which the differential income change is applied. Further refinement (eg. distinguishing tax units and applying different wage increases for manual and non-manual workers in different income groups) would probably not make much difference either. The results may be more sensitive to a differential treatment of persons who depend on different forms of social security. This cannot be checked because of the lack of available data to compute the entitlement to social security benefits in detail. Moreover, this kind of data is also lacking for most countries for which the microsimulation would actually have to be used for updating the poverty statistics (that is, those countries where a HBS is not held every year).

As noted above, the expenditure-based poverty statistics are rather sensitive to the assumptions made about the savings rate. Neither approach (a) nor (b) takes into account the apparent decrease in the actual savings rate between 1985 and 1988 suggested by the National Accounts. The National Accounts figures do not show how this decrease was distributed among different income groups. A proportionate decrease for all households would not affect the poverty rates but that might not accurately describe reality. Using the actual savings rates in different income groups according to the 1988 HBS might give some insight - but this survey actually finds an increase in the average savings rate in the microdata!

Because the increase in disposable income used in the simulations (based on the National Accounts figures) and the increase found by comparing the HBSs of both years differ considerably, the fact that the simulated overall poverty rates move in line with the poverty rates from the actual 1988 HBS is no guarantee of their usefulness. This situation is aggravated by the lack of knowledge about how the savings rate was affected in different income groups. Still, the result that income has risen more for working than for non-working persons, as shown by comparing both HBSs, is corroborated by independent data - and the resulting simulated poverty figures seem quite plausible.

Table 1: Average Equivalent Expenditures and Income (£/year) and Poverty Rates, Based on Actual HBS Data for 1985 and 1988, and Results of Microsimulation*

| | Average Expenditure (1985 prices) | Expenditure poverty rates | | | Average Income (1985 prices) | Income poverty rates | | |
|---|---|---|---|---|---|---|---|---|
| | | 40% | 50% | 60% | | 40% | 50% | 60% |
| | £ | | | | £ | | | |
| Actual data: | | | | | | | | |
| HBS-1985 | 5108 | 4.6 | 14.3 | 25.6 | 5502 | 3.1 | 13.1 | 27.7 |
| HBS-1988 | 5806 | 7.3 | 17.2 | 29.7 | 6630 | 9.0 | 22.4 | 32.2 |
| Reweighted data: | | | | | | | | |
| 1. | 5132 | 4.6 | 14.3 | 25.8 | 5529 | 3.2 | 13.4 | 27.8 |
| 2. | 5224 | 5.0 | 14.6 | 26.0 | 5631 | 3.4 | 14.2 | 27.9 |
| 3. | 5196 | 4.6 | 13.8 | 25.1 | 5616 | 3.4 | 13.6 | 26.9 |
| 4. | 5081 | 4.4 | 14.0 | 25.4 | 5473 | 3.0 | 13.1 | 27.5 |
| 5. | 5254 | 5.0 | 14.6 | 26.1 | 5649 | 3.5 | 14.3 | 27.9 |
| Differential income and expenditure changes | | | | | | | | |
| 6a. | 5243 | 5.6 | 16.0 | 27.8 | 6261 | 6.8 | 21.6 | 31.6 |
| b. | 5791 | 8.4 | 18.7 | 30.0 | | | | |
| 7a. | 5233 | 5.4 | 15.3 | 27.0 | 6107 | 5.4 | 19.7 | 30.4 |
| b. | 5643 | 7.2 | 17.6 | 29.2 | | | | |
| 8a. | 5250 | 5.4 | 15.5 | 27.4 | 6198 | 5.9 | 20.3 | 30.7 |
| b. | 5730 | 7.6 | 17.9 | 29.4 | | | | |
| 9a. | 5293 | 6.0 | 16.5 | 28.2 | 6448 | 6.8 | 20.4 | 30.4 |
| b. | 5967 | 8.3 | 18.5 | 28.9 | | | | |

* Expenditures include imputed rent and housing benefits but exclude mortgage payments. Income is disposable household income and also includes imputed rent and housing benefits. Both are amounts per equivalent adult (modified OECD scale), expressed in 1985 prices.

## 5. Disaggregation

The limitations of the simulated results also translate to the group-specific poverty statistics, which lack a firm basis in view of the aggregate nature of the exogenous information used for the simulations. Still, it is of interest to see whether or not trends in the composition of the poor, as derived from the actual figures, can also be simulated. Only households with and without working members are differentiated. This distinction is highly relevant to poverty incidence and can also be expected to show the largest effects from the differential income changes applied in the simulations.

Table 2 gives the results for the actual HBSs of 1985 and 1988 and of the simulations for households with and without members in paid employment. Starting with approaches 1 to 5, reweighting has small effects not only on the overall poverty rates but also on the composition of the poor population. In all cases, the poverty rates in the two groups are much closer to the figures based on the 1985 HBS than to those based on the 1988 HBS.

Including the effects of differential income changes (approaches 6 to 9), the increase in the income-based poverty rate for households without working members is to a large extent replicated. However, the (much smaller) increase in the poverty rate for households with working members is not reproduced very well for the income

Table 2: Overall Poverty Rates and Poverty Rates for Households With and Without Members in Paid Employment*

| | Expenditure based poverty (50% cut-off) | | | | | Income based poverty (50% cut-off) | | | | |
|---|---|---|---|---|---|---|---|---|---|---|
| | 1 | 2 | 3 | 4 | 5 | 1 | 2 | 3 | 4 | 5 |
| Actual data | | | | | | | | | | |
| HBS '85 | 14.3 | 5.2 | 30.8 | 35.6 | 76.7 | 13.1 | 4.1 | 29.4 | 35.5 | 79.6 |
| HBS '88 | 17.2 | 6.0 | 38.6 | 34.4 | 77.3 | 22.4 | 6.3 | 53.1 | 34.4 | 81.5 |
| Simulated 1988 data - reweighting only | | | | | | | | | | |
| 1. | 14.3 | 5.1 | 31.0 | 35.7 | 77.1 | 13.4 | 4.0 | 30.3 | 35.6 | 80.6 |
| 2. | 14.6 | 5.2 | 32.3 | 34.6 | 76.6 | 14.2 | 4.5 | 32.7 | 34.5 | 79.4 |
| 3. | 13.8 | 5.2 | 31.0 | 33.4 | 75.1 | 13.6 | 4.2 | 32.4 | 33.4 | 79.3 |
| 4. | 14.0 | 4.9 | 30.2 | 35.8 | 77.3 | 13.1 | 3.9 | 29.4 | 35.8 | 80.6 |
| 5. | 14.6 | 5.2 | 32.2 | 34.6 | 76.3 | 14.3 | 4.6 | 32.7 | 34.5 | 79.1 |
| Simulated 1988 data - reweighting and uprating | | | | | | | | | | |
| 6a. | 16.0 | 6.6 | 33.0 | 35.6 | 73.6 | 21.6 | 4.3 | 52.9 | 35.6 | 87.1 |
| 6b. | 18.7 | 4.6 | 44.0 | 35.6 | 84.0 | | | | | |
| 7a. | 15.3 | 6.2 | 32.9 | 34.2 | 73.3 | 19.7 | 4.6 | 48.9 | 34.2 | 85.0 |
| 7b. | 17.6 | 5.4 | 41.3 | 34.2 | 80.0 | | | | | |
| 8a. | 15.5 | 6.5 | 33.0 | 34.2 | 72.5 | 20.3 | 4.6 | 50.7 | 34.2 | 85.2 |
| 8b. | 17.9 | 5.3 | 42.2 | 34.2 | 80.4 | | | | | |
| 9a. | 16.5 | 7.0 | 34.7 | 34.2 | 71.9 | 20.4 | 5.3 | 49.6 | 34.2 | 83.0 |
| 9b. | 18.5 | 6.2 | 42.1 | 34.2 | 77.9 | | | | | |

* Column 1: overall poverty rate
Column 2: poverty rate for households with working members
Column 3: poverty rate for households without working members
Column 4: percentage of households without working members
Column 5: percentage of poor households without working members

poverty line (although many of the results for the expenditure poverty line produced by Simulations 6 to 9 appear very close to the actual estimates). In the case of the income poverty line, this means that the simulated results overestimate the proportion of poor households who have no working members. The overestimation is smallest in approach 9, which uses the most detailed information as input for the simulation.

The expenditure-based results of approaches 6 to 9 again show the sensitivity to the assumption on savings. In almost all cases the poverty rates in the subgroups for 1988 fall between the results according to the two assumptions (a) and (b). However, the differences for households with and without working members have different signs. When a constant relationship between income and expenditures is assumed, poverty is overestimated for households with working members, but underestimated for households without working members. The result is reversed when a constant savings rate is assumed. As a result, the share of poor households with non-working members is underestimated in the first case, but overestimated in the second. Once again, this relationship between income and expenditures appears to be crucial. In this particular case one may assume that the `actual' relationship will be somewhere between the two approaches. However, when the simulation approach is actually used to update poverty statistics, there may not be information on changes in the relationship between expenditures and income. Moreover, there may not be much information about income changes, unlike in the UK. These limitations should be borne in mind when conclusions are drawn for updating poverty statistics in other countries.

## 6. Conclusion

Given the error margins which surround the poverty rates for the base year and the arbitrariness of a number of assumptions used for this microsimulation, considerable care should be taken in interpreting updated poverty statistics obtained in this way. As mentioned earlier, probably the most one can obtain from this type of microsimulation is an indication of whether or not the poverty rates are subject to change as a result of relevant macroeconomic developments and some insight into the direction and the extent of these changes. Despite this, the example of the UK studied in this paper suggests that the simulated poverty statistics should be preferred to those resulting from the assumption of no change. In particular, differential increases in income in different subgroups appear to have affected the poverty rate. Again, that is no guarantee that this would also hold in other countries.

**APPENDIX A**

# Details of the Data Used for the Simulations

## Approach 1: Reweighting based on LFS 1989

The first approach uses household weights which ensure that the reweighted cross-tabulation of *household type* and *region* matches with the 1989 Labour Force Survey. Lacking the 1988 LFS data, it is assumed that the distributions of these variables did not change substantially between 1988 and 1989.

## Approaches 2,3,4: Reweighting based on variables from HBS 1988

Approach 2, 3 and 4 use frequency distribution of variables obtained from the 1988 HBS to compute weights. Because the frequency distribution according to *socioeconomic category* of the head of the households has changed considerably between 1985 and 1988, approach 2 uses the HBS distribution of 1988 for this variable. Likewise, approach 3 uses the *type of tenure*, and approach 4 uses the *age* of the head. Obviously, this approach is not feasible when the simulations will actually be used to update poverty statistics. These approaches are only included to find out to what extent changes in the frequency distributions of these variables are decisive in determining the changes in the poverty rates.

## Approach 5: Reweighting Combining Variables From LFS 1989 and HBS 1988

In approach 5, reweighting ensures that the reweighted cross-tabulation of *region* and *socio-economic category* of the head matches with the 1988 HBS, after the latter has already been reweighted with the weights based on the 1989 LFS cross-classification of *household type* and *region*.

## Approach 6: LFS Weights, Differential Income Increases and Assumptions About Savings

Approach 6 uses LFS weights as in approach 1, but takes into account that incomes changed differently for working persons and persons dependent on social security. As a first approximation, the incomes of households with a non-working head are kept constant (in 1985 prices), and the incomes of households with working heads are increased by 18 per cent. These increases are based on the fact that, according to the National Accounts, social security benefits and other grants from general government have roughly moved with inflation, while the balance of personal disposable income has increased by about 18 per cent (in real terms). This simple approach ignores the fact that households with a working head may also benefit from social security, and that households with non-working heads can have other sources of income. Still, comparing the results with more detailed approaches, the study undertaken here is useful because this approach may have to be used in countries where information on income components is lacking in the HBS.

To obtain expenditures (approach (a)), we estimate the relation between the savings-rate $s$ and the log of income per equivalent adult $y$ for 1985:

$$s = -5.88 + 0.69 \log(y)$$

Substituting the simulated income of 1988 and adding the residual of the regression (assuming that the deviation from the estimated relation remains the same for every household) we compute the savings rate for 1988 and thus obtain expenditures. On average, this results in an increase in the savings rate.

In approach (b) we assume that the savings rate remains constant for every household. Although the 1988 HBS results suggest an increase in the average savings rate, National Accounts figures imply the reverse. Hence, we might also apply decreased savings rates. A proportional decrease would affect neither the shape of the distribution nor the resulting poverty statistics.

### Approaches 7,8: Weights Based on External Information, Differential Income Increases

Approach 7 uses weights derived from the changes in the number of persons in different socioeconomic categories (CSO, 1992), assuming that the number of heads of household in each category changed proportionally to the number of persons (Table A.1). The numbers of manual and non-manual workers are changed in proportion to the numbers of employees in the production and construction industries and in the service industries, respectively. No distinction is made between self-employed in agriculture and non-agriculture. The number of homemakers and others with non-paid activity is assumed constant. Obviously, a more exact basis for these weights would be useful.

The income changes are differentiated according to socioeconomic category of the head. Gross income for households headed by manual and non-manual workers is increased by the average increase in the wage rate, differentiated between these two groups (as obtained from CSO (1992), and corrected for inflation). Gross income for households headed by the self-employed is increased by the increase in the per capita income from self-employment, derived from CSO (1992). The tax rate is kept constant. For households headed by non-working persons income is again assumed to be constant. The figures used in this approach are given in table A.1.

Approach 8 only differs from approach 7 in that income tax is recomputed in view of the decrease in the maximum marginal tax rate from 60 per cent in 1985 to 40 per cent in 1988. The tax free allowance (the part of income over which no income tax has to be paid) in 1985 is approximated on the basis of the actual tax paid by the households in 1985, and the tax rates of 1985. It is increased by 4 per cent for all households to obtain the tax free allowance for 1988. The tax rates for 1988 (averaged from before and after the decrease in the tax rates) are then applied to the resulting taxable income. No attempt is made to distinguish tax units or to differentiate the changes in the tax free allowances.

Table A.1: Weights and Income Increases Used in Simulation Approaches 7 and 8

| | weight | growth factor for gross income |
|---|---|---|
| Households headed by: | | |
| manual workers | 0.995 | 1.085 |
| non-manual workers | 1.078 | 1.157 |
| self-employed | 1.122 | 1.174 |
| unemployed | 0.762 | 1.000 |
| retired persons | 1.036 | 1.000 |
| disabled/sick persons | 1.142 | 1.000 |
| homemakers/others | 1.000 | 1.000 |

### Approach 9: Weights Based on External Information, Differentiated Income Changes Per Income Component

Approach 9 uses the same weights as approaches 7 and 8 but applies different income changes to different components of gross income. Again, the changes are approximations derived from National Accounts figures and from wage rates presented in CSO (1992). Table A.2 gives the growth factors used for the different components of income

Income tax is computed as in approach 8. Obviously, in reality more differences will have occurred in the growth of different income components. Wages have increased more for higher income groups, self-employment income has developed differently for farmers and non-farmers, and social security benefits have been affected by different rules and regulations with respect to entitlements etc. No further attempts have been made to take these differences into account, as information would not be available in such detail and with the same level of reliability in most other EC Member States. Moreover, increasing the precision of the income changes would still leave a crucial uncertainty with respect to the relationship between income and expenditures.

Table A.2: Growth Factors Used for Income Components in Approach 9

| | |
|---|---|
| Wages (hh. headed by manual workers) | 1.085 |
| Wages (hh. headed by non-m. workers) | 1.157 |
| Wages (hh. headed by others) | 1.120 |
| Self-employment income | 1.174 |
| Investment income, pensions, annuities | 1.300 |
| Social security benefits | 1.000 |
| Other incomes | 1.120 |
| Imputed rent | 1.190 |

## References


Bradbury, B. (1990a), 'Modelling the impact of labour market changes on the distribution of family incomes', in: B. Bradbury (1990b).

Bradbury, B., (ed.) (1990b), *Tax-Benefit Models and Microsimulation Methods*, SPRC Reports and Proceedings No. 87, Social Policy Research Centre, University of New South Wales.

Brunner, J.K., and H.G. Petersen, (eds.) (1990), *Simulation Models in Tax and Transfer Policy*, Campus, Frankfurt.

Central Statistical Office (CSO) (1992), *Annual Abstract of Statistics Vol. 128*, HMSO, London.

De Vos, K. (1991), 'Extrapolation of income or expenditure distributions', Discussion Paper EURDP91.08, Department of Economic Sociology and Psychology, Erasmus University, Rotterdam.

De Vos, K. and M.A. Zaidi (1993), 'Trend analysis of poverty in the United Kingdom (1985-1988)', Department of Economic Sociology and Psychology, Erasmus University, Rotterdam/Economics Institute Tilburg.

Eurostat (1991), *National Accounts ESA - Aggregates 1970-1989*, Theme 2, Series C, Office for Official Publications of the European Communities, Luxembourg.

Hagenaars, A.J.M., K. de Vos and M.A. Zaidi (1994), *Poverty Statistics in the Late 1980s: Research Based on Microdata*, Study carried out for Eurostat, Office for Official Publications of the European Communities, Luxembourg.

Haveman, R.H. and K. Hollenbeck, (eds.) (1980), *Microeconomic Simulation Models for Public Policy Analysis*, Academic Press, New York.

Hellwig, O. (1990), 'The overseas experience with microsimulation models', in: B. Bradbury (1990b).

Orcutt, G.H., J. Merz and H. Quinke. (eds.) (1986), *Microanalytic Simulation Models to Support Social and Financial Policy*, North-Holland, Amsterdam.

PART TWO

# Incorporating Behavioural Response

Hans Baekgaard

Ministry of Economic Affairs
Denmark

# Chapter 7: A Microsimulation Approach to the Demand for Day Care for Children in Denmark

## Abstract

*One of the problems with the traditional static microsimulation models is that they typically replicate the impact of changed rules for government programs without allowing for any behavioural response by individuals to the changes. Changes in the price for day care introduced by the Danish Government in 1991 resulted in unanticipated behavioural responses, which meant that the impact of the policy changes had not been accurately estimated by static calculations.*

*This paper presents a method that makes it possible to simulate future changes in the demand for day care as a response to legislative changes in the payment structure. The budgetary consequences when behavioural and demographic effects are taken into account are compared to static law model calculations, and the difference is found to be considerable.*

## 1. Introduction

The background for the investigation presented in this chapter is a recent change in the legislation concerning payment for day care for children in Denmark which, with certain legislative restrictions, is run by the municipalities. In 1991, the Government imposed further restrictions on the freedom of municipal authorities to determine the price of day care.

The result was an immediate reduction in the total parental payment for day care. Subsequently, reports from several municipalities talked about a dramatic increase in the demand for the use of public day care. In the summer of 1992, a group of representatives from the municipalities and the central authorities gathered to investigate the issue. The main object was to analyse the cause of the shift in demand and to propose ways of solving the problem of the growing waiting lists due to capacity shortage.

The method presented in this paper was developed during these investigations, which ended up in a series of proposals to the politicians. The clear message was that the combined effect of increased fertility and lower prices had caused severe pressure on the municipalities - and would continue to do so if no action were taken.

Although warned, the decision makers were not prepared to change the very popular price reduction, of which the main ingredient was the introduction of a mandatory discount to families with more than one child in public day care – the *siblings discount*. On the other hand, the problem created a dilemma in that most of the country before the change already faced rationed demand for day care due to capacity shortage. The problem of growing waiting lists was a touchy subject for many local politicians, but new institutions are expensive to establish. Strong pressure was thus put on the law makers.

A methodology was developed to simulate changes in the demand for day care as a response to demographic and legislative changes in the payment structure. The method was a combination of a static microsimulation day care model and a behavioural response model. Both models rely on data from the cross-section *Model Population*[1] of *the Danish Law Model System.*

The static day care model calculates potential prices for day care for each child in the families represented in the sample, with an indication of the actual users of day care. The model serves as an excellent tool for analysing the *first-round* effects of legislative changes in the payment structure. For any practical purpose, this is sufficient when it comes to making predictions about short-term budgetary or distributive effects of minor changes in the price for day care. However, behavioural responses following major policy changes can make the static law calculations of the budgetary effects unsatisfactory.

To reveal the magnitude of the *second-round* demand effect of price changes, a behavioural response model is used to estimate the relation between the price for day care and the individual choice of whether or not to use public day care. The model used for this purpose is a binary choice logistic regression model. The estimation is based on the potential prices calculated by the day care model for the sample year.

---

1 The Model Population in the Danish Ministry of Economic Affairs is described in detail in the publication *Lovmodel: Overview of the Danish Law Model System*. A brief description of the relevant data is outlined in Section 3.

Based on prices calculated by the static day care model under alternative reforms, the demand model is used to make revised 'grossing-up' factors – that is, to predict the second-round demand effect of the price changes. The predicted demand is then implanted in the static day care model to calculate the budgetary effect of the alternative reforms.

Predictions are made for the years 1992 and 1993, which are one and two years ahead of the sample year in 1991. Taking into account the rapidly increasing birth rates, the grossing-up factors are adjusted with the growth in the population base. An alternative to this procedure is to adopt a dynamic microsimulation approach, where the demographic factors like fertility and changed family composition are handled in a more realistic fashion. However, this is a complicated task when analysing the consequences of the siblings discount. The siblings discount introduces a new element of interdependence between the prices for day care for children within the same family, and this effect is not represented in the data sample. To produce reliable estimates of the budgetary effect of the siblings discount, a behavioural model with the family as the decision unit is essential, but attempts to model the price sensitivity in such a set up have so far been unsuccessful.

Other investigations of the demand for day care have indicated that there is simultaneity between the choice of day care and the labour supply decision of the mother, although no such evidence exists based on Danish data (Garfinkel et al, 1992; Groot and van den Brink, 1992; Gustafsson, 1992; Ribar, 1992). The general result is that the demand is negatively related to the price for day care. Some studies also find a link between the choice of using public day care and the labour supply decision of the mothers. In the present study, the potential simultaneity of the demand for day care and the labour supply of the mother is not taken into account. However, this is considered to be of minor importance to the total demand effect.

In Sweden, where the day care system is similar to the Danish, the main effect of subsidising public day care seems to be a shift in demand from private to public day care (see Gustafsson and Stafford, 1992). Furthermore, the prediction period is short, and the aim of the investigation is to reveal how price changes influence the demand for day care.

The next section describes the existing legislation, as well as the payment structure implemented by the municipalities. A brief description of the data used are outlined in Section 3. Section 4 presents the methodology which produces an aggregate projection of the demographic base and a behavioural microeconomic model of demand. Utilising this method, several implications of alternative reforms are simulated in Section 5. Furthermore, to evaluate the reliability of the method, the section compares subsequent reports about the actual development in the demand for day care with the model predictions. Section 6 concludes the paper.

## 2. Legislation and Payment Structure for Day Care

Although public day care institutions are run by the municipalities, certain legislative restrictions are imposed on the payment for day care for children in Denmark. These restrictions harmonise the quality-price ratio in the field. The purpose of this section is to describe the payment structure.

The legislation has four elements, of which only the first three existed prior to the change in August 1991.

### 2.1 Maximum for parental cost coverage

The municipalities are allowed to charge the parents a *base charge* that covers a certain fraction of the budgetary cost of running a day care institution. In August 1991, this maximum parental cost coverage was reduced from 35 to 32 per cent of the budgetary cost. In 1993 it was further reduced to 30 per cent. The base charge can be less than this maximum, since the municipalities are free to add a voluntary subsidy in order to reduce the base charge below the maximum.

After the base charge is set, there are three types of rebates to eventually reduce the actual payment further. These are described below.

### 2.2 Income tested subsidy

This subsidy is income-tested on parental income, and is calculated as a fraction of the monthly base charge. The fraction starts with 100 percent at an income level of 40,000 Danish Kroner (DKK) per year and goes linearly down to 0 percent at an income level of 135,000 DKK per year. The income that is subject to the income test is reduced by 7,000 DKK per child in the family beyond the first.

## 2.3 Subsidy related to social factors

It is up to the municipality to decide whether special circumstances, such as social factors, shall warrant a reduction irrespective of income level.

## 2.4 Siblings discount

For families with more than one child in public day care, the municipalities are obliged to give a reduction that accounts for one third of the family's total payment. There is a minimum for the total payment, which is the base charge for the most expensive child plus 200 DKK per month for each other sibling using day care.

This mandatory siblings discount was introduced with the 1991 legislation and caused a very considerable reduction in the payment by families with more than one child using public day care.

This system of partial municipal freedom gives rise to a large variation in the across-country payment that confronts the users of day care. Table 1 shows weighted averages, and minimum and maximum base charges in age groups. It is this price variation that provides the possibility of estimating *price elasticity* in a model of behaviour.

Table 1: Monthly Payment (in DKK) for Public Day Care Before Discounts, 1991-93

| Age | 1991 | | | 1992 | | | 1993 | | |
|---|---|---|---|---|---|---|---|---|---|
| | Min | Av | Max | Min | Av | Max | Min | Av | Max |
| 0-2 yrs | 870 | 1,358 | 2,173 | 1,000 | 1,421 | 2,145 | 900 | 1,478 | 2,163 |
| 3-6 yrs | 805 | 999 | 1,480 | 780 | 1,021 | 1,460 | 765 | 1,034 | 1,295 |
| 7-12 yrs | 200 | 658 | 1,130 | 208 | 688 | 1,140 | 200 | 701 | 978 |

# 3. Data

The major data source is the administrative registers in Danmarks Statistik. The Model Population in the Ministry of Economic Affairs is a sample of $3^{1/3}$ per cent of the total population of 5.1 million individuals, counted in about 2.7 million family units. The Model Population is created by merging several registers for incomes, taxes, transfers and, as is relevant in the present context, the use of public day care.

Merging different registers is possible because every individual in Denmark has a personal identifier, the so-called CPR-numbers.[2]

The model population for the year 1990 consists of more than 170,000 individuals grouped in about 90,000 households. From this population, a subsample is selected consisting of all individuals in households with children of relevant ages. This sample contains families of actual users as well as families of non-users of public day care. Together with a detailed set of background information on socioeconomic characteristics of the families, this provides a powerful dataset for the model of individual decision described in the next section.

One particular piece of information – by its nature – is not available from the administrative registers: the potential base charge for non-users. Fortunately this problem is easily overcome, since data are available on the actual prices charged by the municipalities, categorised by institution type and age group.

The estimation data set contains information for all children between 0 and 12 years of age. Only significant variables in relation to the demand for day care are included. These are:

CHARGE : The yearly charge for day care after discounts

*Mother's marital status:*

LONEPAR : 1, if a single parent

MARRIED : 1, if parents are married

COUPLE : 1, if the parents are living together, not married

*Parents' income:*

INCOME : Total parental income the preceding year

[2] CPR refers to a Central Person Registration system which ensures that every individual has a personal identification number used for administrative purposes. Combined with detailed address codes, this system provides a convenient tool for establishing a person/household level model population based on the administrative data.

*Mother's (or single father's) occupation:*

| | |
|---|---|
| HOME : | 1, if mother is a housewife |
| SELFEMP : | 1, if mother is self employed |
| EMPL : | 1, if mother is employed |
| FARMER : | 1, if mother is a farmer |
| STUDENT : | 1, if mother is a student |
| UNEMP : | 1, if mother is unemployed |
| WHOUR : | 1, if mother has full time employment |

*Variables concerning capacity shortage:*

| | |
|---|---|
| WLMRK : | 1, if the child is at high risk of being rationed due to capacity shortage |
| WLFRAC : | the fraction of rationed children to total demand for day care in the municipality/age group. |

*Municipality specific variables:*

| | |
|---|---|
| USERFRAC : | The fraction of users to total number of children in the municipality/age group. |
| WOMEMP : | The fraction of employed women in the municipality/mothers age group. |
| MUNINC : | The average income in the municipality |

The means and standard deviations for all the variables are listed in Table 2.

The variables UNEMP and WHOUR are not part of the exhaustive classification of employment, since they are based on employment during the whole year, whereas the other employment variables represent status in November, 1990.

## 4. The Model

As mentioned in the introduction, there are several good reasons to develop a method that separates the effects of demographic changes from those of the payment structure changes. Methodologically, this means the calculation of revised grossing-up factors. These are evaluated in two steps. In the first step, demographic projections obtained from Danmarks Statistik are used to calculate changes in the population base, to produce a relative change in demand stemming only from changes in the number of children in the relevant age groups. In the second step, the change in demand following the altered payment structure is estimated. Based on the adjusted population base, the change in demand is used to calculate the total change in the demand for day care. These two steps are described in detail below.

Table 2: Mean and Standard Deviation for the Sample Variables

| | Mean | Standard deviation |
|---|---|---|
| CHARGE | 10,104 | 4,714 |
| LONEPAR | 0.1463 | 0.3534 |
| MARRIED | 0.7068 | 0.4552 |
| INCOME | 256,750 | 138,500 |
| HOME | 0.1162 | 0.3205 |
| SELFEMP | 0.0427 | 0.1966 |
| EMPL | 0.8104 | 0.3920 |
| FARMER | 0.0217 | 0.1340 |
| STUDENT | 0.0183 | 0.1456 |
| UNEMP | 0.1242 | 0.3299 |
| WHOUR | 0.6326 | 0.4821 |
| WLMRK | 0.07514 | 0.2636 |
| WLFRAC | 0.0441 | 0.0753 |
| USERFRAC | 0.4793 | 0.2598 |
| WOMEMP | 88.59 | 3.19 |
| MUNINC | 117,280 | 14,205 |

## 4.1 Demographic changes

In this section, the total demand for day care is used, together with projections from Danmarks Statistik of the future population size, to predict the future total demand (under the assumption that a constant fraction of the children will want to use day care).

The first step involves simple calculations of the relative changes in the age groups from the starting year 1991 to the subsequent years of interest. Table 3 summarises the general increase in the population base by age groups and years.

Table 3: Relative Change in the Population Age Groups From the Starting Point in 1991

| Age group | 1991-92 | 1991-93 | 1991-94 | 1991-95 |
|---|---|---|---|---|
| | Growth in per cent | | | |
| 0-2 years | 3.66 | 6.09 | 7.87 | 9.94 |
| 3-6 years | 3.20 | 6.56 | 10.36 | 13.98 |
| 7-12 years | -0.92 | 1.34 | 2.33 | 4.45 |

Table 4 shows the total demand for day care in 1991, augmented to the following years by the changed population. The figures listed in this table are the expected total demand for day care, under the assumption that the fraction of children demanding day care in future years will be the same as in 1991.

The model described in the following section is designed to predict changes in the fraction of demand that originates from sources *other than* the changing population size.

Table 4: The Total Population Base for Day Care Calculated From the Starting Point in 1991

| Age Group | 1991 | 1992 | 1993 | 1994 | 1995 |
|---|---|---|---|---|---|
| | Number of children | | | | |
| 0-2 yrs | 102,428 | 106,177 | 108,666 | 110,489 | 112,609 |
| 3-6 yrs | 170,208 | 175,655 | 181,374 | 187,842 | 194,003 |
| 7-12 yrs | 81,066 | 80,320 | 82,152 | 82,955 | 84,673 |
| All | 353,702 | 362,152 | 372,192 | 381,286 | 391,285 |

* Children on waiting lists in 1991 are included in total demand.

## 4.2 A behavioural model of demand

The model is a *binary choice* model of the logistic regression type with the individual child's use of day care as the response variable:

$$(1) \qquad DAYCARE_i = \begin{cases} 1 & user \\ 0 & non-user \end{cases}$$

The model to be estimated is then:

(2) $P(DAY\ CARE_i=1) = \exp(a+\mathbf{x_i b})/1+\exp(a+\mathbf{x_i b})$

$\mathbf{x_i}$ is a vector of the explanatory variables listed in the preceding section; a and $\mathbf{b}$ are parameters for the constant term and $\mathbf{x_i}$ respectively.

A few comments should be made about the model structure. The model is estimated in thirteen separate groups according to the children's age, from 0 to 12 years. This is done in order to capture the age effect in the price elasticity, which is highly significant.

For each age group, the price variable is split into four separate variables according to the number of children in the family - one, two, three and four or more:

$$\text{CHARGE}_{\#\text{kids}=i} = \begin{cases} \text{CHARGE} & \#\text{kids} = i \\ 0 & \#\text{kids} \neq i \end{cases}$$

This model is a result of reducing the full model of the two classification variables (number of children in the family and the age of the child) and the continuous price variable. The total of 13 times four price variables $CHARGE_{\#kids=i}$ represents the interaction effects of these three variables.

In the model formulation, a detailed description of the structure of price responsiveness had a high priority. This is because the main purpose of the model is to predict the demand response to changes in the price for day care.

For the same reason, the validity of the linearity assumptions implicitly imposed by the logistic model has been verified for the price variables by a so-called smoothing algorithm. Within a relevant price range, the linearity assumption was found to be an acceptable approximation.

## 4.3 Estimation results

Table 5 shows part of the estimation results using data from January 1991 – that is, before the legislative reform introduced in August the same year. As an example, the table includes the full estimation results for the group of children two years of age.

Table 5: Parameter Estimates for Children Two Years of Age

| 1991 | Estimates | | St. dev. |
|---|---|---|---|
| | Children 2 years of age | Typical sign* | |
| $CHARGE_{\#\ kids=1}$ | -0.000057 | - | 0.000029 |
| $CHARGE_{\#\ kids=2}$ | -0.000096 | - | 0.000028 |
| $CHARGE_{\#\ kids=3}$ | -0.000155 | - | 0.000033 |
| $CHARGE_{\#\ kids\geq 4}$ | -0.000172 | - | 0.000055 |
| LONEPAR | 1.23 | + | 0.48 |
| MARRIED | -0.28 | - | 0.27 |
| INCOME | 0.0000057 | + | 0.0000014 |
| HOME | 1.58 | + | 0.96 |
| SELFEMP | 2.63 | + | 1.09 |
| EMPL | 2.48 | + | 0.92 |
| STUDENT | 2.13 | + | 1.04 |
| UNEMP | -0.69 | - | 0.33 |
| WHOUR | 0.67 | + | 0.30 |
| WLMRK | | - | . |
| WLFRAC | -2.77 | - | 1.32 |
| USERFRAC | 3.62 | + | 1.39 |
| WOMEMP | -0.055 | + | 0.042 |
| MUNINC | 0.000011 | + | 0.000014 |
| $R^2$ | 0.459 | | |
| -2 log L | 518.95 | | |

* This the typical sign of parameters in *other* age groups, and is provided so that the usual sign of the parameter is known to the reader.

The variable FARMER is left out to prevent perfect linearity. The parameter estimates for the other occupational variables such as HOME, SELFEMP, EMPL and STUD indicate the level of use of day care among these groups relative to farmers. As can be seen from the positive parameter estimates, all other occupational groups have a stronger tendency to be users than farmers. In the same way, the omitted variable COUPLE is the level variable for marital status. As expected, lone parents are more likely to be users of public day care than couples are.

Table 6 shows the parameter estimates for the price variables in all age groups. All estimates are significantly different from zero at the five per cent level.

The price elasticities[3] exhibit an interesting interaction between the price sensitivity, the age of the child and the number of children in the family. The older the child and the more siblings in the family, the stronger the price sensitivity. This result is not surprising. The younger the child, the more likely it is that he/she needs to be looked after while the parents are away – with little consideration for the cost. Also, the alternative private care is relatively more expensive for small children.

Table 6: Estimated Parameters for the Price Variables

| Age | $CHARGE_{\#\ kids=1}$ | $CHARGE_{\#\ kids=2}$ | $CHARGE_{\#\ kids=3}$ | $CHARGE_{\#\ kids\geq 4}$ |
|---|---|---|---|---|
| 0 | -0.21 | -0.27 | -0.63 | -6.17 |
| 1 | -0.72 | -0.95 | -1.48 | -2.12 |
| 2 | -0.57 | -0.96 | -1.55 | -1.72 |
| 3 | -0.37 | -0.72 | -1.53 | -2.12 |
| 4 | -0.54 | -1.27 | -2.14 | -3.24 |
| 5 | -2.40 | -2.40 | -3.20 | -3.93 |
| 6 | -3.15 | -3.44 | -3.56 | -5.77 |
| 7 | -1.75 | -1.95 | -3.35 | -4.36 |
| 8 | -1.47 | -2.22 | -2.56 | -4.92 |
| 9 | -2.00 | -3.16 | -3.61 | . |
| 10 | -2.74 | -4.19 | -5.24 | -7.36 |
| 11 | -2.97 | -4.19 | -39.3 | -3.21 |
| 12 | -37.2 | -14.0 | -28.8 | -44.6 |

The observed stronger price sensitivity for families with many children is presumably related to the cost of private alternatives to public day care. Consider for example a family where both parents are working and have to choose between engaging a maid or using public day care. In this situation, the bigger family saves more by engaging a maid. The normal cost of a maid is in fact a critical issue, and changes in the payment for day care can make the key difference to the family's decision about day care use.

This point supports the findings by Gustafsson and Stafford (1992) that subsidising public day care shifts demand from private to public day care. It certainly ought to be taken into account in the evaluation of the siblings discount, which subsidises the most price responsive families.

[3] Given the fact that it is a logistic model, we cannot compare the absolute values of the estimated elasticities. For ease of comparison, all estimated parameters in Table 6 are multiplied by 10,000.

# 5. Simulation of Alternative Reforms

During the negotiations that led to the reform in 1991 and the re-evaluation in 1992, several proposals were launched. In 1991 the focus was not on the demand effect of the changed price structure, as it was not considered to be significant. Instead, the static law calculation formed the basis for the decisions. Very shortly after the reform, a dramatic upward shift in the demand for day care was observed by the municipalities. The results presented in this section indicate that ignoring the demand effects of the price reductions was a serious omission.

This section simulates the effects of a series of proposed changes in legislation concerning the price for public day care for children. These proposals were presented by the task force that dealt with the subject in the second half of 1992, one year after the reform. All the results reported appear as they were presented at that time. A more recent survey on total demand is subsequently used to compare the predictions with actual developments in the demand for day care.

## 5.1 Predictions

In Table 7 the predictions under alternative formulations of the siblings discount are compared with the *base line* prediction for the legislation that was introduced in 1991. The first column lists the result from the demographic projections in Table 4 about the expected demand in 1993 - that is, only the change in population size is taken into account. This adjusted demand is a 'no change' baseline for the 1991 legislation, since it can be interpreted as an 'in case nothing is changed' prediction at that time.

The five columns to the right show the demand for day care predicted by the demand model for the five models of siblings discount described in Figure 1. All the predictions are calculated as the percentage change (quoted in parenthesis) predicted by the model from the numbers in the first column.

Each prediction is calculated by using the estimation results to evaluate predicted probabilities under each reform, by substituting the price variables for the parental payment under this particular reform. A new *predicted* day care population is then drawn from the full population using random selection based on calculated probabilities. This prediction routine is iterated several times in order to reduce Monte Carlo uncertainty. The average change in demand for day care from these iterations is presented in Table 7 for each of the alternatives described in Figure 1.

The result for the 'Base line' model shows that there is a predicted total growth of 8.8 per cent in demand due to the legislative price change in August 1991. Note that this change, apart from the introduction of the mandatory siblings discount, also includes a reduced maximum for parental cost coverage. This change is included in the predictions for all the five alternative models.

Table 7: Predicted Total Demand for Day Care in January 1993 Under Alternative Reforms*

| | | Predicted demand – January 1993 | | | | |
|---|---|---|---|---|---|---|
| Age | 1991 demand adjusted to 1993 population size | Base line | 50 per cent rule | $33^{1/3}$ per cent rule | 50 per cent pre school | No mandatory |
| 0-2 yrs | 108,666 | 117,794<br>(8.4) | 113,665<br>(4.6) | 113,339<br>(4.3) | 113,665<br>(4.6) | 113,230<br>(4.2) |
| 3-6 yrs | 181,374 | 197,154<br>(8.7) | 192,619<br>(6.2) | 190,805<br>(5.2) | 188,629<br>(4.0) | 188,266<br>(3.8) |
| 7-12 yrs | 82,152 | 90,121<br>(9.7) | 91,517<br>(11.4) | 87,328<br>(6.3) | 78,784<br>(-4.1) | 81,659<br>(-0.6) |
| Total | 372,192 | 405,069<br>(8.8) | 397,801<br>(6.9) | 391,472<br>(5.2) | 381,078<br>(2.3) | 383,155<br>(2.9) |

* Figures in brackets are percentage changes from 'no change' scenario.

Figure 1: Description of the five alternative models of siblings discount

1. Base line: $33^{1/3}$ per cent discount to children in families with more than one child in public day care. There is a minimum for the family's total payment on the base charge for the most expensive child plus 200 DKK/month for each other sibling.
2. 50 per cent rule: Full base charge for the most expensive child and 50 per cent reduction for the others.
3. $33^{1/3}$ per cent rule: Full base charge for the most expensive child and $33^{1/3}$ per cent reduction for the others.
4. 50 per cent pre school: Full base charge for the most expensive child and 50 per cent reduction for other preschool children.
5. No mandatory: No mandatory discount, but the municipalities are expected to return to voluntary discounts prior to the change, if any.

At the time of making the predictions shown in Table 7, it was still uncertain how the municipalities would implement the reduced maximum for parental cost coverage in the actual prices charged for day care. The predictions are based on an expectation of reduced charges after 1992, but it turned out that the actual prices were somewhat higher than in the initial calculations. Table 1 shows that this was especially the case for the youngest children. The reason was that, in many municipalities, the parental cost coverage was already below the new 30 per cent maximum before the legislative change. A lot of these municipalities increased their prices to cover the extra expenses generated by the siblings discount.

The potential cost of covering the considerable increase in demand caused by the 1991 legislation exceeds the direct budgetary cost related to the lowered prices by a factor of two to three. In this calculation, only the variable cost is included. The cost of establishing new institutions is not. The difference between the first and second-round budgetary effects is dramatic, because payments by parents cover on average only about 22 per cent of the variable costs incurred in running day care institutions.

## 5.2 Comparing model prediction with actual development

In making predictions like the ones reported in Table 7, an assumption of immediate behavioural adjustment is implicitly adopted. When comparing the predictions for 1992 and 1993 with the actual demand, it must be kept in mind that the legislative change was not introduced until August 1991. The full demand effects could not be expected to be observed as soon as January 1992 – only five months later. It is an open question whether the full effects can be observed in 1993 or whether the adjustment process was still continuing beyond that time.

Table 8 compares the model predictions with the results from two separate surveys done in 1992 and 1993 respectively. The model predictions shown in the two 'Predicted Demand' columns in Table 8 are identical to those presented to decision makers when the reform was evaluated in September 1992. The overall impression is that the model predicts fairly well the change in demand due to the price changes. The prediction for 1992 is slightly too high but, as mentioned earlier, the prediction relies on a very fast behavioural response to the new prices, which cannot really be expected to take place.

Table 8: Comparison of Predicted Demand With Surveys on the Actual Total Demand in 1992 and 1993

| | Predicted Demand | | Actual Demand* | |
|---|---|---|---|---|
| | 1992 | 1993 | 1992 | 1993 |
| 0-2 year | 113,928 | 117,794 | 112,025 | 116,000 |
| | (7.3) | (8.4) | (5.5) | (6.7) |
| 3-6 year | 188,127 | 197,154 | 182,839 | 196,000 |
| | (7.1) | (8.7) | (4.1) | (8.1) |
| 7-12 year | 87,067 | 90,121 | 88,560 | 95,300 |
| | (8.4) | (9.7) | (10.3) | (16.0) |
| All | 389,122 | 405,069 | 383,424 | 407,300 |
| | (7.4) | (8.8) | (5.9) | (9.4) |

* The demand for 1992 consists of the actual number of users registered by Danmarks Statistik and the number of children on waiting lists surveyed by 'Bureau 2000'. The 1993 survey is conducted by 'Bureau 2000'. This survey includes only children aged from 0 to 9 year. The total number of actual users in the age group 7-12 years is reported by Danmarks Statistik. Figures in brackets are percentage changes from 'no change' scenario.

The predictions shown in Table 8 are based on assumptions about some factors that were uncertain at the time when the predictions were made. The assumptions were:

- From 1992 to 1993 the *Maximum for Parental Cost Coverage* was reduced from 32 to 30 per cent of the budgetary cost. The municipalities were free to set the price below the 30 per cent maximum and many of them already did so before 1993. In assessing the anticipated price for 1993, an assumption had to be made about how the municipalities would respond to the changed legislation. It was assumed that municipalities with prices above the 30 per cent limit would reduce the prices to this 30 per cent threshold, whereas the municipalities with prices below the threshold were assumed to maintain the same percentage parental cost coverage as in 1992.

- The use of public day care is affected by other factors than the price and the number of children, though these are the only effects accounted for by the model. In the lack of any better alternative, the sum of these effects were assumed to be zero.

As already mentioned, the anticipated response of the municipalities failed to match with observed behaviour. A large number of the municipalities that were giving voluntary price reductions prior to the legislative change actually increased their prices. The average price for public day care before rebates did not decrease as initially expected. As a result, the estimated demand for day care was based on prices that were lower than what actually eventuated.

The main weakness of the model is that it does not take account of the general trend towards growing use of public day care. Even without legislative change, the demand for public day care could be expected to increase. This trend, which is strongest for school age children, is not included in the model, since it was based on one cross-section sample of data. The significance of this shortcoming of the model has been well illustrated since the results presented in this paper were produced, as an improved version of the model has been developed (Finansministeriet et al, 1995). This model is based on two cross-sections of data for the use of public day care. One data set is from January 1991, seven months prior to the change. The other data set is from January 1993, seventeen months after the change. The results from this model suggest that the predictions in Table 7 and Table 8 slightly overstate the price responsiveness of the demand for public day care. The new model decomposes the increase in the share of children using public day care from 1991 to 1993 into three components:

- the change in the number of children on waiting lists;
- the change in the price; and
- the general trend in the demand for day care.

The model provides an estimate of the increase in the share of children using public day care that can be related to each of these three factors. The following table shows how the new model decomposes the observed percentage point increase in the share of children using public day care.

Table 9: Estimated Decomposition of the Change from 1991 to 1993 of the Share of the Children in Age Groups Using Public Day Care

| | ½ - 2 years | 3-6 years | 7-10 years |
|---|---|---|---|
| | | Percentage | |
| 1991 | 56.2 | 70.6 | 32.3 |
| 1993 | 62.6 | 79.0 | 49.0 |
| Difference | 6.3 | 8.4 | 16.7 |
| *comprising* | | | |
| - Waiting lists | 2.7 | 1.8 | 0.6 |
| - Trend | 2.2 | 1.6 | 11.7 |
| - Prices | 1.4 | 4.9 | 4.4 |

Source: Finansministeriet et al, 1995

The table shows that the price-related increase only amounts to a minor part of the total increase in the use of day care. The price responsiveness estimated by the improved model is generally weaker than the results presented in this paper. The estimates from the new model suggest that the percentage increase in the demand for public day care totals 7.2 per cent, rather than the 8.8 predicted in Table 8. As a consequence, the second-round budgetary implications of a price change are estimated to be just over twice the amount estimated by the static calculations.

Finally, it can be noted that another source of error is the way the potential siblings discount is allocated among the children within a family. This is done proportionally to the size of the base charge for each child in the family. This implies that small children, who are the most expensive and the least price responsive, generally get the biggest reductions. This problem arises because the behavioural model of demand has the individual child as the 'decision unit'. The siblings discount is a family rebate, and thus there is no correct way of dealing with it in the model. The evidence from a comparison of the old with the new estimates supports the judgement that the model underestimates demand for the oldest children - with an increase of 13.6 per cent predicted by the new model relative to the 9.7 per cent shown in Table 8. On the other hand, the old model overestimates demand for preschool children, with an increase of 2.5 per cent for the youngest children and 6.9 per cent for 3 to 6 year olds predicted by the new model, relative to the 8.4 and 8.7 per cent shown in Table 8.

## 6. Conclusion

The model of demand for day care exhibits promising qualities concerning price related changes in the demand for day care. In Denmark, such effects have been neglected so far, basically due to the lack of methods for giving any qualified estimates of the size of these effects.

It is important to note that there are serious pitfalls awaiting microsimulation modellers attempting to assess the dynamic impact of legislative changes. Other factors than those directly controlled by the central government may influence the outcome of such changes. This was the case with the unanticipated behaviour of the Danish municipalities, which contributed to an overestimation of the total demand for public day care for children.

The use of a single cross-section of data to estimate behavioural response entails some problems that may be resolved if longitudinal data are available. However, in some cases, the availability of more than one cross-section provides the means of improving the shortcomings of cross-sectional data discovered in this paper. Also, there is little doubt that a more sophisticated model based on the family as the decision unit would improve the variety of legislative changes that could be captured by the demand model.

It may be argued, as it was during the investigation process, that there is a lot of uncertainty attached to the type of statistical methods adopted here. On the other hand, the final budgetary effects of the 1991 legislation should at least have been doubled relative to the first-round estimates. This paper thus illustrates the value of making even uncertain predictions about behavioural and dynamic consequences of legislative changes.

A microsimulation model like the Danish Law Model is the perfect framework for the analysis of first-round effects of changed rules. In many cases, such as the one described in this paper, the second-round effects are very important and should be considered whenever possible. For the results to be used in the political process, they have to be presented in a way that underlines the uncertainties involved. Furthermore, it is essential to achieve a high degree of consensus about the generality of the results and, for this to happen, there must be an efficient exchange of results and experiences between researchers and the administrative system.

The Danish Law Model System is now widely accepted among the ministries in the process of preparing policy initiatives – but only when it comes to static law calculations. Calculations which draw on statistical methods like the one presented here are met, understandably, with a good deal of suspicion.

It took considerable time for static law model calculations to become widely accepted in Denmark. It may be even more difficult to present dynamic results in a way persuasive enough to make them credible as part of the basis for political decision making. Since there is ongoing development of the Danish Law Model System, moving in the direction of a more comprehensive use of dynamic and behavioural methods, it is important that the notion of credibility be taken seriously.

## References

Finansministeriet, Socialministeriet, Økonomiministeriet, Kommunernes Landsforening, Københavns Kommune and Frederiksberg Kommune (1995), *Rapport fra analysearbejdet om børnepasning*, Ministry of Finance, Copenhagen Denmark.

Garfinkel, I., C. Michalopoulos and P. K. Robins (1991), 'A structural model of labour supply and child care demand', *Journal of Human Resources*, Vol. 27, No. 1.

Groot, W. and H. Maassen van der Brink (1992), 'Labour supply, child care and consumption', Paper presented to the Annual Meeting of the European Society for Population Economics, Gmunden, Austria, June.

Gustafsson, S. (1992), 'Public policies and women's labour force participation: A comparison between Sweden, the Netherlands and Germany', Mimeo, University of Amsterdam.

Gustafsson, S., and F. Stafford (1991), 'Child care subsidies and labour supply in Sweden', *The Journal of Human Resources*, Vol. 27, No. 1.

Økonoministeriet (1993), *Lovmodel, overview of the Danish law model system*, Ministry of Economic Affairs, Copenhagen, Denmark.

Let-Sorensen, S. and G. Rohwer (1993), 'Aspects of the female life cycle and labour market participation in Denmark', Unpublished paper, Danmarks Statistik, Copenhagen, Denmark.

Ribar, D. C. (1991), 'Child care and labour supply of married women', *The Journal of Human Resources*, Vol. 27, No. 1.

Elizabeth Symons
University of Keele
United Kingdom

Neil Warren
University of New South Wales
Australia

# Chapter 8: Modelling Consumer Behavioural Response to Commodity Tax Reforms

## Abstract

*This paper examines the importance of incorporating consumer behavioural response when analysing the impact of indirect tax reform. It augments a static simulation model that covers all indirect taxes with a model of consumer demand that is based on an Almost Ideal Demand System (AIDS). In both estimation and simulation, use is made of extensive databases of individual household level data. A selection of simulations are performed using the 1988-89 Australian Household Expenditure Survey. This enables the prediction of consumer expenditure, government revenue and a distributional analysis - both with and without assuming consumer behavioural change. The results suggest that consumer response is significant and any analysis which does not incorporate it needs to carry important qualifications.*

## 1. Introduction

In debates over the need for different tax reforms, surprisingly little empirical analysis is utilised to support the case for change. At best, arguments for tax reform are accompanied by analytically (or theoretically) based assertions about the benefits of change and, sometimes, analysis of the initial impact on equity and the medium to longer term implications for the macro-economy. Very little attention is given to modelling, at the microeconomic level, the behavioural response of various groups in the community to the reforms being considered.

Somewhat paradoxically, in failing to canvass these behavioural responses, the basic case for many tax reforms is not highlighted. Why argue for an increase in tobacco excise based on public health arguments if such a case is not supported by empirical evidence illustrating the response of consumers? Why support an increase in energy (or carbon) taxes if such reforms do not result in the desired energy conservation?

The objective of this paper is to highlight the implications for government revenue forecasting and for the distributional analysis of selected tax reforms of *allowing for the behavioural response of consumers.* The importance of this cannot be underestimated - failure to include consumer response can result in substantial errors in estimates of post-reform government tax revenue and may overstate the distributional impact of reforms. The latter consideration is of importance for any compensation package designed to offset adverse distributional effects. If such effects are not significant, then this omission is probably not important - but such an assumption is not uncontroversial and needs to be evaluated. If the distributional effects are significant, then important qualifications must accompany any tax reform study which does not incorporate the behavioural response of consumers.

This study incorporates a consumer behavioural model into a static microsimulation model of *all* taxes levied in the economy. Previously, models of behavioural response to tax reforms have been restricted to modelling either single taxes in isolation or a small number of related taxes. This paper describes a study which models the implication of one set of commodity tax reforms on *all* other taxes, whether directly related to the reforms being considered or not.

The first problem the study addresses in Section 2 is how to model consumer demand. Section 3 models the impact of taxation reforms on prices and, thus, on consumers' behaviour. Using this modelling framework, Section 4 focusses on revenue-neutral commodity tax reforms which involve replacing a narrow-based commodity tax with one that is broad-based, while Section 5 reviews the case of a revenue-neutral change in the mix of commodity and income taxes. Section 6 examines non-revenue-neutral changes to taxes on petrol, tobacco and alcohol.

## 2. Model of Household Consumption Behaviour

Modelling the behavioural response of consumers to relative and absolute price changes has been the subject of a growing volume of empirical literature. Deaton and Muellbauer (1980) provided a focus with the development of an Almost Ideal Demand System (AIDS), which has been considerably extended over the past decade. One extension, a Quadratic Extended AIDS (QUAIDS), is developed in Blundell et al. (1993). This now forms the basis of the Simulation Program for Indirect Taxation (SPIT) originally outlined in Baker et al. (1990) and used in Pearson and Smith to investigate the effects of environmental taxes in the UK (1991). SPIT is designed to simulate the impact of different indirect tax reforms in the UK using the 1988 Family Expenditure Survey. This involves two effects: first, how UK households respond to changes in the absolute and relative price of commodities, and, second, how the environmental taxes affect aggregate tax revenue and how the tax burden is distributed across different groups of households. SPIT does not model the direct tax system, implicitly assuming total income is fixed.

QUAIDS assumes households allocate their income in two stages. First, total income is divided between savings, expenditure on durable goods (including housing) and non-durable goods. In the simulation, durable expenditure is treated as 'fixed' or 'rationed' in terms of quantity consumed. Thus, households maintain their consumption of these goods at the observed level. At the second stage, non-durable or unrationed expenditure is allocated to broad categories of consumption goods, with the shares determined using QUAIDS. The categories covered are: Beer, Wine, Spirits, Food, Fuel, Clothing, Household Services, Personal Goods and Services, Leisure Goods, Travel Fares, Tobacco, Motoring, Petrol and Leisure Services. Jointly, these account for approximately 80 per cent of total expenditure.

QUAIDS relates household expenditure shares for each of the commodity groups to the prices of these groups and to the household's level of real non-durable expenditure, the latter variable entering as a quadratic. This extension to the standard AIDS model allows goods to be 'luxuries' in some income ranges but 'necessities' in others, thereby providing a better 'fit' for the data. Thus the share of expenditure for the *ith* good out of total non-durable expenditure for household $h$ in time period $t$ is given by:

$$W_{it}^{h} = \alpha_i + \sum_{j=1} \gamma_{ij} \ln P_{jt} + \beta_i \ln x_t^h + \lambda_i \left(\ln x_t^h\right)^2$$

where: $W_{it}^{h}$ budget share of good $i$ from total unrationed expenditure for household $h$ at time $t$

$P_{jt}$ price of good $j$ at time $t$

$x_t^h$ total unrationed real expenditure by household $h$ at time $t$, ie. $y_t^h / P_t^h$ where $y_t^t$ is total unrationed expenditure

$\ln P_{jt}^h$ household specific Stone price index evaluated at observed shares ie. $\sum_{j=1} W_{jt}^{n} \ln P_{jt}$

and $\alpha_i, \gamma_{ij}, \beta_i$ and $\lambda_i$ are preference parameters which can vary with the household characteristics and other conditioning variables.

The expenditure equations are estimated as an overall demand system in order to impose the restrictions required by economic theory on the parameters. Blundell et al. (1993) found that the parameters $\alpha$ and $\beta$ were not constant across different types of households and periods. Part of this inconstancy can be incorporated in estimation by allowing the intercept, $\alpha$, to vary with household characteristics (eg. age of head of household, number of children, region, housing type, employment status etc.) and deterministic variables (eg. seasonal dummies and time trend). The real expenditure coefficient, $\beta$, is allowed to vary with a limited number of characteristics (eg. season, children, occupation of head).

The introduction of household-varying intercept and expenditure coefficients allows for considerable variation in the preference parameters across households. Despite this, problems of parameter stability arise when estimating across the full sample. The most important distinctions are whether or not the household includes a smoker and/or whether or not the household possesses a car. To accommodate this, the sample was split into four sub-groups according to the presence or absence of a car and whether or not recorded tobacco expenditure was positive. This allows the full set of parameters to vary across the four groups so that the price response, in addition to the intercept and income response, can differ across household type (for a more detailed discussion of estimation see Baker et al. 1990).

## 3. Modelling Australian Commodity Tax Reforms

Estimating a QUAIDS model of consumer behaviour for Australia using Australian data is difficult. Australia has conducted national household expenditure surveys (HES) for 1975-76, 1984 and 1988-89. Each survey adopts a methodology similar to the UK FES and contains detailed information on the income and expenditure of individuals within some 7,225 households. However, household data are only available from the latter two surveys. (In contrast, household data from the UK FES are available annually from 1957.) However, even if household data from the 1975-76 HES were available, this would not provide sufficient price variation to estimate the consumer price effects in QUAIDS.

Various techniques have been developed to approximately estimate QUAIDS. These have focussed on the use of intertemporal National Accounts data on Private Final Consumption Expenditure (PFCE). This technique is inferior to that of working with intertemporal household data for a number of reasons. First, grossed-up FES type surveys result in aggregates different to those in National Accounts PFCE. Secondly, PFCE may change over time because of factors other than prices. Thirdly, PFCE consumption categories may differ markedly from those in FES. Finally, PFCE does not enable differences in household behavioural responses to be modelled.

This study has taken an alternative approach that, while a compromise, is not unreasonable. The QUAIDS estimates from the annual UK FES between 1974 and 1988 (across 100,000 households) have been applied to Australian households. This assumes that Australian and UK consumers respond similarly to price changes, have similar goods and services available and are surveyed using similar methodology with comparable commodity categories. The most important assumption is the degree of similarity between consumption patterns in the UK and Australia. This has potential implications for the own-price and cross-price elasticities of consumers and, therefore, their response to commodity tax changes. Table 1 details the budget shares of household expenditure derived from the 1988 UK FES and the Australian 1988-89 HES.

There are clear similarities between the way households in the two countries distribute their total expenditure. A major source of difference can be attributed to climatic differences: in the UK a larger share of the household budget is spent on

Table 1: A Comparison of Australian and UK Household Expenditure Patterns

| Item | Australia | | UK | |
|---|---|---|---|---|
| | % of total expenditure | % of unrationed expenditure | % of total expenditure | % of unrationed expenditure |
| Beer | 2.3 | 3.0 | 2.8 | 3.7 |
| Wine | 0.6 | 0.8 | 1.0 | 1.3 |
| Spirits | 0.6 | 0.7 | 0.7 | 1.0 |
| Food | 18.4 | 24.0 | 18.6 | 25.0 |
| Fuel | 2.5 | 3.2 | 5.1 | 6.8 |
| Clothing | 5.4 | 7.0 | 7.0 | 9.5 |
| Household Services | 6.6 | 8.6 | 4.8 | 6.4 |
| Personal Goods and Services | 4.9 | 6.3 | 4.0 | 5.3 |
| Leisure Goods | 4.9 | 6.4 | 4.7 | 6.3 |
| Travel Fares | 1.2 | 1.5 | 2.4 | 3.2 |
| Tobacco | 1.7 | 2.1 | 2.2 | 2.9 |
| Motoring | 9.8 | 12.7 | 9.2 | 12.4 |
| Petrol | 4.7 | 6.1 | 3.2 | 4.3 |
| Leisure Services | 13.5 | 17.6 | 8.9 | 11.9 |
| *All Unrationed Expenditure* | *77.1* | *100.0* | *75.5* | *100.0* |
| Household Goods (Durables) | 8.8 | | 7.3 | |
| Housing | 14.1 | | 18.1 | |
| *ALL EXPENDITURE* | *100.0* | | *100.0* | |

heating fuel and clothing while, in Australia, a higher proportion is spent on leisure activities. The more widespread availability and use of public transport (travel fares) in the UK is demonstrated by the results. The other major difference is the cost of housing, which takes a larger share of the UK budget. This is partly explained by the smaller owner-occupier sector in the UK than in Australia. Thus, in Australia there are more outright owners who will have no nominal rent or mortgage payments attributed to them and so have lower housing costs. As real housing expenditure is assumed fixed in the simulation, this is probably not an important difference.

The model of consumer behaviour discussed above must be incorporated into a model of indirect tax reform. The tax model used is STATAX (Warren, 1987,1990) which is a static microsimulation model of the household sector. STATAX is structured around the 1988-89 HES and can be used to examine the impact on Australian households of *all* taxes levied by Federal, State and local governments, and of the Federal social security system.

The Personal Income Tax and Social Welfare modules are relatively straightforward and simply apply official rules on tax liability and welfare entitlements to the household, using data on the income and circumstances of each individual within the household.

The indirect tax module within STATAX models over 60 different taxes and is based on an Input-Output price model constructed using inter-industry (or National Accounts Input-Output) data (Warren, 1987). The UK SPIT model focusses only on VAT and excise duties and does not, in general, attempt to model those taxes which are imposed on inputs into the production and distribution of goods and services. The exception to this is an investigation into the effects of a carbon tax on fossil fuels (including production inputs) on the price of consumer goods (Symons et al. 1994).

A basic Input-Output price model is relatively simple but in practice a broad range of indirect taxes are levied both on inputs into the production process and on use by final consumers. These indirect taxes could be specific (per unit of output), ad valorem (on units of value) or lump sum; they could be levied on primary inputs, produced (or intermediate) inputs or on final demand; and they could be levied ex-factory (producer prices), on the wholesalers or on the retailers. An indirect tax module which ultimately accommodates all these factors is inevitably quite complex, a task which STATAX addresses but is detailed elsewhere (Warren 1987,1990). It should be noted that, in the analysis of this chapter, no account is taken of producer's response, the response of households to personal income tax or welfare payment changes, or the purchase of durable goods. Thus, this can be considered a short to medium-term model.

Modelling QUAIDS in SPIT within STATAX involves several steps. First, STATAX is used to estimate the impact of the current (direct and indirect) tax system and the welfare system on households. From this estimates are derived of:

- commodity prices;
- effective indirect tax rates (adjusting for National Accounts and HES differences);
- Personal Income Tax (PIT) and Social Welfare payments (SW); and
- rationed and unrationed expenditure.

From this one can identify for each household, ignoring the $h$ superscript:

$$Y_m = SW_m + I_m^f - S_m - PIT_m \tag{2}$$

where:

| | |
|---|---|
| $SW_m$ | government transfers in period $m$ |
| $I_m^f$ | factor income in period $m$ |
| $Y_m$ | total expenditure in period $m$ |
| $PIT_m$ | personal income tax in period $m$ |
| $S_m$ | savings in period $m$ |

and where $m = 0$ indicates pre-reform values and $m = 1$, post-reform values.

With this knowledge of income and expenditure from the survey data, one can calculate the following identities:

Pre-reform

$$y_0 = Y_0 - R_0^D - R_0^D \tag{3}$$

Post-reform

$$y_1 = Y_0 + (SW_1 - SW_0) + (I_1^f - I_0^f) - (S_1 - S_0) - (PIT_1 - PIT_0) - (P_1^D / P_0^D) * R_0^D - (P_1^H / P_0^H) * R_0^H \tag{4}$$

where: $y_m$ unrationed expenditure in period $m$ (non durables and non housing)

| | |
|---|---|
| $R_m^D$ | durable expenditure in period $m$ |
| $R_m^H$ | housing expenditure in period $m$ |
| $P_m^D$ | housing expenditure Price Index in period $m$ |
| $P_m^H$ | durable expenditure Price Index in period $m$ |

If it is then assumed factor income and savings are constant in nominal terms such that:

$$S_1 = S_0$$

$$I_1^f = I_0^f$$

then:

$$y_1 = Y_0 + (SW_1 - SW_0) - (PIT_1 - PIT_0) - (P_1^D / P_0^D) * R_0^D - (P_1^H / P_0^H) * R_0^H \qquad (5)$$

The budget share, (1), can then be calculated for each period using $y_m$ deflated by the appropriate Stone price index. Allowance has to be made for the 'share prediction errors' present at the estimation period for each household. This is interpreted as the unrationed expenditure for a household which QUAIDS cannot explain, because of the specific characteristics which are not included in the model but which do not change with prices. The predicted budget shares include these household-specific errors, assuming that they are independent of the price and income changes that the household faces. The predicted shares are set at a minimum of zero. If equation (1) predicts a negative share it is set to zero and the other shares are adjusted according to their unrationed expenditure to ensure that they sum to unity. The budget shares are then converted to expenditure for each commodity, group which then allows the estimation of tax revenue.

This process is repeated for all households and summed across all households to provide government tax revenue estimates which can then be grossed up to national accounts level.

## 4. Simulation of Revenue-neutral Commodity Tax Reforms

The above framework can be used to examine the response by consumers to tax reforms; the implications for tax revenue; and the distributional impact of the reforms. Three types of tax reforms in Australia are examined in the following sections: (i) revenue-neutral tax reforms; (ii) revenue-neutral tax mix reforms; and

(iii) non-revenue-neutral commodity tax reforms. This section focusses on changes in the structure of Australian commodity taxes, replacing the narrowly based Wholesale Sales Tax (WST) with a broad-based Goods and Services Tax (GST). Section 4.1 considers the replacement of WST with a 5.2 per cent GST, while Section 4.2 examines the replacement of WST with a 6.3 per cent GST which excludes food from the tax base.

## 4.1 A 5.2 per cent broad based GST replacing current WST

Consider the case of a revenue-neutral substitution of the Australian Wholesale Sales Tax by a broad-based, single rate GST.[1] Under the GST, all private final consumption expenditure is assumed taxable, with the only exceptions being those associated with banking and non-banking financial institutions, insurance, residential property, welfare and religious institutions, health, education and gambling. However, the residential building and construction industry, banking and non-banking financial institutions, insurance and gambling industries are assumed to be input taxed.

### Revenue effects

In 1990-91, the WST raised $8,780 million. If no behavioural response takes place, a GST levied at a single rate of 5.2 per cent on all goods and services would raise $8,628 million (or $152 million less). After allowing for the behavioural response of domestic consumers, total GST revenue falls by a further $57 million to $8,571 million.

As Table 2 shows, a first-round analysis of the impact of replacing WST with a 5.2 per cent GST would overstate total tax revenue (from all taxes) for domestic consumers by $207 million. After also taking account of the building industry and foreign households, total tax revenue collected under this reform option would be $208 million less after taking account of the behavioural responses of consumers.

---

[1] The wholesale tax (WST) is levied at three rates - 10 per cent, 20 per cent and 30 per cent - across a narrow range of manufactured commodities (excluding food and clothing). In February 1992 the 20 per cent rate on non-luxury passenger motor vehicles was reduced to 15 per cent. Whether the GST is single stage (retail sales tax) or multiple stage (value added tax) is not important when modelling this tax. However, it might be when considering administrative issues in regard to the tax or the avoidance and evasion of the tax.

Table 2: Total Tax Revenue Under Various Tax Reforms: 1990-91 ($M)

| Reform option | Domestic Household Consumers | | | Building | | Foreign House-holds | Total | |
|---|---|---|---|---|---|---|---|---|
| | Without behavioural reponse | After behavioural response | Difference | Residential | OTHER | | Without behavioural response | After behavioural response |
| *Replacement of WST with a:* | | | | | | | | |
| 5.2% GST | 108207.2 | 108000.3 | -206.9 | 500.2 | 186.9 | 6398.4 | 115292.7 | 115084.9 |
| 6.3%GST exc. food | 108048.0 | 107722.0 | -326.0 | 603.5 | 225.6 | 6416.3 | 115293.4 | 114967.4 |
| *20% Increase in Excise Duty on:* | | | | | | | | |
| Petrol | 109531.0 | 109347.6 | -183.4 | - | - | 7027.9 | 116558.9 | 116375.5 |
| Tobacco | 108796.3 | 108714.3 | -82.0 | - | - | 6902.1 | 115699.0 | 115617.0 |
| Alcohol | 108762.3 | 108567.3 | -195.0 | - | - | 6899.1 | 115617.0 | 115466.4 |
| *Tax Mix Change:* | | | | | | | | |
| 10% GST Less WST | 116140.6 | 115895.9 | -244.7 | 958.7 | 358.4 | 6478.0 | 123935.7 | 123691.0 |

As Table 3 shows, the majority of the decline in total tax revenue comes from taxes other than the GST. These results illustrate the importance of using a model that incorporates all taxes rather than a limited range (Lee and Pashardes, 1988; Jones and Savage, 1989) as a change in one tax can affect the revenue from what, at first sight, seems to be an unrelated tax. This broad-ranging impact on other taxes is caused by the change in consumption patterns resulting from the tax reform.

The biggest positive effect is on State taxes, through the registration and purchase of motor vehicles (some $125m), with smaller increases in the revenue collected from alcohol. A substantial fall is recorded for petrol excise, because people substitute out of this and into other commodities. This is not inconsistent with increased motoring expenditure, as the purchase of new cars could be accompanied by less driving of these motor vehicles due to petrol price increases. There is a sizeable fall in gambling tax revenue, one category of expenditure which consumers find relatively easy to substitute away from in order to maximize welfare from a given level of real expenditure. The overall change in consumer expenditure also appears to have affected revenue from payroll taxes.

Table 4 shows the changes in real and nominal expenditure driven by the tax induced changes in prices after the tax reforms. The largest price changes are

Table 3: Tax Revenue Collected From Domestic Households Before and After Allowing for Behavioural Responses to Commodity Price (Assuming Total Real Expenditure by Households is Constant) ($M)

| Type of tax | 1990-91 revenue | 5.2% GST | | | 6.3% GST excluding Food | | |
|---|---|---|---|---|---|---|---|
| | | Before behavioural response (NBR) | After behavioural response (ABR) | Difference (ABR less NBR) | Before behavioural response (NBR) | After behavioural response (ABR) | Difference (ABR less NBR) |
| Federal-Corporation | 9131.0 | 9131.0 | 9111.9 | -19.1 | 9131.0 | 9107.7 | -23.3 |
| GST | 0.0 | 8627.5 | 8570.8 | -56.7 | 8464.6 | 8399.9 | -64.7 |
| Wholesale Sales | 8780.4 | 0.0 | 0.0 | 0.0 | 0.0 | 0.0 | 0.0 |
| Liquor Excise | 501.2 | 501.2 | 524.6 | 23.4 | 501.2 | 518.8 | 17.6 |
| Petrol Excise | 6572.8 | 6572.8 | 6460.4 | -112.4 | 6572.8 | 6460.7 | -112.1 |
| Tobacco Excise | 1354.0 | 1354.0 | 1344.4 | -9.6 | 1354.0 | 1336.1 | -17.9 |
| Beer Excise | 878.5 | 878.5 | 908.8 | 30.3 | 878.5 | 894.9 | 16.4 |
| Primary Production | 506.9 | 502.4 | 498.4 | -4.0 | 502.5 | 501.2 | -1.3 |
| Customs Duties | 2598.6 | 2598.6 | 2603.9 | 5.3 | 2598.6 | 2598.8 | 0.2 |
| Personal Income Tax | 50168.0 | 50168.0 | 50168.0 | 0.0 | 50168.0 | 50168.0 | 0.0 |
| Fringe Benefit Tax | 1181.8 | 1181.8 | 1180.3 | -1.5 | 1181.8 | 1179.5 | -2.3 |
| Other Indirect | 1273.2 | 1272.0 | 1250.9 | -21.1 | 1272.4 | 1254.6 | -17.8 |
| **SUB TOTAL** | *82946.5* | *8787.8* | *82622.2* | *-165.6* | *82625.5* | *82420.0* | *-205.5* |
| State-Land Tax | 1704.7 | 1704.7 | 1701.4 | -3.3 | 1704.7 | 1700.9 | -3.8 |
| Motor Vehicle | 2689.0 | 2689.0 | 2813.6 | 124.6 | 2689.0 | 2764.6 | 75.6 |
| Stamp Duties | 3502.6 | 3502.8 | 3491.7 | -11.1 | 3502.8 | 3489.5 | -13.3 |
| Payroll Tax | 5099.9 | 5099.9 | 5065.8 | -34.1 | 5099.9 | 5066.9 | -33.0 |
| Gambling Taxes | 2034.8 | 2034.6 | 1973.4 | -61.2 | 2034.7 | 1954.2 | -80.5 |
| Franchise Taxes | 3145.8 | 3118.9 | 3109.7 | -9.2 | 3119.3 | 3093.7 | -25.6 |
| Other Indirect | 2581.1 | 2579.7 | 2545.6 | -34.1 | 2580.6 | 2549.1 | -31.5 |
| **SUB TOTAL** | *20757.9* | *20729.7* | *20701.2* | *-28.5* | *20731.0* | *20619.1* | *-111.9* |
| Local-Rates | 4689.6 | 4689.6 | 4676.9 | -12.7 | 4689.6 | 4682.6 | -7.0 |
| ***GRAND TOTAL*** | *108393.9* | *108207.2* | *108000.3* | *-206.9* | *108045.8* | *108048.0* | *-326.0* |

reductions for motoring, leisure goods and beer (11, 7.2 and 7.5 per cent, respectively) and increases in travel and clothing (5.8 and 4.6 per cent). All other price changes are under 4.5 per cent in absolute value. The price changes for these broad commodity groups are aggregate indices, reflecting the price changes of goods within each group. The most important changes in the quantity of goods consumed due to these price changes are increases in wine, personal goods and services and motoring (27 per cent, 6.3 per cent and 5.5 per cent) and a reduction in fuel of 6.2 per cent. Again, all other changes are under 4.5 per cent.

The increase of 6.3 per cent in expenditure on personal goods and services is partly induced by a fall in price of 4.3 per cent. This significant price fall is to be expected, because many of the items were previously taxed at either 20 per cent (electric shavers, razors, and shaving accessories, combs, brushes, toilet and dressing sets) or 30 per cent (jewellery, cosmetics). The resulting increased consumption of these goods is augmented by the increase in the price of substitutes (fuel, clothing and leisure goods) and the fall in the price of complements. The offsetting price effects are relatively small in aggregate.

For leisure goods there is a 7.2 per cent reduction in price, with only a 4.4 per cent increase in quantity consumed. This is less than implied by the own-price elasticity of around -0.8. The effect of the price changes of other goods reduces the own-price effect for this good. The main factors contributing to this less-than-expected effect are the reduction in the price of substitutes (notably, personal goods and services) and increases in the price of complements (notably, travel).

Table 4: Index of Nominal and Real Expenditure Under Different Indirect Tax Reform Packages (Assuming Total Real Expenditure by Households is Constant)*

| Commodity | 5.2% GST | | 6.3% GST excl Food | |
|---|---|---|---|---|
| | Nominal | Real | Nominal | Real |
| Beer | 0.959 | 1.037 | 0.957 | 1.022 |
| Wine | 1.237 | 1.272 | 1.266 | 1.287 |
| Spirits | 0.996 | 1.025 | 0.983 | 1.002 |
| Food | 1.003 | 0.980 | 0.993 | 1.002 |
| Fuel | 0.980 | 0.938 | 1.012 | 0.959 |
| Clothing | 1.030 | 0.985 | 1.030 | 0.974 |
| Household Services | 1.017 | 0.995 | 1.013 | 0.984 |
| Personal Goods and Services | 1.017 | 1.063 | 1.030 | 1.065 |
| Leisure Goods | 0.969 | 1.044 | 0.984 | 1.047 |
| Travel Fare | 1.074 | 1.015 | 1.047 | 0.975 |
| Tobacco | 1.037 | 0.993 | 1.040 | 0.987 |
| Motoring | 0.939 | 1.055 | 0.936 | 1.039 |
| Petrol | 1.010 | 0.974 | 1.019 | 0.973 |
| Leisure Services | 0.992 | 0.969 | 0.988 | 0.962 |
| Household Goods (Durables) | 0.958 | 1.000 | 0.967 | 1.000 |
| Housing | 1.002 | 1.000 | 1.003 | 1.000 |
| All Goods and Services | 0.994 | 1.000 | 0.994 | 1.000 |

* For example, column 2 shows real expenditure on beer increasing by 3.7 per cent after the 5.2 per cent GST reform

One commodity that has a large price change is motoring. This is not unexpected given the abolition of a 20 per cent WST on cars (and of 30 per cent on luxury cars). The own-price elasticity is -0.82 but quantity only increases by 5.5 per cent. This appears to be due to the impact on motoring of other price effects (as indicated by cross-price elasticities). One factor contributing to this low quantity response is the impact of reduced beer prices on motoring. With a cross-price elasticity of 0.3 (implying drinking and driving are substitutes), the large reduction in beer prices leads to reduced motoring expenditure.

## Distributional effects

One advantage of working with household data is that it facilitates a comprehensive distributional analysis of the effects of tax changes. It is possible to break down the impact for different parts of the income distribution or by other household characteristics. Figure 1(a) shows the average gain in income (measured in dollars per week) from the 5.2 per cent GST reform across adult equivalent gross income deciles[2]. Figure 1(b) shows this gain as a proportion of income. Before allowing for behavioural response the majority of the population would be net losers from the tax reform. Figure 1(b) shows that those at the bottom end of the income distribution would suffer most. Once behavioural response is allowed for (termed ABR), consumption patterns change due to the relative price changes and the gain is increased (or the loss is reduced). The tax saving due to behavioural response is given by the difference before and after response, shown as blocks in Figure 1. A comparison of Figures 1(a) and 1(b) shows that this absolute tax saving for lower income deciles is fairly constant across income deciles but, when taken as a proportion of income, falls substantially for higher deciles. This is because of the smaller level of savings by the lower income deciles.

This result is made more explicit if households are ranked by adult equivalent consumption rather than income (shown in Figures 1(c) and 1(d)). Then the tax saving of higher consumption deciles is much larger than that of the lower ones, indicating that they are able to switch away from goods that have become relatively expensive more easily, because a lower proportion of their consumption is devoted

---

[2] These decile groups are obtained by applying an equivalence scale where the first adult has a weight of one, second and subsequent adults a weight of 0.7 and all children a weight of 0.5. Therefore a household with 2 adults and 2 children is assumed equivalent to 2.7 adults. If the household has an income of $540pw, their adult equivalent income is $200pw and this income is used to rank the household.

to necessities. However, as a proportion of consumption (Figure 1(d)), the tax saving is more evenly spread.

As this discussion indicates, it is very important to include behavioural response of consumers in any analysis of indirect tax reform for several reasons. First, the response may affect taxes other than those immediately related to the tax reform package. Second, the distributional impact of the response is not uniform across different groups and, third, higher income groups have a greater opportunity to switch away from goods that have become relatively expensive.

## 4.2 6.3 per cent broad-based GST (excluding food) replacing WST

In the debate over a GST it is often stated that, for reasons of economic and administrative efficiency, the base of the GST should be kept broad and the tax should be levied at a single rate. While this may be true for administrative efficiency it need not be the case for economic (resource allocation) efficiency. Economic efficiency involves minimising the deadweight loss from taxes and, from Ramsey's rule, this implies that the tax rate for each good should be inversely related to the elasticity of demand for the good. Thus, a multiple rather than single rate GST is more efficient. In practice, issues other than just economic efficiency determine the tax rate - but it should be noted that exemptions from the GST base or a multiple rate can be supported on the grounds of economic efficiency. This section investigates the implications of such exemptions, again considering the importance of consumer behavioural response.

Table 2 indicates how the revenue from a 6.3 per cent GST which excludes food is distributed between domestic and foreign households. The impact on revenue collected due to behavioural response is marked and greater than that seen above with the broad-based single rate tax ($326 million compared with $207 million). Table 3 shows that the reduction in GST revenue caused by behavioural response is slightly more when food is exempt than with the broad base and lower tax rate ($64.7 million compared with $56.7 million). The higher GST rate (6.3 per cent rather than 5.2 per cent) means that the prices of some commodities have not fallen as much with the abolition of WST, notably motor vehicles and alcohol (generally reductions of one percentage point less). In consequence, the revenue from alcohol excise duties and State tax on motor vehicles does not increase as much. Overall, however, the over-riding conclusion is that the behavioural response by consumers to a revenue-neutral exemption of food from the GST base creates only a minor change in the revenue from GST.

Figure 1: Distributional Impact of Replacement of Wholesale Sales Tax by a 5.2 Per Cent Goods and Services Tax

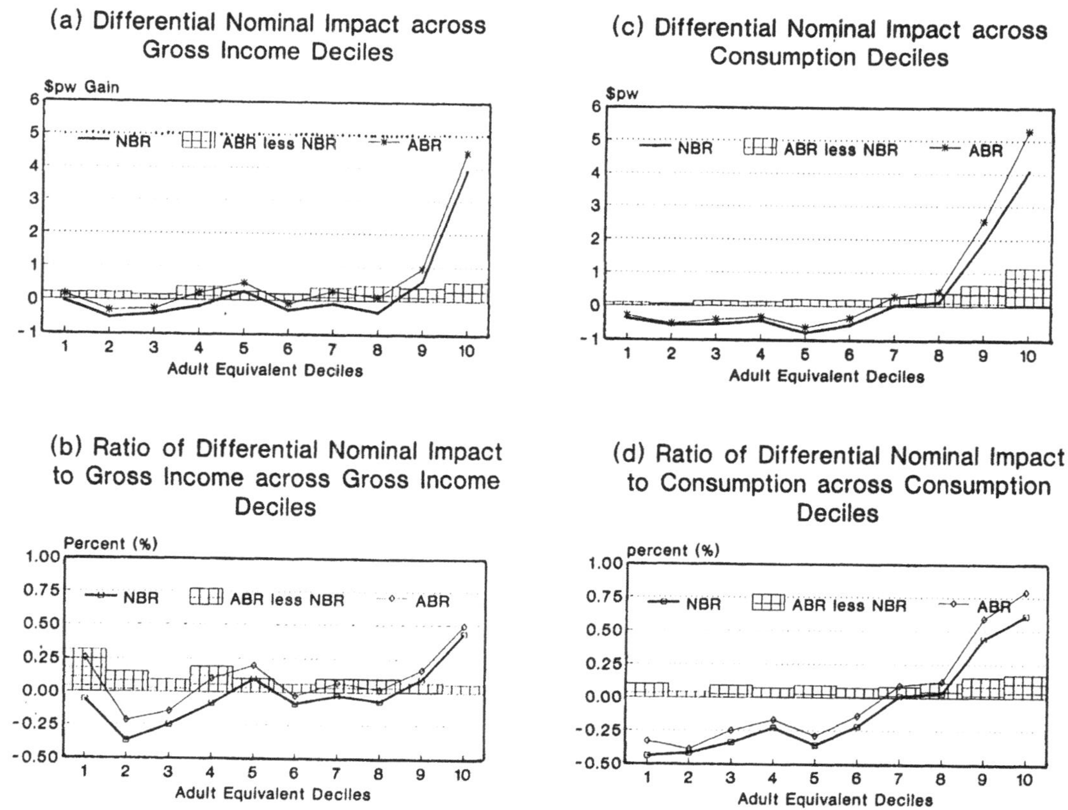

The exemption of food from the 6.3 per cent GST base yields some interesting changes in the distributional impact of the reform. Figure 2 indicates that this tax has a rather different impact than the 5.2 per cent broad-based GST on the lower adult equivalent income deciles before any behavioural response is considered.

Figure 2: Distributional Impact of Replacement of Wholesale Services Tax by a Goods and Services Tax

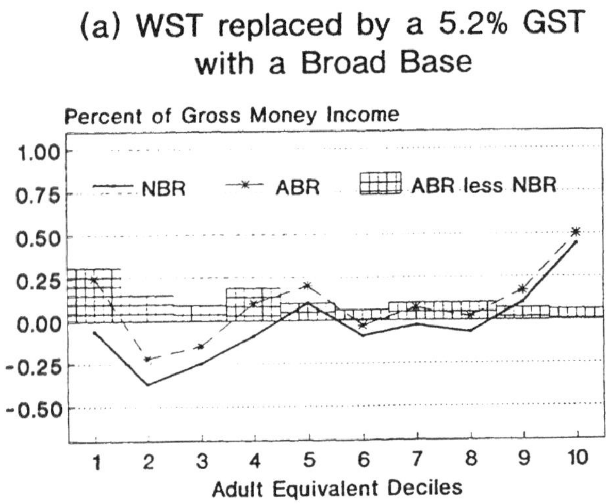


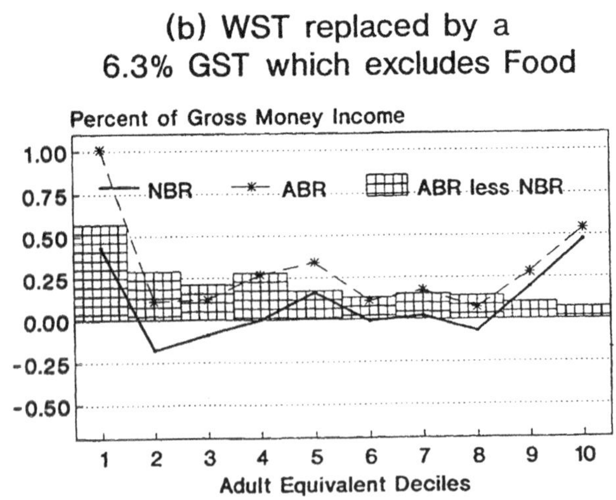

The exemption of food would appear to result in a greater level of commodity substitution by consumers compared to the case where food is included in the GST base. This is probably the result of greater opportunities for consumers to adjust their consumption in response to the greater opportunities for commodity substitution resulting from the relative price differential between food and non-food consumption. The now higher GST rate (6.3 per cent as against 5.2 per cent) will also further encourage substitution between the taxed commodities and those which were already untaxed commodities under the 5.2 per cent GST.

The average adult equivalent gain from the tax reform package increases from $0.28 to $0.53 a week as a result of the exemption of food and consumer behavioural response. A comparison of Figure 2(a) and Figure 2(b) shows that a majority of this gain accrues to the lower and middle income groups, who gain most from substituting into untaxed goods such as food.

The net effect of exempting food is that from an initial (pre-response) position where five deciles in Figure 2(b) were net losers from the tax reform package, their subsequent behavioural response has resulted in all of these groups becoming net gainers. However, the price of this benefit for households is that the tax reform package (where a 6.3 per cent GST replaces the WST) is now some $328.8 million in deficit.

## 5. Simulation of Revenue-neutral Tax Mix Changes

Tax reforms which have attracted much interest in recent years are those which involve a change in the mix of personal income and consumption taxes - so-called 'tax mix' reforms. Such reforms have been implemented in the UK in 1979, in New Zealand in 1986, in Canada in 1991 and were mooted for Australia in 1985, to name but a few.

This section will not review the case for or against such a tax reform but examines the sensitivity of revenue forecasts to consumer behavioural response. No attention will be given to the response by factor owners to lower personal income taxes or to increases in social security payments which might accompany such reform. In this sense, the study will only consider part of the total response by households to tax mix reforms.

For illustrative purposes, a 10 per cent GST is modelled on the same tax base as the 5.2 per cent GST examined in Section 4, with the same assumption that the government acts to maintain each household's real consumption expenditure. The revenue raised from domestic consumers by a 10 per cent GST replacing the WST is detailed in Tables 2 and 5. As with the revenue neutral 5.2 per cent GST, the impact of consumers' responses on GST revenue is but a minor part of the complex changes that occur in revenue collected following the introduction of a GST.

Table 6 indicates the source of the changes that occurred. The increase in the demand for alcohol stems from the replacement of a 20 per cent WST by a 10 per cent GST. Likewise, the increased State revenue from motor vehicle taxes stems from the fall in the price of motor vehicles, which were previously taxed at 20 per cent (and 30 per cent for luxury cars). An interesting result is that State governments have become net beneficiaries from the Federal tax reforms, although this is the result of the revenue from different State taxes moving in different directions. For the Federal Government, the impact of behavioural response on revenue is just as complex. Significant falls in petrol excise and, to a lesser extent, GST revenue, are offset by increases in the revenue from alcohol excise duties.

The distributional impact of the commodity tax reforms without consideration of any personal income tax or social welfare increases are shown in Table 7. It is apparent that the reforms impact quite severely on the lower income groups and that much of the compensation would need to be directed towards these groups. Probably the most difficult task for those designing the compensation package would be how to target each household group simply and directly. It was shown in Section 4 that excluding commodities consumed by lower income groups is not an efficient way of targeting the compensation dollar.

## 6. Non-revenue-neutral Excise Duty Changes

The two packages of reforms examined above have assumed real expenditure remains constant through compensation by the government. However, many reforms are implemented in a framework where governments do not undertake any compensatory measures. This is particularly the case with excise duties on petrol, tobacco and alcohol, which are often increased for reasons other than simply to raise general revenue. Without further pursuing the rationale for such tax increases, this section examines the consequences for consumers and government revenue of a 20 per cent increase in petrol, tobacco and alcohol excise duties.

Table 5: Tax Revenue Collected from Domestic Households Before and After Replacement of the WST with a 10 Per Cent GST (Assuming Total Real Expenditure by Households is Constant) ($m)

| Type of Tax | 1990-91 | 10% GST REPLACES WST | | |
|---|---|---|---|---|
| | | Before behavioural response (NBR) | After behavioural response (ABR) | Difference (ABR less NBR) |
| FEDERAL-Corporation Tax | 9131.0 | 9131.0 | 9108.1 | 22.9 |
| - GST | 0.0 | 16553.1 | 16365.1 | 187.5 |
| - Wholesale Sales Tax | 8780.4 | 0.0 | 0.0 | 0.0 |
| - Liquor Excise | 501.2 | 501.2 | 535.5 | -34.3 |
| - Petrol Excise | 6572.8 | 6572.8 | 6422.8 | 150.0 |
| - Tobacco Excise | 1354.0 | 1354.0 | 1372.8 | -18.8 |
| - Beer Excise | 878.5 | 878.5 | 951.5 | -73.1 |
| - Primary Production Taxes | 506.9 | 502.7 | 497.8 | 4.9 |
| - Customs Duties | 2598.6 | 2598.6 | 2585.5 | 13.0 |
| - Personal Income Tax | 50168.0 | 50168.0 | 50168.0 | 0.0 |
| - Fringe Benefits Tax | 1181.8 | 1181.8 | 1180.1 | 1.7 |
| - Other Indirect Taxes | 1273.2 | 1273.8 | 1241.8 | 32.0 |
| **SUB TOTAL** | *82946.5* | *90715.5* | *90429.6* | *285.9* |
| STATE-Land Tax | 1704.7 | 1704.7 | 1700.6 | 4.1 |
| - Motor Vehicles Taxes | 2689.0 | 2689.0 | 2863.8 | -174.8 |
| - Stamp Duties | 3502.6 | 3503.0 | 3495.2 | 7.8 |
| - Payroll Taxes | 5099.9 | 5099.9 | 5048.2 | 51.7 |
| - Gambling Taxes | 2034.8 | 2034.8 | 1994.1 | 40.6 |
| - Franchise Taxes | 3145.8 | 3120.4 | 3154.6 | -34.3 |
| - Other Indirect Taxes | 2581.1 | 2583.7 | 2536.7 | 47.0 |
| **SUB TOTAL** | *20757.9* | *20735.5* | *20793.2* | *-57.8* |
| LOCAL-Rates | 4689.6 | 4689.6 | 4672.8 | 16.8 |
| ***GRAND TOTAL** | *108393.9* | *116140.4* | *115895.9* | *244.7* |

## 6.1 Petrol excise

Table 8 illustrates the impact of a 20 per cent increase in the petrol excise duty on domestic consumers before and after allowing for their behavioural response to the consequent change in relative prices. The change in consumption patterns evident in Table 9 highlights the broad impact on expenditure on petrol of a change in petrol prices induced by a petrol price increase. The change in real expenditure is most marked for petrol, although there is a significant increase in expenditure on travel fares. Obviously, consumers are substituting away from the use of petrol towards travelling by other means, particularly public transport.

Table 6: Index of Nominal and Real Expenditure when WST Replaced with a 10 Per Cent GST (Assuming Total Real Expenditure by Households is Constant)

| Item | 10% GST REPLACES WST | |
|---|---|---|
| | Nominal | Real |
| Beer | 1.006 | 1.030 |
| Wine | 1.304 | 1.283 |
| Spirits | 1.053 | 1.035 |
| Food | 1.044 | 0.975 |
| Fuel | 1.031 | 0.944 |
| Clothing | 1.080 | 0.987 |
| Household Services | 1.054 | 1.000 |
| Personal Goods and Services | 1.069 | 1.063 |
| Leisure Goods | 1.015 | 1.037 |
| Travel Fares | 1.114 | 0.988 |
| Tobacco | 1.086 | 1.000 |
| Motoring | 0.984 | 1.047 |
| Petrol | 1.056 | 0.971 |
| Leisure Services | 1.033 | 0.981 |
| HH Goods (Durables) | 1.007 | 1.000 |
| Housing | 1.006 | 1.000 |

Table 7: Distributional Impact of a 10 Per Cent GST Replacing WST

| Adult Equivalent Deciles | Adult Equivalent Cash Gain ($p.w) | | | Adult Equivalent Cash Gain as a Percentage of Money Income % | | |
|---|---|---|---|---|---|---|
| | Before behavioural response | After behavioural response | Difference | Before behavioural response | After behavioural response | Difference |
| 1 | -8.89 | -8.63 | 0.26 | -13.78 | -13.37 | 0.40 |
| 2 | -7.24 | -7.03 | 0.21 | -5.11 | -4.96 | 0.15 |
| 3 | -7.62 | -7.43 | 0.19 | -4.47 | -4.36 | 0.11 |
| 4 | -9.14 | -8.72 | 0.42 | -4.42 | -4.22 | 0.20 |
| 5 | -9.55 | -9.25 | 0.30 | -3.71 | -3.60 | 0.12 |
| 6 | -11.33 | -11.07 | 0.26 | -3.64 | -3.55 | 0.08 |
| 7 | -12.22 | -11.80 | 0.43 | -3.26 | -3.15 | 0.11 |
| 8 | -15.66 | -15.14 | 0.52 | -3.42 | -3.30 | 0.11 |
| 9 | -16.97 | -16.48 | 0.49 | -2.96 | -2.87 | 0.09 |
| 10 | -18.88 | -18.23 | 0.65 | -2.10 | -2.02 | 0.07 |
| ALL | -11.74 | -11.37 | 0.37 | -3.40 | -3.29 | 0.11 |

The impact of the changes in nominal and real expenditure on the tax revenue shown in Table 8 again highlights the importance of adopting a modelling approach which considers the impact on all indirect taxes, not just those specifically under study. In the case of the petrol excise, the $79 million revenue fall after modelling consumer behavioural response is less than half of the impact on total tax collections ($183.4 million). The revenue from the WST is also affected, as is that from State taxes on motor vehicles and payroll tax.

It would appear that the petrol excise tax also encourages increased consumption of beer and tobacco - the consumer has substituted drinking and smoking in the local hotel for using their car to undertake other types of leisure activities. The distributional impact of these changes is shown in Table 10 and Figure 3(a). Lower income groups appear more sensitive to changes in the price of petrol in both absolute and relative terms. Admittedly, this is in a context in which the petrol excise impacts disproportionately heavily on the first decile. Nonetheless, the lower income deciles respond much more to this tax than do the higher income deciles.

## 6.2 Tobacco excise

A 20 per cent increase in tobacco excise results in a very limited response by consumers. Where a consumer response does occur, it is to commodities other than tobacco - with the apparent objective of enabling nominal expenditure on tobacco to be increased so as to maintain real expenditure on tobacco. To do this consumers reduce their real consumption of a range of other commodities, particularly spirits, fuel, leisure goods and travel fares. This provides the rather interesting result that, while there appears to be almost no change in the revenue from tobacco excise after including behavioural response ($0.6 million), tax collections overall are reduced by $82 million. This result appears to be attributable to smokers maintaining their habit by reducing their consumption of other commodities (which results in petrol excise collections falling along with WST revenue).

The overall distributional impact of the tobacco excise increase is shown in Table 10 and Figure 3(b). The response of the higher income groups is insignificant, but it is not for the lower income groups. An important conclusion from this result is that consumers' behavioural response acts to lesson the regressivity of the tax increase (and the revenue raised overall). The tax does, however, still appear to impact disproportionately on the lower income groups.

Table 8: Tax Revenue From Domestic Households Following a 20 per cent Increase in Excises - $m

| Type of Tax | 1990-1991 | Petrol Excise | | | Tobacco Excise | | | Alcohol Excise | | |
|---|---|---|---|---|---|---|---|---|---|---|
| | | Before behavioural response | After behavioural response | Difference | Before behavioural response | After behavioural response | Difference | Before behavioural response | After behavioural response | Difference |
| FEDERAL-Corporation | 9131.0 | 9131.0 | 9115.9 | -15.1 | 9131.0 | 9124.5 | -6.5 | 9131.0 | 9123.3 | -7.7 |
| - GST | 0.0 | 0.0 | 0.0 | 0.0 | 0.0 | 0.0 | -0.0 | 0.0 | 0.0 | 0.0 |
| - Wholesale Sales | 8780.4 | 8791.9 | 8767.7 | -24.2 | 8780.4 | 8762.6 | -17.8 | 8829.9 | 8778.5 | -51.4 |
| - Liquor Excise | 501.2 | 501.2 | 499.9 | -1.3 | 501.2 | 494.1 | -7.1 | 601.5 | 560.1 | -41.4 |
| - Petrol Excise | 6572.8 | 7614.4 | 7535.2 | -79.2 | 6572.8 | 6560.9 | -11.9 | 6572.8 | 6568.2 | -4.6 |
| - Tobacco Excise | 1354.0 | 1354.0 | 1363.1 | 9.1 | 1624.8 | 1624.2 | -0.6 | 1354.0 | 1358.3 | 4.3 |
| - Beer Excise | 878.5 | 878.5 | 887.0 | 8.5 | 878.5 | 886.7 | 8.2 | 1054.2 | 995.0 | -59.2 |
| - Primary Production | 506.9 | 507.0 | 506.4 | -0.6 | 506.9 | 506.8 | -0.1 | 506.9 | 506.7 | -0.2 |
| - Customs Duties | 2598.6 | 2598.6 | 2589.2 | -9.4 | 2598.6 | 2594.3 | -4.3 | 2598.6 | 2591.6 | -7.0 |
| - Personal Income Tax | 50168.0 | 50168.0 | 50168.0 | 0.0 | 50168.0 | 50168.0 | -0.0 | 50168.0 | 50168.0 | 0.0 |
| - Fringe Benefits Tax | 1181.8 | 1181.8 | 1180.2 | -1.6 | 1181.8 | 1181.3 | -0.5 | 1181.8 | 1180.7 | -1.1 |
| - Other Indirect | 1273.2 | 1273.2 | 1265.9 | -7.3 | 1273.2 | 1267.9 | -5.3 | 1273.2 | 1272.1 | -1.1 |
| **SUB TOTAL** | *82946.5* | *83999.7* | *83878.8* | *-120.9* | *83217.3* | *83171.1* | *-46.2* | *83272.1* | *83102.5* | *-169.6* |
| STATE-Land Tax | 1704.7 | 1704.7 | 1702.2 | -2.5 | 1704.7 | 1703.4 | -1.3 | 1704.7 | 1702.7 | -2.0 |
| - Motor Vehicle Taxes | 2689.0 | 2689.0 | 2673.5 | -15.5 | 2689.0 | 2679.6 | -9.4 | 2689.0 | 2713.2 | 24.2 |
| - Stamp Duties | 3502.6 | 3502.6 | 3499.3 | -3.3 | 3502.6 | 3500.6 | -2.0 | 3502.6 | 3502.3 | -0.3 |
| - Payroll Tax | 5099.9 | 5099.9 | 5082.0 | -17.9 | 5099.9 | 5089.8 | -10.1 | 5099.9 | 5090.0 | -9.9 |
| - Gambling Taxes | 2034.8 | 2034.8 | 2032.0 | -2.8 | 2034.8 | 2032.0 | -2.8 | 2034.8 | 2037.9 | 3.1 |
| - Franchise Taxes | 3145.8 | 3229.7 | 3226.7 | -3.0 | 3277.7 | 3277.6 | -0.1 | 3188.7 | 3153.4 | -35.3 |
| - Other Indirect | 2581.1 | 2581.2 | 2568.8 | -12.4 | 2581.1 | 2573.1 | -8.0 | 2581.1 | 2579.6 | -1.5 |
| **SUB TOTAL** | *20757.9* | *20841.9* | *20784.5* | *-57.4* | *20889.7* | *20856.1* | *-33.6* | *20800.9* | *20779.1* | *-21.8* |
| LOCAL-Rates | 4689.6 | 4689.6 | 4684.7 | -4.9 | 4689.6 | 4687.2 | -2.4 | 4689.6 | 4685.8 | -3.8 |
| ***GRAND TOTAL** | *108393.9* | *109531.0* | *109347.6* | *-183.4* | *108796.3* | *108714.3* | *-82.0* | *108762.3* | *108567.3* | *-195.0* |

Table 9: Index of Expenditure Following a 20 Per Cent Increase in Excise Duties (Assuming Total Nominal Expenditure by Households is Unchanged)

| Commodity | Petrol Excise | | Tobacco Excise | | Alcohol Excise | |
|---|---|---|---|---|---|---|
| | Nominal | Real | Nominal | Real | Nominal | Real |
| Beer | 1.016 | 1.016 | 1.013 | 1.013 | 1.000 | 0.944 |
| Wine | 0.990 | 0.990 | 0.991 | 0.991 | 0.971 | 0.911 |
| Spirits | 1.003 | 1.003 | 0.967 | 0.967 | 1.005 | 0.942 |
| Food | 0.997 | 0.997 | 0.999 | 0.999 | 1.001 | 1.001 |
| Fuel | 0.975 | 0.972 | 0.986 | 0.986 | 1.006 | 1.006 |
| Clothing | 0.991 | 0.991 | 1.000 | 1.000 | 0.984 | 0.984 |
| Household Services | 0.997 | 0.997 | 1.002 | 1.002 | 0.999 | 0.999 |
| Personal Goods and Services | 0.993 | 0.993 | 0.998 | 0.998 | 0.989 | 0.989 |
| Leisure Goods | 0.988 | 0.988 | 0.985 | 0.985 | 1.006 | 1.006 |
| Travel Fare | 1.044 | 1.043 | 0.989 | 0.989 | 1.033 | 1.033 |
| Tobacco | 1.007 | 1.007 | 1.093 | 1.000 | 1.003 | 1.003 |
| Motoring | 0.997 | 0.996 | 0.998 | 0.998 | 1.004 | 1.004 |
| Petrol | 1.046 | 0.981 | 0.999 | 0.999 | 0.999 | 0.999 |
| Leisure Services | 1.000 | 0.998 | 0.999 | 0.999 | 1.002 | 1.002 |
| HH Goods(Dur) | 1.000 | 1.000 | 1.000 | 1.000 | 1.000 | 1.000 |
| Housing | 1.000 | 1.000 | 1.000 | 1.000 | 1.000 | 1.000 |
| All Goods and Services | 1.000 | 0.996 | 1.000 | 0.999 | 1.000 | 0.998 |

While the finding that tobacco smoking is only marginally affected by the level of excise on tobacco does not correlate with anecdotal evidence that, in Australia, an increase in excise on tobacco is correlated with a decline in smoking, it should be recognised that taxes are only one of the factors which influence consumer behaviour. In the case of tobacco, it is heavily influenced by anti-smoking campaigns, workplace restrictions by employers and community attitudes.

## 6.3 Alcohol excise

The revenue impact of a 20 per cent increase in alcohol (beer and liquor) excise duty is shown in Table 8. While the tax increase led to an estimated pre-behavioural response increase in revenue of some $368.4 million, consumers' response to the increase resulted in over half (52.9 per cent) of the initial estimate of increased revenue being eroded (or $195.0 million). This contrasts with the case of the petrol excise where the reduction was only 16.1 per cent and, for tobacco, 20.4 per cent. A large part of the decline in revenue from the response to alcohol excise comes from the change in WST revenue. This is because most alcohol is also taxed at a WST rate of 20 per cent - and any decline in alcohol consumption induced by the excise increase will thus flow over to a reduction in revenue from the WST.

Table 10: Distributional Impact of 20 Per Cent Increase in Excise Duties (Assuming Total Nominal Expenditure by Households is Constant)

| Adult Equivalent Deciles | Adult Equivalent Cash Gain ($pw) | | | Adult Equivalent Cash Gain as a Percentage of Money Income (%) | | |
|---|---|---|---|---|---|---|
| | Before behavioural response | After behavioural response | Difference | Before behavioural response | After behavioural response | Difference |
| **PETROL** | | | | | | |
| 1 | -1.41 | -1.11 | 0.30 | -2.19 | -1.72 | 0.46 |
| 2 | -0.96 | -0.77 | 0.19 | -0.68 | -0.54 | 0.13 |
| 3 | -1.01 | -0.83 | 0.18 | -0.59 | -0.49 | 0.11 |
| 4 | -1.23 | -0.99 | 0.24 | -0.60 | -0.48 | 0.12 |
| 5 | -1.39 | -1.18 | 0.21 | -0.54 | -0.46 | 0.08 |
| 6 | -1.64 | -1.36 | 0.28 | -0.53 | -0.44 | 0.09 |
| 7 | -1.78 | -1.49 | 0.28 | -0.48 | -0.40 | 0.07 |
| 8 | -2.19 | -1.81 | 0.38 | -0.48 | -0.39 | 0.08 |
| 9 | -2.48 | -2.12 | 0.36 | -0.43 | -0.37 | 0.06 |
| 10 | -3.16 | -2.81 | 0.35 | -0.35 | -0.31 | 0.04 |
| ALL | -1.72 | -1.45 | 0.28 | -0.50 | -0.42 | 0.08 |
| **TOBACCO** | | | | | | |
| 1 | -0.45 | -0.36 | 0.09 | -0.70 | -0.56 | 0.14 |
| 2 | -0.53 | -0.35 | 0.18 | -0.37 | -0.25 | 0.13 |
| 3 | -0.52 | -0.34 | 0.18 | -0.30 | -0.20 | 0.11 |
| 4 | -0.55 | -0.35 | 0.20 | -0.27 | -0.17 | 0.10 |
| 5 | -0.69 | -0.48 | 0.21 | -0.27 | -0.19 | 0.08 |
| 6 | -0.62 | -0.53 | 0.09 | -0.20 | -0.17 | 0.03 |
| 7 | -0.70 | -0.58 | 0.13 | -0.19 | -0.15 | 0.03 |
| 8 | -0.72 | -0.63 | 0.09 | -0.16 | -0.14 | 0.02 |
| 9 | -0.66 | -0.61 | 0.05 | -0.11 | -0.11 | 0.01 |
| 10 | -0.66 | -0.63 | 0.02 | -0.07 | -0.07 | 0.00 |
| ALL | -0.61 | -0.49 | 0.12 | -0.18 | -0.14 | 0.03 |
| **ALCOHOL** | | | | | | |
| 1 | -0.41 | -0.18 | 0.22 | -0.64 | -0.28 | 0.34 |
| 2 | -0.29 | -0.13 | 0.16 | -0.20 | -0.09 | 0.11 |
| 3 | -0.29 | -0.14 | 0.15 | -0.17 | -0.08 | 0.09 |
| 4 | -0.32 | -0.19 | 0.13 | -0.15 | -0.09 | 0.06 |
| 5 | -0.45 | -0.20 | 0.25 | -0.17 | -0.08 | 0.10 |
| 6 | -0.47 | -0.23 | 0.24 | -0.15 | -0.07 | 0.08 |
| 7 | -0.52 | -0.25 | 0.27 | -0.14 | -0.07 | 0.07 |
| 8 | -0.82 | -0.38 | 0.44 | -0.18 | -0.08 | 0.10 |
| 9 | -0.93 | -0.41 | 0.52 | -0.16 | -0.07 | 0.09 |
| 10 | -1.10 | -0.51 | 0.58 | -0.12 | -0.06 | 0.06 |
| ALL | -0.56 | -0.26 | 0.30 | -0.16 | -0.08 | 0.09 |

Figure 3: Distributional Impact of Increases in Excise Duties

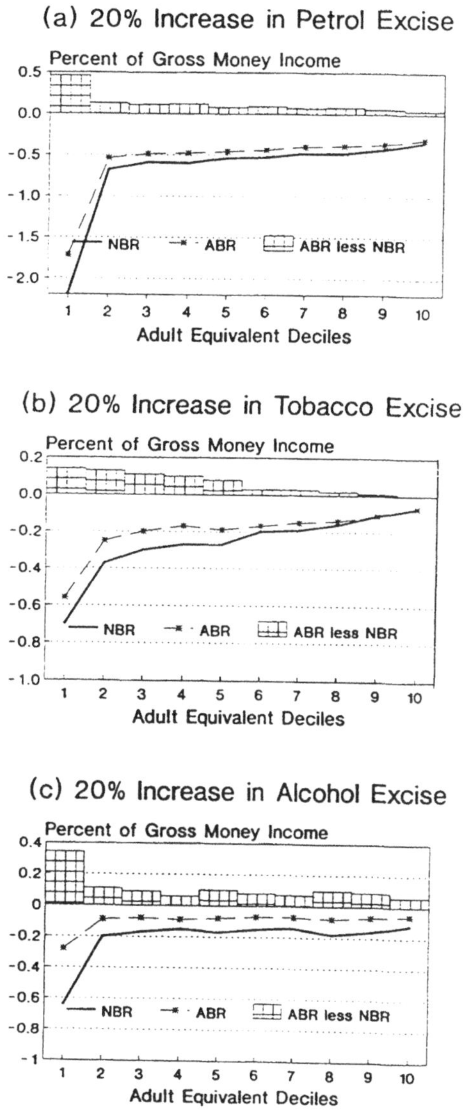

The distributional impact of the alcohol excise duty increase is shown in Table 10 and Figure 3(c). Excluding the lowest income decile[3], the excise change is almost proportional to income after consideration of consumer behavioural response. However, if consumption of alcohol is (as expected) a declining proportion of increasing income, the results in Figure 3(c) indicate a major adjustment by the higher income groups. The net effect of this behavioural response is to almost cut the tax increase in half. In the case of the other excise duties, it has been the lower income deciles that have been most responsive to the tax increases.

The reason for this is most likely the forms in which alcohol is consumed and the different patterns of consumption for each type of alcohol. For example, beer consumption has a different distribution across income deciles to that of wine or spirits. Combined, the expenditure on these different types of alcohol is closer to being proportional to income than is, say, beer consumption alone.

The modelling in this section of various increases in excise duties (with no government compensation to maintain real expenditure constant) has shown how consumers respond by substituting away from now relatively expensive products.

## 7. Conclusion

The findings in this paper contain several important lessons for those considering the behavioural response of consumers to GST-based tax reforms. First, the impact on tax revenues of excluding commodities from the GST base are not additive because the interaction between taxes and their tax bases is not linear, due to the very different consumption patterns of different households and the different way combinations of taxes impact on households. Second, the cross-price elasticities of goods and services are as important as own-price elasticities to the analysis of the impact of tax changes on tax revenue collection. Third, taking account of behavioural response when assessing the distributional impact of commodity tax reform appears to make most difference to the lower income groups. Fourth, adding exemptions to the GST base, and subsequently adjusting the GST rate to ensure the

[3] The distributional results for the lowest income decile should be treated with some caution as this group comprises a range of individuals whose experience could be argued as being transitional. Included here are the unemployed and the self-employed (such as farmers whose income is temporarily low due to factors such as drought). In some studies, these groups are excluded since their experience is not indicative of their normal position (as in Warren 1991).

tax reform package (before behavioural response) is revenue-neutral, has little impact on the revenue from the GST. The main effect is on the revenue from taxes other than the GST. Any study into the behavioural response of households to changes in one tax should therefore give considerable attention to modelling the potential impact of the reforms on the revenue from the full range of taxes.

The major conclusion from this paper is that any failure to consider the behavioural response of consumers to tax-induced changes in relative prices is likely to lead to inaccurate estimates of tax revenue and a misleading picture of the distributional impact of commodity tax reforms. Moreover, it is clear that different groups respond differently depending upon their circumstances.

However, this paper considers only one of the potential responses that could occur following the introduction of tax reforms. No consideration has been given to the behavioural response of factor owners (labour and capital), to the response of producers through the production techniques adopted, or to the response by foreigners. Even so, the study has highlighted the importance of considering such factors, through a case study of the response of consumers to commodity tax reforms in Australia.

## References

Baker, P., S. McKay and E. Symons (1990), 'The simulation of indirect tax reforms: The IFS simulation program for indirect taxation (SPIT)', Working Paper W90/11, Institute for Fiscal Studies, London.

Blundell, R., P. Pashardes and G. Weber (1993), 'What do we learn about consumer demand patterns from microdata', *American Economic Review*, Vol. 83, No. 3.

Blundell, R.W., A.S. Duncan and C. Meghir (1992), 'Taxation in empirical labour supply models: Lone mothers in the UK', *Economic Journal*, Vol. 102, No. 411.

Cullis, J. and P. Jones (1992), *Public Finance and Public Choice*, McGraw-Hill, London.

Deaton, A. and J. Muellbauer (1980), *Economics and Consumer Behaviour*, Cambridge University Press, Cambridge, England.

Dilnot, A. and A. Duncan (1992), 'Lone mothers, family credits and paid work', *Fiscal Studies*, Vol. 13, No 1.

Jones, G. and E. Savage (1989), 'Modelling Australian indirect tax reforms: A welfare consistent approach', in: J.G. Head (ed.), *Australian Tax Reforms in Retrospect and Prospect*, Conference Series 8, Australian Tax Research Foundation, Sydney.

Lee, C. and P. Pashardes (1988), 'Who pays indirect taxes?', IFS Report Series R32, Institute for Fiscal Studies, London.

Lee, C., M. Pearson and S. Smith (1988), 'Fiscal harmonization: An analysis of the European Commission's proposal', IFS Report Series R28, Institute for Fiscal Studies, London.

Pearson, M. and S. Smith (1991), 'The European carbon tax: An assessment of the European Commission's proposal', IFS Report Series R39, Institute for Fiscal Studies, London.

Symons, E., and N.A. Warren (1993), 'Australian commodity tax reforms and consumer behavioural response', Discussion Paper 93/13, University of New South Wales, July.

Symons, E., J. Proops and P. Gay, (1994), 'Carbon taxes, consumer demand and carbon dioxide emissions: A simulation analysis for the UK', *Fiscal Studies*, Vol. 15, No. 2.

Warren, N.A., (1987), 'The distributional impact of a change in the tax mix in Australia', Research Study No. 6, Australian Tax Research Foundation, Sydney.

Warren, N.A., (1990), 'A goods and services tax for Australia: An evaluation of the National Farmers' Federation outlay tax reform proposal', Research Study No. 11, Australian Tax Research Foundation, Sydney.

Warren, N.A., (1991), 'The changing incidence of federal indirect taxes: 1975-76 to 1988-89', Background Paper No 13, Economic Planning Advisory Council, AGPS, Canberra.

Joachim Merz

University of Lüneburg
Germany

# Chapter 9: Market and Non-Market Labour Supply and the Impact of the Recent German Tax Reform - Incorporating Behavioural Response

## Abstract

*This chapter analyses the market and non-market labour supply effects of the 1990 German tax reform. The theoretical microeconomic model explicitly takes into account taxes, transfer payments, and the socioeconomic characteristics of the individual and the household. Activity-specific participation probabilities, relevant wages and hours of work are estimated with selectivity correction. The microdata base is the representative Sfb 3-Secondary Occupation Survey 1984, enlarged by taxation, regional economic and labour demand information. The results of a combined dynamic and static microsimulation show that there are second-round effects, on both market and non-market activities, which differ greatly by socioeconomic group.*

## 1. Introduction

Individual labour supply and the incentive or disincentive effects of governmental tax and transfer programs are an ongoing topic of economic and social-political debate. There are numerous studies on labour supply in general (with and without taxes), which concentrate on market activities in the formal economy. However, because of changed opportunity costs and other market circumstances, tax and transfer reforms might affect not only labour supply in paid work but also *unpaid* non-market economic activities within a household. In addition, female labour

* Financial support by the German National Science Foundation (DFG) for my former project 'Market and Non-market Activities of Private Households', a research project of the Sonderforschungsbereich 3 (Sfb 3) 'Microanalytic Foundation of Social Policy' at the Universities of Frankfurt and Mannheim and for my recent DFG-project on 'Time Use in Employment and Household Production - Dynamic Microanalyses with Paneldata' at the University of Luneburg is gratefully acknowledged. The tax computations within the Socio-Economic Panel data (1st wave, 1984) were done by the researchers of the Sfb 3 tax project (C-8) (v. Essen, et al. 1988). I gratefully acknowledge their assistance and support.

supply is strongly related to female household activities. Therefore, a combined empirical investigation of paid and unpaid labour supply is required to comprehensively analyse the impact of tax reforms.

Income taxes, in principle, can lead to increases or reductions in the supply of market labour. Reduced taxes raise net wages and thus raise labour supply (the substitution effect). At the same time, however, reduced taxes raise total income. As a result, demand for goods, including leisure time activities, will rise. Increased consumption of leisure leads to decreasing labour supply (the income effect). Whether the substitution or the income effect finally dominates the supply of labour can only be analysed on an empirical basis. Moreover, the effect of taxes on unpaid non-market labour supply - that is, on household production activities - is, so far, little known and little understood or investigated (although see Flood, 1988). The effects of tax reform on market and non-market labour supply is the focus of this microanalysis. The study employs a microeconomic model, microeconometric estimates describing behavioural response, and microsimulation of second-round impacts of the 1990 German tax reform. To open the box of modelling behavioural response in a microsimulation model, this study enhances earlier work (Merz, 1990) via a more detailed theoretical and econometric discussion and by additional microsimulation results.

In this chapter, a microeconomic model of multiple market and non-market labour supply activities within the time allocation/household production framework is first formulated. Second, a microeconometric model is used to estimate activity-specific participation, endogeneous (marginal) wage rates and hours of work (taking into account the selectivity bias problem of non-negative wages and hours of work). Third, a microsimulation of the German 1990 tax reform package is run to show the incentive and disincentive effects on market and non-market labour supply. The microanalysis of tax reform effects is based on a representative sample for the Federal Republic of Germany (FRG), the Sfb 3-Secondary Occupation Survey 1984, enlarged by regional, economic and labour demand information.

## 2. A Microeconomic Labor Supply Model of Multiple Market and Non-Market Activities

The first part of the analysis involves a time allocation microeconomic model of multiple labour supply activities.

## 2.1 Time allocation of market and non-market labour supply

The microeconomic time allocation model assumes a rational decision-maker who optimally allocates time and commodities according to a utility maximising process. In the sequel of the household production model (Becker, 1965; Lancaster, 1966; Gronau, 1977; and Gronau 1980), an individual receives utility, u, not from the purchased market commodities themselves but from the produced output of goods and services **z** in the household:

$$u = u(\mathbf{z}) = u(z_E, z_F) \tag{1}$$

Given the empirical data at hand (Section 4) the household goods and services produced may be divided into specific economic non-market products $z_E$, and all other leisure household products combined in $z_F$. Thus $\mathbf{z} = (z_E, z_F)$. The efficient production of all non-market goods on the basis of the household technology at one's disposal is described by the household production function:

$$\mathbf{z} = \mathbf{z}(X, t_C) = \mathbf{z}(X, t_E, t_F), \tag{2}$$

where the Hicksian composite purchased commodity X is combined with consumption time $t_C$ producing $\mathbf{z} = (z_E, z_F)'$, where $z_E = z_E(X, t_E)$ and $z_F = z_E(X, t_F)$.

The inputs in the household production process are restricted by a time constraint and a budget constraint. Total consumption time $t_C$ is divided into time spent on non-market production $t_E$ and time spent on all other household production activities $t_F$, together simply called 'leisure'. Thus $t_C = t_E + t_F$. Time spent in paid market activities $t_M$ is divided into time spent in a primary occupation $t_H$ and time spent in a secondary occupation $t_N$. Thus $t_M = t_H + t_N$. Total time is then allocated within the time constraint to:

$$t = t_M + t_C = t_H + t_N + t_E + t_F. \tag{3}$$

The disposable income available for consumer expenditures PX (P is the price of the Hicksian 'composite commodity') is made up of non-labour income and labour income. Within a family/household unit, the activities of other household members influence the individual decision maker. The household's non-labour income ($V^b$), in addition to the individual's non-labour income ($V^b_{individual}$), might also include other household members' labour ($I^b_{others}$) and non-labour income ($V^b_{others}$):

$$V^b = V^b_{individual} + I^b_{others} + V^b_{others} \tag{4a}$$

There might be further intra-household interdependencies (via joint utility, household production etc.) but the above function, at least, is in accordance with our empirical data. With multiple market activities, labour income might be subdivided into income from a primary occupation $w_H{}^b t_H$ and income from a secondary occupation $w_N{}^b t_N$, where $w_H{}^b$ and $w_N{}^b$ are the respective gross wages per time unit (eg. hourly wage rates). The budget restriction is then:

$$PX = V^b + w_H{}^b t_H + w_N{}^b t_N \tag{4b}$$

with gross non-labour income and gross individual labour income as defined above.

## 2.2 Taxes, transfers and socioeconomic characteristics

Taxes paid either on labour or non-labour income reduce the corresponding gross terms (the gross non-labour income $V^b$ and the gross labour income $w_H{}^b t_H$ + $w_N{}^b t$) by the tax liability $T(V^b, w_H{}^b t_H, w_N{}^b t_N)$. Considering the taxes explicitly, the budget restriction with taxes is:

$$PX = V^b + w_H{}^b t_H + w_N{}^b t_N - T(V^b, w_H{}^b t_H, w_N{}^b t_N) \tag{5a}$$

or

$$PX = I = V + w_H t_H + w_N t_N \tag{5b}$$

where

$$w_H = (1 - \tau_A)\, w_H{}^b \ , \ w_N = (1 - \tau_A)\, w_N{}^b \tag{5c}$$

$$V = V^b - \{T(V^b, w_H{}^b t_H, w_N{}^b t_N) - \tau_A [w_H{}^b t_H + w_N{}^b t_N]\} \tag{5d}$$

$$\tau_A = \partial T(V^b, w_H{}^b t_H, w_N{}^b t_N) / \partial (w_H{}^b t_H + w_N{}^b t_N]. \tag{5e}$$

With the marginal tax rate of wage income, $\tau_A$, the gross terms (with an upper index b in (5c)) are transformed into marginal net wage rates $w_H$ and $w_N$. V is the transformed so-called 'virtual' non-labour income and marks that kind of non-labour income that is assigned to the net wage rate of a certain segment of the kinked budget curve.[1] As equation (5d) shows, the virtual non-labour income is not only directly influenced by the total tax liabilities T(.) but also, and in particular, by the marginal tax liability of labour income $\tau_A [w_H{}^b t_H + w_N{}^b t_N]$.

Transfers are a part of gross non-labour income $V^b$ and may include property income, subsidies of third private parties and, in particular, public and social

---

[1] Burtless and Hausman (1978) describe V as 'virtual' because it is assumed that the decision maker acts 'as if' its non labour income is $V_i$ when being on the budget segment i.

security payments. With regard to other household members, the individual's tax and transfer function only includes those parts of total household labour and non-labour income which are critical for the individual under investigation.

The time restriction (3) and the budget restriction (5b) can be integrated in an extended full-income equation following Becker (1965). The 'full-income' S is total time income, allocated to the consumption expenditures PX and valued consumption time ($t_C = t_E + t_F$), valued by the wage from secondary occupation $w_N$:

$$PX + w_N(t_E + t_F) = V + w_H t_H + w_N(t - t_H) = S. \tag{6}$$

In the case of multiple paid jobs, the time spent on the secondary occupation wage is regarded as opportunity cost. This is based on the assumption that a certain desired income is only achievable if, in addition to a *rationed primary occupation*, a secondary occupation is pursued[2]. Hence, labour supply in the secondary occupation $t_N$ becomes the decision variable in the case of multiple (here two) market labour supply activities.

The formulation of the time allocation problem so far is based on a representative individual decision-maker with some intra-household interdependencies. In empirical applications, different individuals will have different preferences and make different allocation decisions, even if they face the same non-labour income and the same market wage rates. Different preferences can be explicitly taken into account via individual observed variables - that is, socioeconomic characteristics (such as gender, age, household situation etc.). If they are unobservable they may be included in the error term of the microeconometric specification. If they are observable, they consequently enter the utility function $u = u(\mathbf{z};\mathbf{s})$ as preference factors described by a vector s. The following derivation of the optimal market and non-market labour supply time allocation equations takes into account these different individual preferences, via socioeconomic characteristics.

## 2.3 Optimal time allocation equations

To derive the optimal time allocation decision, the individual maximises his/her utility (1) with regard to the output of household goods and services described by the household production function (2). This decision is influenced by the time

---

[2] Shisko and Rostker 1976 follow a similar rationing approach. However, they do not include taxes and informal activities.

constraint (3), the budget constraint (5) and the available household production technology:[3]

$$\max \{u = u(z;s) = u(z_E,z_F;s), \tag{7a}$$

$$\text{s.t.} \quad t = t_M + t_C = t_H + t_N + t_E + t_F \quad \text{and}$$

$$PX = V^b + w_H{}^b t_H + w_N{}^b t_N - T(V^b, w_H{}^b t_H, w_N{}^b t_N)\}$$

The corresponding Lagrangian with respect to the input terms is:

$$L(X,t_E,t_F,\lambda_I,\lambda_t;s) = u[f(X,t_E,t_F);s] - \lambda_I(PX - V - w_H t_H - w_N t_N) - \tag{7b}$$

$$\lambda_t(t_H + t_N + t_E + t_F - t)\,.$$

The first order conditions are $u_X = \lambda_I P$ , $u_{tF} = u_{tE} = \lambda_t$ , $w_N = \lambda_t/\lambda_I$ and after transformations via $u_{tF}/u_X = \lambda_t/(\lambda_I P) = G^r$ to $(G^r - f') = 0$ with $f' = u_{tF}/u_X$ respectively:

$$t_E(G^r - f') = 0, \tag{7c}$$

as well via $w_N = \lambda_t/\lambda_I$ with $G = \lambda_t/\lambda_I = G^r P$ to $(G - w_N) = 0$ and

$$t_N(G - w_N) = 0. \tag{7d}$$

In order to achieve an interior solution (given a convex budget set and a quasi concave utility function) for the optimal home production time $t_E$ ($t_E > 0$), the optimum condition is $G^r = f'$. That is, the shadow price of leisure ($G^r$) has to be equal to the shadow price of the household production ($f' = f'_{tE}/f'_X$ is the ratio of marginal productivities). As long as the shadow price of household production is lower than that of leisure ($f' < G^r$), there will be no time invested in household production of non-market economic activities (that is, $t_E = 0$).

In order to achieve an interior solution regarding the optimal labour supply in secondary occupation $t_N$ ($t_N > 0$), the optimum condition is $G = w_N$. That is, the shadow price of leisure (G = marginal rate of substitution between consumption and leisure) must correspond to the marginal net wage ($w_N$). As long as the marginal net wage is lower than the shadow price of leisure ($w_N < b$), there will be

---

[3] A similar model was formulated by Flood (1988) who, with respect to Gronau (1977, 1980), takes account of taxes; besides taxes and transfer payments, this study also includes multiple activities.

no secondary labour supply (that is, $t_N = 0$). The Marshallian allocation equations derived from these optimum conditions for the consumption goods X and the leisure time $t_C$ are:

$$X = X(P,w_N,S,s) = X(P,w_H,w_N,V,t_H,t,s) \tag{8a}$$

$$t_C = t_C(P,w_N,S,s) = t_C(P,w_H,w_N,V,t_H,t,s) \tag{8b}$$

The relevant optimal time allocation equations for labour supply in the secondary occupation $t_N$ and in the household production $t_E$ are, therefore:

$$t_N = t - t_H - t_C = t_N(P,w_H,w_N,V,t_H,t,s) \tag{8c}$$

$$t_E = t_E(P,w_N,S,s;f'_{tE}) = t_E(P,w_H,w_N,V,t_H,t,s;f') \tag{8d}$$

where $t_H$ again is given rationed primary occupation hours of work.

Note that the optimal household production, in particular, results from the ratio of the marginal productivities of the household production function $f' = f'_{tE}/f'_X$. The above input-oriented multiple time allocation problem and its solution is formulated taking account of taxes and transfer payments, via the marginal net wages $w_H$ and $w_N$ and the virtual non-labour income V. For further details, and for examination of an output oriented formulation with explicit household commodity shadow prices arising from a three-stage activity model and with regard to duality, see Merz (1989).

## 3. Microeconometric Specification of Participation, Wage Rates and Labour Supply

Further results support the hypothesis that labour supply in the primary occupation is widely independent of further multiple activities. The modelling consequence is a block-recursive microeconometric time allocation specification with primary labour supply in the first block, followed by the block of secondary occupation and household production activity hours.

The first block (explaining time spent in a primary occupation) is determined by the model's non-labour income, the estimated primary (marginal net) wage rate, socioeconomic characteristics, and the (Heckman, 1979) selection correction for primary occupation. Time allocated to the second market occupation as well as to household production could be influenced by primary market activities. Following the microeconomic model, the respective hours equation in the second block encompasses the estimated primary and decision-driving secondary (marginal net) wage, time spent in primary occupation, non-labour income including other household member income, socioeconomic characteristics, and the respective selection correction.

In the sense of the seemingly unrelated regression (SUR) argument relating to complete demand system estimation (that is, that a simultaneous estimation of all system equations with the same set of explanatory variables could similarly be estimated consistently by single equation OLS), secondary occupation and household production hours will be estimated here, equation by equation, by COLS.

The relevant (marginal) wage rates are endogenized and then considered explicitly as explanatory variables in the hours of work equations. The market wage rates for primary and secondary occupation may, on the one hand, depend on individual abilities, such as schooling, vocational training, work experience (human capital terms) and, on the other hand, on labour market conditions which are the elements of an explanatory vector $\mathbf{x}_{wi}$ of an individual i. An error term in the wage equation describes, among other things, non-observable variables such as motivation, innate abilities etc. Within each block the microeconometric time allocation model is then specified in three stages:

***Stage 1:*** Estimation of the participation probability (PROBIT) of an activity j; calculation of the 'selectivity bias' correction term $\lambda_{ji}$ for all persons with an activity j (j=primary occupation (H), secondary occupation (N), household production (E) respectively);

***Stage 2:*** 'selectivity bias' corrected estimation (COLS) of the (marginal) wage rate in the activity j (j=H,N);

***Stage 3:*** 'selectivity bias' corrected estimation (COLS) of weekly hours of work with endogenized (marginal) wage rates from stage 2 for all formal and informal activities j (j=H,N,E).

One problem, at least, remains - that of the kinked budget curve when non-linear taxes and transfers tariffs are regarded with the consequence of non-linear estimation. Fortunately, the 1990 German tax reform structure results in an almost overall linear marginal tax rate. It seems therefore to be appropriate to approximate the in-principle kinked budget curve by a respective linear specification of the model. Hence, and finally, the block-recursive equation system with a linear approximation of the respective relations will be specified as follows:

**Block 1: Primary occupation**

*Participation:* $N^{-1}[p_j(t_{ij}>0)] = \mathbf{a}_j'\mathbf{x}_{pji} + \varepsilon_{pji}$ $(i \in A_j, j=H)$ (9a)

*Wages:* $\log w_{ji} = \beta_j'\mathbf{x}_{wji} + \delta_{wj}\lambda_{ji} + \varepsilon_{wji}$ $(i \in A_j, j=H)$ (9b)

*Hours:* $t_{Hi} = \gamma_{wHH}(\hat{w}_{Hi}) + \gamma_{VH}V_i + \gamma_{sH}'\mathbf{s}_i + \delta_{tH}\lambda_{Hi} + \varepsilon_{tHi}$
$(i \in A_H)$ (9c)

**Block 2: Secondary occupation and household production**

*Participation:* $N^{-1}[p_j(t_{ij}>0)] = \alpha_j'\mathbf{x}_{pji} + \varepsilon_{pji}$ $(i \in A_j, j=N,E)$ (9d)

*Wages:* $\log w_{ji} = \beta_j'\mathbf{x}_{wji} + \delta_{wj}\lambda_{ji} + \varepsilon_{wji}$ $(i \in A_j, j=N)$ (9e)

*Hours:* $t_{ji} = \gamma_{Nj}(\hat{w}_{Ni}) + \gamma_{wHj}w_{Hi} + \gamma_{tHj}t_{Hi} + \gamma_{Vj}V_i + \gamma_{sj}'\mathbf{s}_i + \delta_{tj}\lambda_{ji} + \varepsilon_{tji}$
$(i \in A_j, j=N,E)$ (9f)

$N^{-1}$ is the transformation of the dependent 0,1-variable based on the normal distribution (PROBIT-approach); $p_j(\mathbf{x}_{pji})$ is the participation probability; $\mathbf{x}_{pji}$ is the vector of the explanatory variables of stage 1; $\mathbf{x}_{wji}$ is the vector of the explanatory wage rate variables (human capital terms) with a log specification to allow for non-linear wage relations (the imputed wages ($\hat{w}_{ji}$) then are logarithms of the estimated net marginal wage rates ($\log w_{ji}$)); $V_i$ is the virtual non-labour income (transfer payments and property income of the individual and other household members); $\mathbf{s}_i$ is the vector of socioeconomic characteristics; $\lambda_{ji}$ is the respective individual selection correction variable; $\varepsilon_{pji}$, $\varepsilon_{wji}$ and $\varepsilon_{tji}$ are normally distributed error terms and $\alpha_j$, $\beta_j$, $\gamma_{s.}$, $\delta_j$ are parameters to be estimated. $A_j$ is the set of persons who are involved in the activity j(j=H,N,E). A is the set of all persons in the sample.

The theoretical model shows that time spent in household production is dependent on the ratio of marginal productivities. With the empirical data at hand the marginal productivities reflecting the individual skill has to be approximated by socioeconomic variables (probably not the worst approximation). Finally, in order to correctly determine the significance of the estimated parameters, the correct asymptotic covariance matrix of the parameter vectors is calculated and shown in the following Tables.

# 4. Microdata Base

The microanalysis is based on the Sfb3-Secondary Occupation Survey 1984 of the Special Collaborative Program 3 (Sonderforschungsbereich 3, Sfb3) 'Microanalytic Foundations of Social Policy' of the Frankfurt and Mannheim Universities designed by Helberger et al. (1985). The Sfb3-Secondary Occupation Survey aimed at obtaining, for the first time, representative microdata for the FRG on primary occupation, secondary occupation, household production and combined activity patterns. Questions concerning illicit work were included in those parts dealing with secondary occupation. For these main topics, the survey included questions about the kind of activity, the amount of time spent therein, and the evaluation of the monetary/non-monetary income of secondary occupation and near-market economic activities in the household production. In addition, information on involvement in so-called 'social networks' and other socioeconomic data were obtained. The representative survey conducted by MARPLAN was based on the West German residential population aged 14 years and older. There were 7,826 interviews within four seasonal subsamples.

The definition of primary occupation in the survey follows traditional concepts, but there is no standard definition of secondary occupation. Secondary occupation is not only undertaken by those with primary employment. A secondary activity is of particular importance for non-gainfully employed persons as well. Thus, in addition to multiple job-holding in primary and secondary occupation, the self-reported secondary activities of housewives/housemen, unemployed, pupils, students and pensioners were included in the sample.

For household production, this study concentrated on near-market goods and services, which are particularly open to market substitution (Merz and Wolff, 1993). The open formulation of the question about household production allows a self-reported statement on the corresponding activities. The given answers describe typical 'Do it yourself' (DIY) activities (like renovation, repairs, craft or gardening etc.), but do not include full-time housework or child care. This method uses a more restricted - but at the same time more precise - selection criterion than other surveys on household production.

The following analysis and the microsimulation of the 1990 tax reform is based on the fourth representative subsample (4th quarter 1984), as this subsample provides particularly detailed information on taxes and transfer payments. Extensive regional microdata about the economic and labour demand situation in 1984 from the Federal Research Institute on Regional Planning was merged onto the primary Sfb3 database. Finally, when simulating the impact of the 1990 tax reforms, the original Sfb3 data were reweighted to forecast 1990 demographic totals. More detailed and comprehensive microdata about the individual tax situation in the Socio-Economic Panel of the Sfb3/DIW, individual marginal taxes and the tax paid by the household were also merged with the Sfb3 Secondary Occupation Survey.

# 5. Multiple Labour Supply Estimation Results

This chapter examines possible primary and/or secondary occupation. The secondary activities can be carried out either in addition to a paid main occupation or in addition to a main (but not paid) activity - as for pupil/student, housewife/houseman or pensioner. A secondary occupation would be illicit work when taxes and/or social welfare contributions are evaded.[4] In addition, unpaid activities in household production are further classified as non-market economic activities and leisure. Using the model and data described above, the respective determinants of participation in each activity, the endogeneous wage rates in primary and secondary occupations, and the selectivity bias-corrected market and non-market labour supply are now analysed and quantified.

## 5.1 Participation

The first stage of the three-stage model explains the probability of participation in market and non-market activities. To save space only the main results of the PROBIT estimations[5] are discussed; the single estimated coefficients are in Merz (1993, pp. 13-15). The main explanatory variables used are: *personal characteristics* (gender, family status, age, schooling, occupational status, social self-assessment, secondary occupation activities, costs, household production activities, social network, other activities, hours of work, wage rate, tax burden, satisfaction with household income, transfers); *household characteristics* (residential property, number of employed persons, size, children, income); and *regional characteristics* (community size, tertiary job holders, unemployment quota, gross national income per capita).

---

[4] According to the studies by Wolff (1991) and Merz and Wolff (1993) based on the same database, 40 per cent of secondary occupation could be classified as illicit work.

[5] All estimates are computed by the author's TOBIT/COLS program package.

The probability of *primary labour force participation* significantly depends on gender, non-linearly integrated age, school education, personal transfer payments and household characteristics. Transfer payments, such as pensions, grants from public funds for pupils/students and private subsidies, significantly reduce the probability of a primary occupation. The more persons in the household are gainfully employed (with an income of >100 Deutschmark - DM a month), the higher the probability that the interviewed person will have a primary occupation

The formulated theoretical model assumes that the primary occupation is carried out regardless of both secondary occupation and engagement in household production. In order to check this hypothesis, secondary occupation and household activity (in different activity categories respectively), as well as the corresponding time of activity and remuneration, were included in the participation equation of primary labour supply. The resulting non-significant coefficients indicate that participation in primary occupation is indeed independent of further economic activities.

It can be expected that labour market conditions (eg. the possibility of being gainfully employed at all), as well as individual characteristics, determine participation in a primary occupation. The enlarged microdata base incorporated such labour demand influences to a certain extent. The results show that the regional and labour demand indicators (community size, share of job holders in the tertiary sector, ratio of vacancies to unemployed, number of long-term unemployed and economic strength of a region measured as the gross added value per capita) in the participation equation are, however, of no significant importance. Other-person and household-specific factors, such as a specific school-leaving certificate, seem to be of greater importance for primary labour force participation.

For labour force participation in a *secondary occupation* (with paid secondary activities like crafts, construction, cleaning, selling etc.), specific informal activities in household production are important. DIY-activities such as renovating, house construction and car repair are positively correlated with a higher probability of secondary occupation. Thus, a person who carries out an unpaid activity in this category is also likely to have a paid secondary occupation.

Unpaid activities in social networks - in particular, help for relatives and friends - increase the probability of a paid secondary occupation. Active social contact in this area, with all its information processes, makes a paid secondary activity more likely. It also indicates an especially active and multiple-engaged part of the population which is also engaged in household production. Compared with other professional statuses, being a pensioner significantly increases the probability of a secondary occupation, as does a high household income. In contrast to the case of primary occupation, the labour market situation does play a role - the more unfavourable the ratio of vacancies to the number of unemployed, the higher the probability of having a secondary occupation. An economic need to earn additional money becomes apparent in those regions which have a tight labour market.

A number of factors affect the likelihood of participation in *household production*, including active social networks and the cost of secondary occupation. Higher commuting and other secondary occupation costs make a secondary occupation unattractive and home production more attractive. Women, in particular, are engaged in non-market economic activities. Ownership of a flat and/or house increases DIY participation. Regional and labour market indicators are of importance for DIY participation, in as much as they indicate a higher activity level in smaller towns/communities. A higher quota of regional long-term unemployment reduces the engagement in both secondary occupation and household production significantly. A generally decreasing activity pattern seems to appear together with a lasting unemployment quota in regional areas.

All in all, it is clear that different factors affect participation in market and non-market labour supply. Individual tax liabilities are not very important in the decision to undertake a secondary occupation or household activity. However, if only those undertaking two jobs had been analysed, the influence of overall tax liabilities on the participation decision might have been greater.

## 5.2 Marginal net wages

In our microeconomic model the influence of taxes is formulated by way of two variables - the marginal net wage rate and the virtual property income. The endogenous wage rate is thus a marginal net wage rate which, according to equation system (9), has to be estimated for a primary and a secondary occupation. When estimating the tax-dependent marginal net wage rate, there is a simultaneity problem with respect to the final estimated hours of work. The problem is that individuals have, in principle, non-linear budget sets, so that the marginal wage rate varies with the number of hours the individual works. In general, there are three procedures to account for this - a full information approach, a linear approximation and the instrumental variable method (see Wales and Woodland, 1979; Wales 1973).

To deal with the simultaneity problem of endogeneous marginal net wages and hours the instrumental variable approach is followed. The marginal tax rate $\tau$ is individually computed with the data of the Socio-Economic Panel. It is then merged via socioeconomic variables to each person in the Sfb 3-Secondary Occupation Survey. Then, for each person in the sample, the net marginal wage rate is calculated by $\hat{w} = (1-\tau)w^b$. In this three-stage-model the respective net marginal wage rate (calculated by the individual marginal tax rate) is estimated endogenously and then entered into the working hours equation.

The marginal net wages parameter estimates (PROBIT/COLS approach) include the selection correction variable (the hazard rate) computed from the participation equation. To deal with possible non-linearities the dependent variable is the logarithm of the marginal net wage rate. The single coefficients are provided in Merz (1993, p. 14)

When predicting the marginal net wage in a *primary occupation*, variables relating to human capital (like years of work experience, school-leaving certificate, position and part time employment) prove to be highly significant. Women have a significantly lower marginal net wage rate in the primary occupation than men. Labour market indicators are not significant, but the selection correction variable is. Thus, an estimation of the marginal net wage rate in the primary occupation with OLS alone would have given biased results. In other words, the theoretical model and estimation approach correspond to the empirical facts observed.

The explanation of the marginal net wage rate in a *secondary occupation* reveals a different structural picture. The respective school-leaving certificates are significant in a similar way as for primary occupation. However, years of experience and present professional status are not significant. As expected, illicit work increases the marginal net wage rate and a clear tax effect might well be linked with it. The negative significant coefficient of the gross added value per capita is an indicator of the character of an alternative economy: the worse the economic situation of a region the higher the marginal net wage rate, characterising a special engagement in secondary occupation. The $R^2$ values, which are quite high when compared with cross-section results in general, show a relatively good fit with the model approach.

## 5.3 Labour supply in market and non-market activities

Estimated marginal net wages enter the working hours equations in market and non-market activities, as explanatory equations in the form of 'imputed wages'. To avoid multicollinearity, the labour supply equations do not include the variables which have already been used for explaining marginal net wages. Those variables encompass gender, work experience, school leaving certificate, occupational status, secondary occupational characteristics (including illicit work) and regional and labour market information. With the model specified by equation (9) the own-log marginal primary net wage is the imputed wage for supplied hours of work in the primary occupation. The own-log marginal secondary net wage is the imputed wage for labour supply in the secondary occupation and in household production. As the estimates show (Merz, 1993, p. 15) higher remuneration reduces working hours.

The (non-linear specified) age and number of gainfully employed persons in the household play a significant role in supplied hours of work in a *primary occupation*. In addition, there is a time interdependency between market and non-market labour supply - regular secondary occupation, handicraft and gardening activities in household production significantly reduce the supply of primary occupation. If, however, the DIY-activity is carried out on a regular basis, it complements the market activities. *Taxes* on market labour income are captured via the own-log marginal primary net wage and the virtual non-labour income. The signs of both coefficients are different and virtual non-labour income is both highly significant and positively correlated with hours of primary work.

Looking at working hours in a *secondary occupation*, human capital variables are taken into account via the imputed wage. The remaining variables, particularly, show an association between DIY-activities and unpaid social network activities and positively support secondary occupation. In addition, secondary activities are particularly important in low income households. Working hours in *household production* are significantly determined by the nature of the activity. Unpaid neighbourly help reduces the own-time spent in household production. Thus, social networks are important within the market and non-market activity profiles. As expected, owners of flats or houses invest more time in repairs and maintenance etc.

# 6. Market and Non-market Labour Supply Effects

## 6.1 Characteristics of the 1990 German tax reform

Since the beginning of the 1980s, wages and income taxation have been repeatedly modified in the Federal Republic of Germany. Besides the re-introduction of the allowance for dependent children in 1983, the overall 1986-1990 fiscal law modifications are taken account of in this study. The fiscal policies of 1986-88 aimed to reduce the tax burden on all wage and income taxpayers and on supporting families with children. There was an increase in the basic tax-free threshold for single (married) taxpayers from DM 4,212/(8,424) to DM 4,752/(9,504) - and a reduction of the marginal tax rate from 28 per cent to 22.5 per cent (starting in 1988) in the lower progression range (DM 60,000/(120,000). In addition, there was a reduction in progressivity from the third income tax bracket onwards. For families, there was an increase in the allowance for dependent children to DM 2,484; a doubling of the allowance for children's education; and an increase in the household allowance. The 1986-88 changes resulted in a tax reduction amounting to about DM 24 billion.

The 1990 German tax reform, with an estimated gross reduction for the taxpayer of about DM 32.2 billion, surpassed in terms of volume all fiscal law modifications so far. After reducing tax subsidies amounting to DM 19 billion, a net total reduction of about DM 20 billion was expected. The most important characteristics of the 1990 tax reform were:

- reduction of the marginal tax rate in the lowest marginal tax rate zone from 22 to 19 per cent;
- reduction of the top rate from 56 to 53 per cent;
- introduction of a linear progressive marginal tax schedule;
- modification of the progression zone from DM 18,036 - 130,031 to DM 8,154 - 120,041
- increased allowance for dependent children from DM 2,484 to DM 3,024; and
- reduction of the corporation tax rate from 56 to 50 percent.

When the 1990 tax rates are compared to the old tax rates, the tax reform is of benefit of all taxpayers. However, the individual tax reduction - in absolute terms - increases with income.

## 6.2 Microsimulation results

Traditional impact analyses investigate microeconomic substitution and income elasticities at sample means of the explanatory variables. The disadvantage of this approach is the artificial mean person/household concept, with no distributional results at all. Distributional results, however, are a key outcome of microsimulation analyses. The original 1984 baseline microsimulation uses the microeconometric wages and hours equations, allied with respective individual characteristics and before tax reform characteristics, adjusted to 1990 demographic totals provided by the Dynamic Sfb3 Microsimulation Model (Galler and Wagner, 1986). The reformed new tax schedule then changes the respective parameters in the behavioural equations. The key question to be answered is how the 1990 reform reductions stimulate or reduce market and non-market labour supply.

Taxes enter the time allocation equations of activities in the formal and informal economy by way of marginal net wages and virtual non-labour income. A modification of the fiscal system must first and foremost start at these two terms. For all those who were active in a primary occupation, secondary occupation or household production, respectively, weekly working hours were computed using individual marginal net wages and labour income from the pre-reform tax regulations in force after 1984. This is the *before tax reform* or baseline simulation.

To analyse the 1990 tax reform, new marginal net wages and new virtual non-labour income were computed (again for each active person respectively), applying the new 1990 tax schedules. This is the *after tax reform* simulation.[6] These new tax variables, in combination with the other socioeconomic terms within the estimated behavioural equations, yield the new (after tax reform) weekly working hours for each individual's formal and informal activity. The analysis focuses only on those persons who are active, so the impacts of a changed participation structure are neglected here.

Figure 1 summarises the *overall effects* of the 1990 German tax reform on time allocation in market and non-market labour supply[7]. The tax cuts produced relatively small overall effects, but distinct group specific effects. They reduced the weekly working hours for men in primary occupations by about -0.2 per cent, whereas female labour supply in primary occupation increased by about 0.3 per cent on average. It should be noted that all following Figures respectively correspond to a male or female interviewed person, with only one person interviewed in the household. For secondary occupations there was an increase in hours of work for both men (0.2 per cent) and women (0.9 per cent).

---

[6] In order to follow the concept of imputed wages as an instrument for those without actual respective wages, after-tax-reform net marginal secondary ln wages are estimated by COLS in a similar way as before-tax-reform wages and then computed for all persons in the sample.

[7] The underlying tables for all following figures are provided in Merz (1993).

Figure 1: Overall Percentage Change in Weekly Market and Non-Market Hours of Work Due to 1990 German Tax Reform

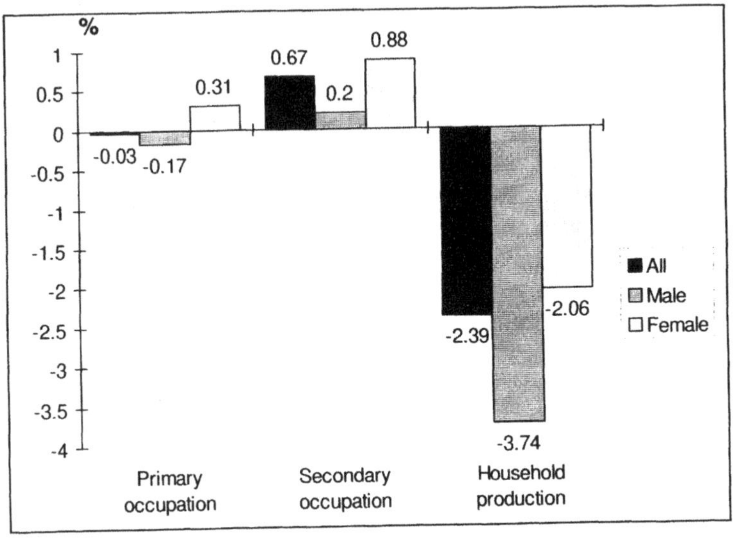


Paid market work is often the focus of interest in debate about tax reforms. In the discussion of the formal and informal economy, however, it is often argued that higher marginal taxes imply less work in the market and more work outside the market - that is, work at home. This would suggest that the 1990 tax cuts should lead to more market work but less non-market work.

The microsimulation results of the 1990 tax reform, with reduced marginal tax rates but increased marginal net wages, show that economic activity in closed market household production, the so-called DIY-activities, will be diminished in favour of 'pure' leisure activities and other paid activities - by -3.7 per cent for men and by -2.1 per cent for women. There are specific tax effects on non-market production and the effects, considered absolutely and relatively, are greater than in paid activities.

The tax reductions cause an increase in the marginal net wage. In accordance with microeconomic theory, the price of leisure then increases, which results in a lower demand for leisure and hence a higher demand for working hours. With the hours of work increasing in secondary occupation, this substitution effect outweighs the opposing income effect in this empirical analysis. The results establish that for men, on average, the 1990 tax reform will increase hours of work only in paid secondary occupations. With an additional overall reduction in male primary occupation, the 1990 tax reform thus changes working patterns. There is a shift from a (single) primary job to paid secondary activities. Women are increasingly engaged in paid economic activities at the cost of unpaid non-market activities. There is an incentive for additional paid primary and secondary hours. Hence, there is an overall tendency in favour of paid secondary market labour supply for men and women.

While average effects are interesting, one of the strengths of microsimulation is its ability to analyse the impact of tax reform upon sub-groups of the population. Figure 2 shows the changes in working hours by *occupational status* due to the tax reform. For male white collar workers and self-employed, there is a small average decrease in hours of work in primary occupations as a result of the tax reforms. For both male and female civil servants and blue collar workers there is little or no impact on primary paid labour supply (Figure 2). For secondary jobs, however, there is an overall average increase in male working hours for blue-collar workers and civil-servants, but disincentive effects for white-collar workers and self-employed.

Women show a different profile. There is a small overall average increase in female hours in primary work, especially for the self-employed. Incentives to increase hours in a secondary job are strongly experienced by white-collar working women.

Figure 2 shows clearly how a tax reform like the German one, which focuses on paid work in the market, nonetheless produces remarkable behavioural impacts for household production activities. Unpaid work is reduced in favour of paid work, which becomes more attractive because of the reduced tax burden.

For those not engaged in a primary occupation, the tax reform reduces the incentive to hold secondary jobs for male and female students, whereas housewives will be more likely to be engaged in a paid secondary occupation (see Merz, 1993). For all non-occupied persons, regardless of their gender, unpaid DIY-activities in the household will be diminished (but to a different extent for men and women).

In the tax reform debate the increased child and other family allowances played an important role. Figure 3 shows the impact of the tax reforms by *family status*, and indicates that the tax reform will reduce male and female labour supply in primary and secondary occupations for singles. However, the opposite effect is apparent for lone parents with children (with the exception of males' reduced secondary activity).

Figure 2: Percentage Change in Weekly Market and Non-Market Hours of Work Due to 1990 German Tax Reform for Primary Occupied Persons

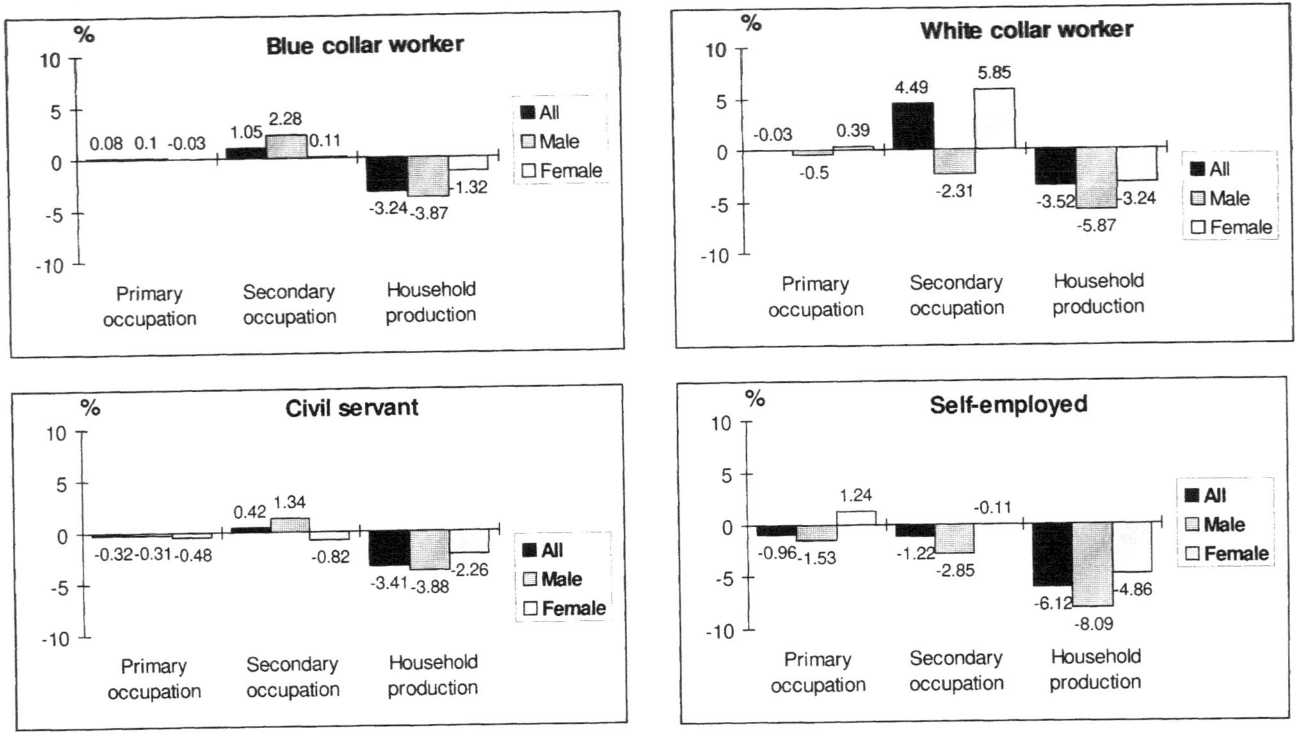

In economic terms, the substitution effect of a reduced tax burden dominates the income effect when there are children. For couples with and without children a similar effect is apparent. In particular, the existence of children will not diminish but increase paid female secondary activities, regardless of whether the woman is single or married. For females living in couple with children families, the tax reform will lead to about a 1.8 per cent increase in hours worked.

Another way of assessing the impact of the tax reform is to examine the changes in hours by the *level of family income*. Figure 4 shows the market and non-market labour supply effects according to the following categories of pre-tax-reform family/household net income: low <1500 DM; mean 1500-3000 DM; high >3000 DM. The tax reform will, above all, increase labour supply in primary and secondary occupations for families with higher incomes. For families with higher income, the substitution effect of an increased marginal net wage will be dominant, particularly for a secondary occupation. Conversely, the income effect dominates the behaviour of families with medium or lower household income. This general pattern, however, differs by the gender of the household head and kind of paid activity. Thus, all the 1990 tax scheme regulations together seem to stimulate the labour supply of richer households. Remarkably, a specific improvement in labour supply incentives for low and middle income families cannot be observed.

Unpaid DIY-activities among families with higher incomes (>3000 DM net) will reduce, relatively speaking, their level of unpaid household production activities. This reduction seems to be translated into further increases in primary and secondary paid work.

## 7. Conclusion

This study on market and non-market labour supply and taxes is based on a theoretical microeconomic model with multiple labour supply in market and non-market activities in the formal and informal economy. The multiple time allocation model, which explicitly takes into account taxes, transfer payments and socioeconomic characteristics, is specified and estimated in a multiple three-stage approach explaining the participation probability, the relevant wages and the hours of work in different activities.

The microsimulation of the 1990 German tax reform overall shows clear gender- and activity-specific labour supply results: the working time of men in their primary

Figure 3: Percentage Change in Weekly Market and Non-Market Hours of Work Due to 1990 German Tax Reform, by Family Type

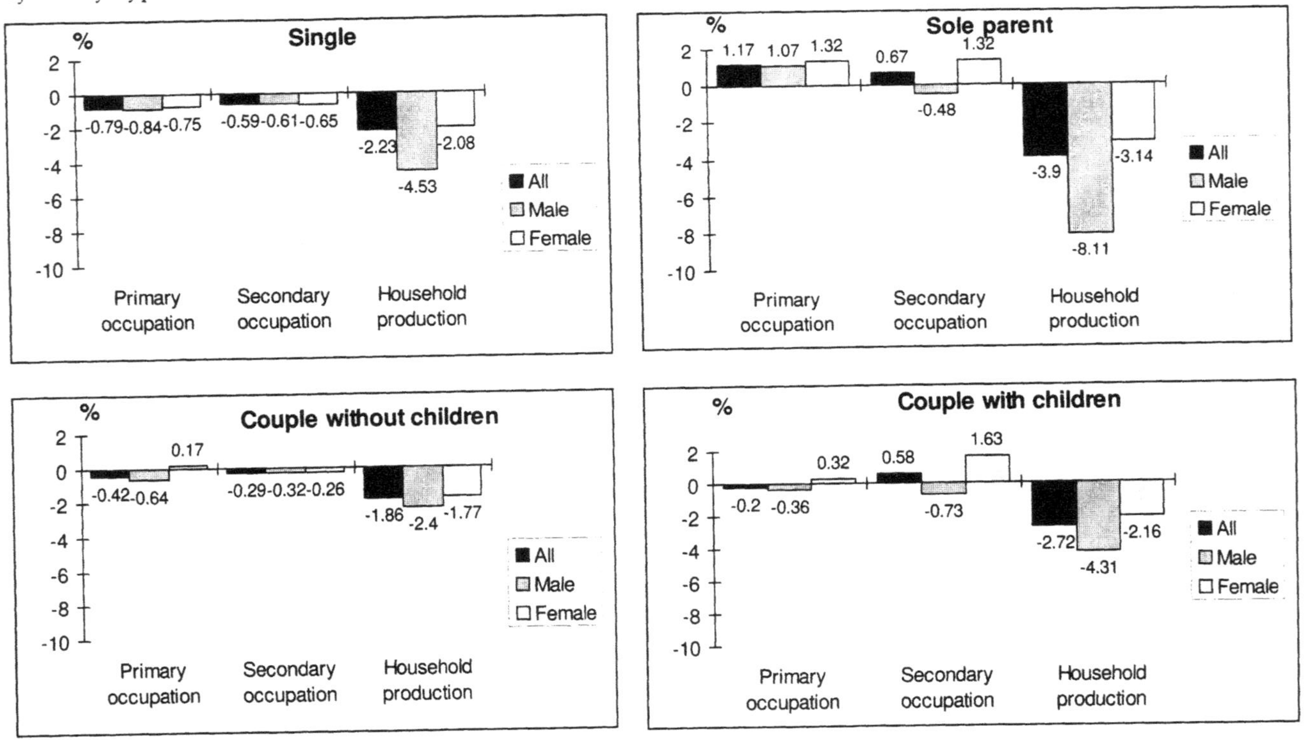

Figure 4: Percentage Change in Weekly Market and Non-Market Hours of Work Due to 1990 German Tax Reform, by Household Income Level

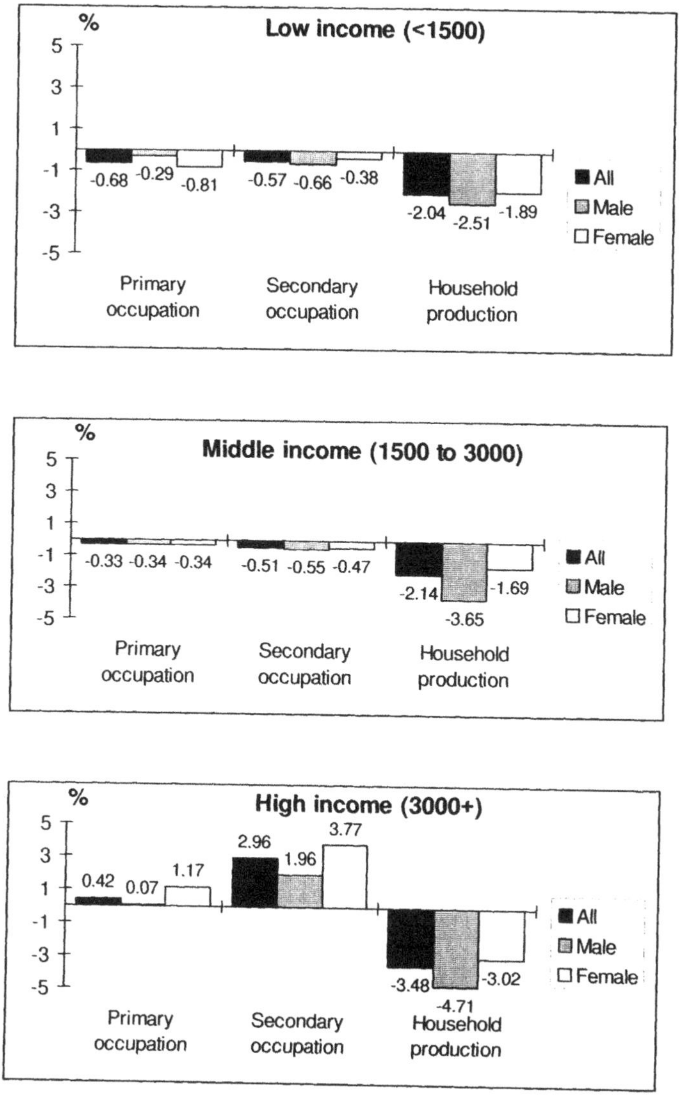

occupation will be somewhat reduced ('disincentives') while the working time of women will increase ('incentives'). The hours of work in a secondary occupation (including illicit work) will, in particular, increase for women. Unpaid household production activities will be affected by the tax reform. Market opportunities, thus, play a role in affecting participation in the non-market sphere. The tax reductions, in general, will reduce non-market activities for men - and, more particularly for women - in favour of paid market activities.

Besides these overall average results, the 1990 tax reform yields quite different results for socioeconomic groups - categorised by occupational status in a primary occupation, by the existence of children in the family, and by family income. Different labour supply profiles for men and women become apparent for these groups. So, for example, the existence of children in the families of singles and couples is an incentive for additional earnings in a secondary occupation. For families with a higher income, paid secondary working hours are stimulated and the substitution effect is dominant (suggesting a forward bending labour supply curve). Many further results show the relative importance of various socioeconomic characteristics in a multivariate approach of multiple paid and unpaid activities.

Whether the stated effects are intended or not by the political process, the results show that the impact of tax reform has to be discussed, not only in relation to the traditional first-round money effects, but also with regard to behavioural impacts like those discussed above.

These microsimulation tax reform results are empirically founded on microeconometric estimates, which themselves are based on a theoretical microeconomic model of activities. With this procedure and representative microdata, individual behavioural response effects and their importance could be made apparent in a microsimulation approach. These results could not have been achieved by such traditional approaches as the analysis of typical cases or tax schedules.

It is obvious that for real policy advice on further tax changes the microdata base must be enhanced. Nevertheless, a microsimulation analysis like the one discussed above promises a fairly adequate analysis of real world distributional patterns - incorporating not only institutional regulations but also the behavioural reactions of those who are affected.

## References

Aaron, H. and J. Pechman, (eds.) (1981), *How Taxes Affect Economic Behavior*, Brookings Institution, Washington DC.

Becker, G.S. (1965), 'A theory of the allocation of time', *Economic Journal*, Vol. 75, No. 299.

Burtless, G. and J.A. Hausman (1978), 'The effect of taxation on labour supply: Evaluating the Gary negative income tax experiment', *Journal of Political Economy*, Vol. 86, No. 5.

v. Essen, U., H. Kaiser and P.B. Spahn (1988), Verteilungswirkungen der Einkommensteuerreformen 1986-1990: Eine Simulationsstudie fur die Bundesrepublik Deutschland auf der Basis des Sozio-okonomischen Panels, *Finanzarchiv*, Vol. 46.

Flood, L. (1988), 'Effects of taxes on non-market work: The Swedish case', *Journal of Public Economics*, Vol. 36.

Galler, H.P. and G. Wagner (1986), 'The microsimulation model of the Sfb3 for the analysis of economic and social policies', in: G.H. Orcutt, J. Merz and H. Quinke (eds.), *Microanalytic Simulation Models to Support Social and Financial Policy*, North Holland, Amsterdam.

Gronau, R. (1977), 'Leisure, home production and work: The theory of the allocation of time revisited', *Journal of Political Economy*, Vol. 85.

Gronau, R. (1980), 'Home production: A forgotten industry', *Review of Economics and Statistics*, Vol. 62, No. 3.

Hausman, J.A.(1981), *Labor Supply*, Aaron and Pechman, New York.

Heckman, J.J. (1979), 'Sample selection bias as a specification error', *Econometrica*, Vol.47, No. 1.

Helberger, C., J. Merz and H Schneider (1985), Nebenerwerbstätigkeitsumfrage - Documentation elaborated by Klaus Wolff, Frankfurt.

Killingsworth, M.R. (1983), *Labor Supply*, Cambridge University Press, Cambridge, Mass.

Lancaster, K. (1966), 'A new approach to consumer theory', *Journal of Political Economy*, Vol. 74, No. 2.

Merz, J. (1989), *Markt- und nichtmarktmäbige Aktivitäten privater Haushalte - Theoretischer Ansatz, repräsentative Mikrodaten, mikroökonometrische Analyse und Mikrosimulation wirtschafts- und sozialpolitischer Malbnahmen für die Bundesrepublik Deutschland*, Habilitationsschrift, Frankfurt.

Merz, J. (1991), 'The 1990 German tax reform - microsimulation of time allocation effects in the formal and informal economy', in: J.K. Brunner and H.G. Petersen (eds.), *Simulation Models in Tax and Transfer Policy*, Campus, Frankfurt.

Merz, J. (1993), 'Market and Non-market labour supply and recent German tax reform impacts', FFB Discussion Paper No 6, University of Lüneburg, Germany.

Merz, J. and K. Wolff (1993), 'The shadow economy: Illicit work and household production - a microanalysis of West Germany', *The Review of Income and Wealth*, Vol. 39, No. 2.

Shisko, R. and B. Rostker (1976), 'The economics of multiple job holding', *American Economic Review*, Vol. 66, No. 3.

Wales, T.J. (1973), 'Estimation of a labor supply curve for self-employed business proprietors', *International Economic Review*, Vol. 14, No. 1.

Wales, T.J. and A.D. Woodland (1979), 'Labour supply and progressive taxes', *Review of Economic Studies*, Vol. 2.

Anders Klevmarken

Uppsala University
Sweden

Paul Olovsson

Gothenburg University
Sweden

# Chapter 10: Direct and Behavioural Effects of Income Tax Changes - Simulations with the Swedish Model MICROHUS

## Abstract

*A dynamic microsimulation model of the Swedish household sector is used to evaluate the direct and indirect (behavioural) effects on the income distribution of the 1991 Swedish tax reform. The direct effect is a relatively high increase in income inequality while the indirect effects appear to be very small.*

## 1. Introduction

The building of a microsimulation model for the Swedish household sector was motivated partly by the need to have a tool with which to analyse both the direct and indirect effects of tax and transfer changes on the distribution of income and wealth - that is, including behavioural adjustments to policy changes. This model was also to act as an 'umbrella' under which results obtained from analysis of the HUS-project database could be successively collected (Klevmarken and Olovsson, 1993). These data, a random sample of Swedish households, provide the sample of individuals and households on which the simulation model operates. They have also, with few exceptions, been the empirical base for the estimation of behavioural relations. The attempt to build a microsimulation model which includes the tax and transfer systems, as well as behavioural relations, is a never-ending one, as there are always improvements that can be made. In order to get something which can run before all research funds have been used up, it has occasionally been necessary to sacrifice the ambition of front-line research for less sophisticated and quicker solutions. In this sense the model, now christened MICROHUS, should be considered work in progress!

There are of course major differences between a behavioural model and the type of 'accounting' models which are used to simulate the first-order effects of tax policy changes. The latter type of model typically includes all or almost all details of the tax and transfer rules and, depending only on the accuracy of the data used and coding detail, it can reproduce the distribution of income with almost no error. In contrast, any attempt to model behavioural factors will necessarily imply sizeable specification errors and, as a result, it might not be necessary to include the tax and transfer system with the same degree of detail as in the accounting models. The key goal of this study was to get some indication of the *relative size* of the behavioural adjustment effects. If they are small, then accounting models will suffice. But if the adjustment effects are not negligible, more effort has to go into modelling behaviour.

After a few general comments on microsimulation in Section 2, a description of the general structure of MICROHUS is presented in Section 3. The results from some simulations of the distributional effects of the 1990-91 Swedish income tax reform are outlined in Section 4.

## 2. Microsimulation

Microsimulation can be viewed as an attempt to model and simulate the distribution of policy target variables, not only their mean values. For instance, in a microsimulation study one might analyse the impact of an income tax change on the distribution of income - who loses and who gains. The microsimulation approach is thus primarily designed for studies of the distributional effects of economic policy, and one of its main advantages is that it permits assumptions of heterogeneous behaviour. Every individual, household or firm does not necessarily behave as the average economic agent. When economic relations are highly nonlinear, when tax laws and rules of transfer programs introduce censoring and truncation, and when subpopulations differ in behaviour, models of average behaviour inadequately evaluate the impact of policy changes. In such cases, a microsimulation model can do better. A good example is one of the first microsimulation models actually used for policy evaluation, namely the model of the Swedish supplementary pension scheme (Eriksen, 1973; Klevmarken, 1973).

Many microsimulation models identify subgroups or subpopulations, each of which are assumed homogeneous in behaviour. In these models it is only necessary to simulate the behaviour of each subgroup. Population totals and means are then obtained by weighting each group by its relative size. An alternative, and usually more general and flexible, approach is to simulate each individual, household or firm. In this case, the simulation model usually operates on a real sample of individual units. Gradually, through the simulation process, the sample values of each unit are updated. An advantage of this approach is that the analysis is not limited to certain preselected subgroups of units. Rather, the analyst can choose any mode of analysis of the updated sample. If this sample were a probability sample from some population, for example, the sampling weights could, in principle, be applied to the updated sample to infer the population.

A limitation of many microsimulation models is that they include the rules which determine the outcome of economic policy (such as the tax rules and tax schedules) but exclude behavioural relations. These models can thus only be used to simulate the 'first-order effects of policy changes. MICROHUS operates on a sample of individuals and households from the 1984 HUS-surveys (Klevmarken and Olovsson, 1989, 1993). It includes behavioural relations which model both demographic changes and adjustments in the housing and labour markets. It is dynamic in the sense that the outcomes of previous simulation runs affect those of later runs. There is, however, no macroeconomic feedback.

## 3. The General Structure of MICROHUS

MICROHUS is structured in model 'blocks' each of which consist of one or more modules (Klevmarken et al., 1992). 'Preparations' is a block which adjusts the age distribution of the 1984 HUS-sample to conform with that of the Swedish population. It compensates for a mildly selective non-response in the 1984 HUS-wave. This block also imputes weekly work hours for those employed in 1984 and who did not respond to the question about work hours.

The 1984 HUS-sample only included individuals below the age of 75. For this reason MICROHUS does not simulate the behaviour and income paths of the oldest cohorts. In the first module of the 'Demographics' block, every sample member becomes one year older and, when a household head becomes 75, the whole household is removed from the sample and put in a file which is later used in the block 'Inheritance'. Every remaining individual is then exposed to the risk of dying using conventional life tables. Every individual for whom death is simulated is deleted from the population and the block 'Inheritance' is used to determine how the wealth of the deceased should be allocated among his or her heirs.

In the block 'Demographics' there are also modules which simulate the birth of new children, the formation of consensual unions and marriages, and separations. All these modules change the composition of existing households, form new ones and delete old ones. Fertility is modelled by a Box-Cox type quadratic duration model which has the hourly wage rate and disposable income as well as several socioeconomic variables as arguments. This model was estimated on HUS-data. Continuous time duration models are also used to model the formation of consensual unions and dissolutions of unions. The matching of spouses is based on a canonical correlation analysis of the characteristics of spouses. Matching is done randomly by age, schooling and labour market experience.

In the 'Demographic' block, economic incentives influence the decisions to form a union and find a partner through the 'schooling' and 'years of work experience' variables. The more schooling a woman has, the higher the probability she will enter a consensual union and marry. Work experience also increases the probability of forming a union. The probability of cohabiting is an inversely u-shaped function of years of work experience with a peak at about 15 years, while the probability of marrying is a continuously increasing function of work experience. The probability of having a child also depends on these variables and on the woman's wage rate and the disposable income of the household. Females with relatively long schooling postpone their first child but tend to have their next child (children) quite soon thereafter. Females with just a few years of work experience have a smaller probability of having a child than women with both less and more experience. The wage rate has a negative effect on fertility, while there is no consistently positive or negative income effect.

The block 'Geographic Mobility' simulates household moves between the big cities and the rest of the country using a constant transition matrix. Households which are simulated to move are randomly allocated to municipalities. The probability of moving to a particular municipality is proportional to the number of respondents from that municipality in 1984.

The 'Housing' block simulates changes in tenure and consequent changes in market values, tax-assessed values and the expenditure on housing. The probability of moving to a new house or apartment is modelled by a Probit relation, while tenure choice relating to a decision to move is simply simulated by a constant transition matrix. The decision to move is assumed dependent on economic incentives. The Probit function includes among its explanatory variables labour income, net wealth and the housing costs of the household, as well as the marginal tax rate of the head of the household. Separate models are used for couples and singles. The wealthier a household and the higher the marginal tax rate, the less likely the household is to move. High labour incomes increase the probability of moving for couples but decrease the probability for singles.

The number of observed changes of tenure in the HUS-data was not large enough to permit the estimation of a model which made tenure choice a function of economic incentives. For this reason we have used a constant transformation matrix. For each household simulated to move the cost of housing (rent for tenants), interest payments on mortgages and the market and tax-assessed values of owner-occupied homes are assigned by a minimum distance method - that is, the values of these variables are copied from an existing household which, as closely as possible, resembles the receiving household in size, number of children, age and sex of the head, population density of the housing area, and the earnings and wealth of the household.

The 'Labor Market' block simulates entries into and exits from the labour market, hours worked, absence from work because of sickness, unemployment and wage rates. This block also includes modules which simulate the length of schooling and the choice of occupation. In MICROHUS every individual is assumed to decide about his/her future education at the age of 15. A model for the total length of theoretical education and vocational training was estimated from HUS-data. Due to the major changes in the educational system in post-war Sweden, this model was estimated for three groups of birth cohorts (those born between 1935 and 1944; 1945 and 1954 and from 1955 on) and also by gender. In simulating the behaviour of future cohorts the estimates for the youngest cohorts are used. The length of schooling depends on the father´s and the mother´s schooling, the father´s occupation and whether the simulated individual grew up in a big city, abroad or somewhere else. A general result is that the parents' background is of less importance for the young cohorts than it is for the older cohorts. The model also includes measures of national unemployment and the relative wage rate of white collar workers compared with blue collar workers. This represents an attempt to capture economic incentives on the schooling decision, but none of these variables turned out to be significant.

Transitions into the labour force are currently simulated only by age and sex-dependent constant transition probabilities. People who are simulated to join the labour force are randomly allocated to a full-time (40 hours per week) or a half-time (20 hours per week) job, using constant transition probabilities. The absence of economic incentives to join the labour force is a weakness of the model which will be removed in future model revisions. This deficiency might, however, be less serious for a country like Sweden - with its high labour force participation rates for both genders - than it would be for a country with lower participation rates.

Those who enter the labour market are assigned a wage rate by a conventional earnings function. The model assumes that all educational activities take place immediately after compulsory schooling is finished. It thus cannot handle repeated spells of continuing education. Each individual who joins the labour force is assigned one of four occupations - blue collar worker, white collar worker, entrepreneur/manager or executive. The assignment is done by constant transition probabilities which depend on the father´s occupation and schooling. The model does not currently allow for changes in occupation.

The model which simulates annual hours of market work is of the Hausman type with random heterogeneity and measurement errors. In addition to the marginal wage rate and virtual income, the following explanatory variables were used - age, number of children less than seven years old, number of household members, whether or not the household owns their home, and whether they live in a big city. The model was estimated separately for males and females living in *de facto* relationships. The spouse´s labour income was included in the virtual income, but in no other way does the model captures any jointness in the decisions of two spouses. The model was also estimated for singles but, in this case, data did not permit a model for each gender. Estimates were obtained from the 1984 HUS cross-section of employed and self-employed, using income data from 1983. The budget sets computed were determined by the 1983 income tax system, and income dependent transfers and subsidies did not influence them. The Slutsky conditions were enforced, and turned out to be binding only for females.

The elasticities shown in Table 1 were obtained, when evaluated at 2,100 hours, a marginal wage rate of 21 Swedish Kroner (SEK) and a virtual annual income of 56,000 SEK. Most previous estimates for Sweden are limited to males aged 25-55 years. There is a range of results depending on data set, model and estimation method (see the surveys in Gustafsson and Klevmarken (1993) and Aronsson et al (1994)). Most uncompensated and compensated elasticities are close to 0.1 and the corresponding income elasticities are very small. The few estimates of compensated wage rate elasticities available for females are higher than those for men, ranging from 0.3 to 1.3. The income elasticities are also relatively small for females. Compared to these results, our estimated compensated elasticity for males of 0.131 is of the same magnitude, while the estimate for females of 0.112 is much smaller. Our estimates also differ from previous results as to the relative importance of the income and substitution effects. For couples, the income and substitution effects are almost of the same magnitude and the result is that any change in marginal tax rates will have almost no effect on the labour supply of couples. Singles, however, are more responsive to tax changes. Both the income and the substitution effects are relatively large, but the latter dominates the former - and a decrease in marginal tax rates will thus result in an increased labour supply.

When the model is simulated there is no constraint which prevents a non-positive outcome. This is interpreted as a transition out of the labour force. Another way out of the labour force is to retire. Everyone is assumed to take an old age pension at the age of 65. Early retirement is simulated using age- and time-specific transition probabilities obtained from the Swedish National Insurance Board. An individual can choose between full-time or part-time early retirement. This choice is simulated with age-specific but constant transition probabilities from the same source. There are thus no economic incentives influencing the retirement decision. People may also be temporarily absent from the labour market, either because they are sick or caring for their children or because they are unemployed. When a child is born the mother is assumed to use the maximum number of months of paid leave. No months of parental leave are allocated to the father.

Short spells of absence of less than 11 weeks due to sickness are simulated by a Tobit model. The marginal income loss due to sickness, after compensation from sickness insurance but before tax, has a small negative effect on the duration of absence. The more income a person loses, the shorter time he/she is likely to stay away from work. Females have longer spells of sickness than males and young and old people longer than the middle-aged. Families with children and families who live in the big cities also have longer spells. The Tobit model does not simulate well very long spells of sickness. To get a good representation of spells exceeding 10 weeks, the Tobit model is supplemented by random drawings from a rectangular distribution.

Table 1: Labour Supply Elasticities for Full-time Working Males and Females by Marital Status

| | Singles | Couples | |
|---|---|---|---|
| | Males and Females | Males | Females |
| Wage rate | 0.143 | -0.013 | -0.006 |
| Income | -0.581 | -0.181 | -0.149 |
| Compensated | 0.601 | 0.130 | 0.112 |

The model which simulates months of unemployment is also a Tobit model. The general level of unemployment is exogenously given by the national unemployment rate. Individual differences are explained by differences in gender, experience, schooling, seniority and age. Well-educated, middle aged individuals and females have shorter spells of unemployment.

The annual work hours simulated by the labour supply model are reduced in proportion to the simulated number of weeks sick and months unemployed. The corresponding compensation from national sickness insurance is computed, as well as unemployment compensation. Finally, the Labor Market block also includes a module which updates the wage rates of people in the labour force. The model is a combination of an autoregressive structure and a conventional Mincer type earnings function. Separate models are used for males and females.[1]

MICROHUS distinguishes between a few assets and liabilities - these are real estate (owner-occupied house or condominium), mortgages, financial assets, other loans, and other assets. The latter category includes consumer durables. For taxation purposes, the model assumes that household wealth is divided equally between two married spouses. Cohabiting but not married couples keep the assets they brought into the union as their personal property.

In the current version of MICROHUS market values of real estate, financial assets and other assets are adjusted exogenously by average national rates obtained from official statistics. Mortgages and loans are assumed amortised by three per cent each year. For households which change housing or experience a marriage or a separation there are consequential changes in assets, while other households are assumed not to have any changes in assets.

A household which moves from one house/apartment to another will get a new real estate market value and new mortgages in the 'Housing' block. Any residual balance between new and old market values and mortgages is picked up as a change in financial assets. Similarly, a household which moves from a rented apartment to an owner-occupied house will have a decrease in financial assets equal to the difference between the market value of the house and the mortgage.

---

[1] The model was estimated from the 1984 and 1986 waves of the HUS panel. The log wage rate is thus explained by its lagged value with a lag of two years. The estimated autoregression parameter for females was estimated to be 0.637. The predictions from the model, however, implied an unreasonable decrease in the average female wage rate. For this reason this parameter was increased to 0.785. The estimate for males was 0.801.

If two spouses separate each spouse will bring his/her share (50 per cent) of the household wealth to the new households. This implies that a former wife, who keeps the couple's house, will pay to her former husband his share in the house. If two individuals marry their assets will be added and then divided into equal shares.

Inheritance is another source of wealth changes. The transfer of assets between generations and between deceased and surviving husbands and wives is controlled by the block 'Inheritance'. A separate file keeps track of all children-parents and husband-wife relations. Bequests to more distant heirs are not considered. If a married person dies and the total wealth exceeds a certain relatively low threshold the spouse is assumed to inherit 50 per cent (the legal share) and any children 50 per cent. Otherwise the spouse inherits everything. When a parent dies and there is no living spouse the children who are alive will share the wealth left by their parent in equal parts. In order not to run the total wealth of the simulation population towards zero, the wealth of a dead person who has no spouse or children is also redistributed among sample households.

The block 'Incomes and taxes' collects information about labour supply, wages, wealth and demographic matters in order to compute labour income, transfer payments and other non-labour income components which generate taxable income. Income and wealth taxes are then computed. The tax scales, tax bases and taxation rules for each year in the period 1984-1992 are applied. Transfer payments are also computed using the rules for each year. This block finally computes disposable income.[2]

The behavioural models used in most modules have been estimated in constant prices (that is, monetary variables are in constant prices). The tax system has tax bases, thresholds etc. in current prices. Monetary values thus have to be converted from fixed prices to current prices when the simulation enters the tax modules and back again to fixed prices when the simulation leaves the tax modules. The conversion is done by CPI, which is given exogenously.

Finally, MICROHUS also includes a few modules which assign identification numbers and household numbers and store lagged values from previous years.

---

[2] A more detailed description of the treatment of each income component is given in Klevmarken and Olovsson (1994).

There are also modules which help in extracting output data from MICROHUS as the simulations proceed.

Each simulation consists of a number of runs and each run is a full sequence, involving all or a selected number of modules and simulating the changes in one year. The simulation is done recursively - that is, each module is called sequentially (approximately in the order of presentation above) and a later module cannot influence the outcome of an earlier module until the next run. This implies that all demographic changes are determined before changes in housing and labour supply, and geographical moves are simulated before changes in housing, which in turn precede changes in labour supply. Joint decisions about, for instance, housing and jobs are thus ruled out. This is primarily done for reasons of simplicity. A model of joint housing and job decisions would become rather complex.

## 4. Simulated Effects of Swedish Income Tax Changes 1984-1992

In 1991 a major tax reform was implemented in Sweden. Income tax rates were reduced, in particular in high income brackets, and the tax base was broadened. Income from capital was no longer taxed jointly with labour income and a separate 30 per cent flat-rate tax was introduced. Jointly with changes in income taxes there was an increase in value-added tax and a broadening of its base. The changes in the relative prices caused by the new value-added tax are, however, not implemented in MICROHUS. The simulation model only includes the old and the new income tax systems.

The old income taxes were highly progressive and, until 1991, spouses were jointly taxed for capital incomes. The new 1991 tax system introduced separate taxation of all incomes and almost 90 per cent of taxpayers had only to pay a proportional local tax, the size of which varied from one municipality to another but with an average close to 30 per cent. Above a certain threshold, high income earners had to pay an additional 20 per cent state tax. In principle, the new tax system had only two tax brackets, one with a tax of about 30 per cent and one with 50 per cent. In practice as a result of a political compromise, a few additional tax brackets with a marginal income tax rate higher than 30 per cent were introduced for low incomes. Technically this was brought about by allowing low income earners certain deductions which decreased with increasing income.

The major effect of the income tax reform was a substantial reduction in marginal tax rates. In 1983 the progressivity of the tax system was at its peak. There had been minor adjustments of the tax system before 1991 to reduce the degree of progressivity, but the major change came in 1991.[3] Figure 1 gives the evolution of average marginal tax rates between 1984 and 1992 by gender, as simulated by the full MICROHUS model. Males have experienced a considerable decrease in marginal tax rates, while the decrease for women has been more moderate. The explanation is, of course, that women on average have lower incomes than men.

The tax reform was preceded by studies of its distributional effects and the anticipated labour supply adjustments. These studies predicted only small distributional effects. The income tax reduction, primarily for high income earners, was balanced by increased transfer payments to families with children and low income families. It was also predicted that a reduction in interest deductions, which was part of the new tax system, would increase the tax of many high income earners. One might also note that many households included a full-time working, and sometimes well paid, male and a part-time working, and less well paid, woman - which would tend to even out the tax effects within the household. A potential weakness of these distributional studies was their limitation to first-order effects on the income distribution. No attempt was made to estimate any adjustment effects on the income distribution.

In the public debate which preceded the tax reform much attention was, however, given to so-called dynamic effects. One of these effects was an expected increase in labour supply as a result of the decreased marginal taxes. The reform was not fully funded, as 7.6 per cent of the expected decrease in public revenues was assumed to be offset from higher taxes resulting from increased labour supply and other behavioural adjustments. The decision not to fund the tax reform fully was supported by empirical studies of labour supply, but such studies did not address distributional issues. Very little is known about the effects of behavioural adjustments on income inequality.

---

[3] For a more detailed analysis of the evolution of the Swedish tax and transfer system in the last few decades see Gustafsson and Klevmarken (1993).

Figure 1: Simulated Swedish Average Marginal Tax Rates by Gender, 1984-1992

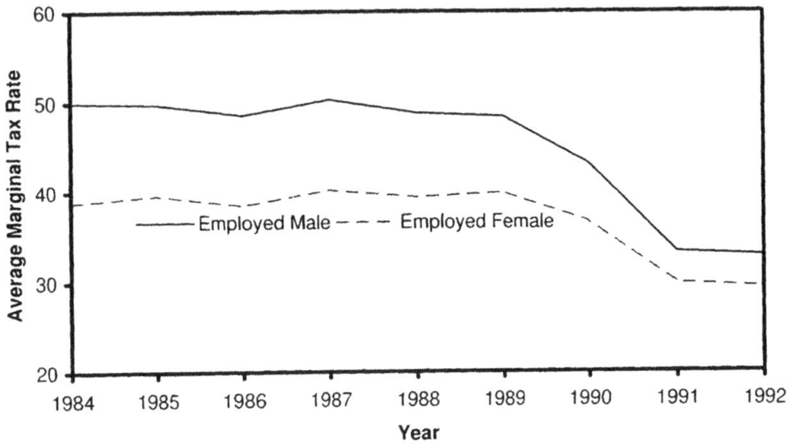


To evaluate the effects of the 1991 tax reform a reference case is needed against which alternative hypothetical runs can be compared. The case where the full model is simulated (including the tax/transfer systems of each year) for the period 1984-1992 is used. This is also the case which should trace the development of the Swedish economy. Appendix A gives results from this simulation and Table 2 compares aspects of the simulated development with those in official statistics (the following characterisation of the reference case economy draws upon these).

The simulated population is an ageing population.[4] The proportion of children below 18 years of age decreases, as the large cohorts of the 1940s move into the upper middle-age brackets. The proportion of recently retired (that is, those in the age bracket 65-74) has not, as of 1992, started to increase. These changes in the age composition of the simulated population agree well with those of the Swedish population. For the period 1984-1992 changes in the marital status of the population are also simulated. The proportion of married persons decreases, while the proportion of individuals in consensual unions and living as singles increases. These

[4] The simulated population size decreases from 5,250 individuals in 1984 to 4,866 in 1992. This is a decrease which has no correspondence in the Swedish population. The explanation is that MICROHUS does not allow for migration.

trends can also be found in the Swedish population, although the share of singles does not increase as strongly as in MICROHUS.

Comparisons with official estimates of Swedish labour force participation rates and hours of work are difficult, because MICROHUS does not use exactly the same variable definitions and population concepts as official statistics. The labour force

Table 2: Comparison Between Official Statistics and the Simulations of MICROHUS

| | 1984 | 1985 | 1986 | 1987 | 1988 | 1989 | 1990 | 1991 | 1992 |
|---|---|---|---|---|---|---|---|---|---|
| **A. LABOR FORCE PARTICIPATION RATES** | | | | | | | | | |
| **Males** | | | | | | | | | |
| Labor force survey | 85.6 | 86.0 | 85.9 | 85.7 | 86.2 | 86.8 | 87.0 | 86.0 | 84.0 |
| Index 1984=100 | 100 | 100.5 | 100.3 | 100.1 | 100.7 | 101.4 | 101.6 | 100.5 | 98.0 |
| Simulation 1 | 72.3 | 73.0 | 73.4 | 72.2 | 71.8 | 69.0 | 69.1 | 70.2 | 68.9 |
| Index 1984=100 | 100 | 101 | 102 | 100 | 99 | 95 | 96 | 97 | 95 |
| **Females** | | | | | | | | | |
| Labor force survey | 78.2 | 79.2 | 80.0 | 81.1 | 81.8 | 82.2 | 82.6 | 81.7 | 79.9 |
| Index 1984=100 | 100 | 101.3 | 102.3 | 103.7 | 104.6 | 105.1 | 105.6 | 104.4 | 102.2 |
| Simulation 1 | 65.4 | 66.3 | 66.6 | 64.0 | 64.3 | 63.7 | 63.8 | 62.9 | 63.9 |
| Index 1984=100 | 100 | 101 | 102 | 98 | 98 | 97 | 98 | 96 | 98 |
| **B. HOURS OF WORK (EMPLOYED)** | | | | | | | | | |
| **Males** | | | | | | | | | |
| Labor force survey (weekly) | 40.6 | 40.7 | 40.7 | 41.4 | 41.4 | 41.5 | 41.5 | 41.2 | 41.2 |
| Index 1984=100 | 100 | 100.2 | 100.2 | 102.0 | 102.0 | 102.2 | 102.2 | 101.5 | 101.5 |
| Simulation 1 (annual means) | 1858 | 1902 | 1880 | 1933 | 1937 | 1930 | 1909 | 1944 | 1932 |
| Index 1984=100 | 100 | 102 | 101 | 104 | 104 | 104 | 103 | 105 | 104 |
| **Females** | | | | | | | | | |
| Labor force survey (weekly) | 31.7 | 32.1 | 32.3 | 32.7 | 33.2 | 33.5 | 33.7 | 33.6 | 33.5 |
| Index 1984=100 | 100 | 101.3 | 101.9 | 103.2 | 104.7 | 105.7 | 106.3 | 106.0 | 105.7 |
| Simulation 1 (annual means) | 1391 | 1443 | 1418 | 1430 | 1445 | 1451 | 1429 | 1440 | 1376 |
| Index 1984=100 | 100 | 104 | 102 | 103 | 104 | 104 | 103 | 104 | 99 |
| **C. INEQUALITY IN HOUSEHOLD DISPOSABLE INCOME PER ADULT EQUIVALENT (GINI COEFFICIENT)** | | | | | | | | | |
| Statistics Sweden* | 0.220 | 0.221 | 0.230 | 0.221 | 0.221 | 0.223 | 0.231 | 0.261 | 0.253 |
| Index 1984=100 | 100 | 100.4 | 104.5 | 100.4 | 100.4 | 101.4 | 105.0 | 118.6 | 115.0 |
| Simulation 1 | 0.171 | 0.181 | 0.171 | 0.180 | 0.173 | 0.179 | 0.188 | 0.220 | 0.207 |
| Index 1984=100 | 100 | 106 | 100 | 105 | 101 | 105 | 110 | 129 | 121 |

Sources: Statistiska Meddelanden Be21SM9501, Income distribution survey in 1993 - Statistics Sweden. Labor force surveys, Statistics Sweden. * The tax reform in 1991 made various benefits taxable which had previously been exempt from taxation. If these benefits had been taxable in 1990 and included in the income concept the Gini coefficient would have increased from 0.231 to 0.246.

survey estimates of participation rates are higher than those of MICROHUS, because the age group 66-74 is included in the latter estimates but not in the former. The official estimates show a weakly increasing rate of participation in the labour force for males until 1990 and then a declining rate. For females they also increase to a peak in 1990, after which the official series shows a decline. The simulated series reaches a peak earlier and the subsequent decline is stronger, in particular for men.[5] As shown in Appendix A the retired component of the population increases which, as a general trend, agrees with what can be observed. Early retirement in particular increases. The model simulates an initial decrease in the rate of unemployment and, after 1988, an increase.[6] This pattern approximately agrees with that observed. The Swedish rate of unemployment declined until 1990, after which it increased to levels not previously experienced in Sweden since the 1930s.

The labour force surveys show that average weekly hours of work for men increased slightly until 1990 and then levelled off. Women increased their hours of work quite substantially until 1990, after which their work effort declined too. Given the difficulty in comparing weekly hours with the annual hours simulated by MICROHUS, the slightly increasing trend for males is a little underestimated by MICROHUS, while the more strongly increasing trend for females is reproduced even less well. A closer look at the distribution of annual work hours for females as simulated by MICROHUS reveals that the dispersion of hours increases. Women who work part-time decrease their hours, while those who work full-time or almost full-time increase their hours. This increase in dispersion is particularly strong in 1991-1992 and might be the result of the tax reform. There is no similar increase in dispersion for males.

The model simulates an increase in the inequality of wage rates for both males and females. Wage rate differentials have increased in Sweden in the second half of the 1980s, but the model probably exaggerates the magnitude of the increase. The relative pay of women increased in Sweden at least until the end of the 1980s. The model, however, does not show any trend-like increase but makes the relative female wage rate fluctuate between 82 and 85 per cent of male rates.

---

[5] It should be noted that these comparisons are based on only one single simulation of MICROHUS. Because the model is stochastic, replications of the same simulation will not give exactly the same results. Comparisons with official statistics should ideally be based on several replications. An indication of the built-in stochastic variability of MICROHUS can be obtained from Klevmarken and Olovsson (1994), Appendix B.

[6] The definition of being unemployed is at least 12 months of unemployment.

The income concept used is disposable income (ie. the sum of all income of a household less income taxes and wealth taxes). A household equivalence scale is also applied in which children below 13 years of age are given the weight 0.5, teenagers aged 13-17 and adults other than the household head the weight 0.7, and the household head the weight 1.0. The resulting disposable income per equivalent adult is assigned to each household member. The unit of observation used when computing measures of central tendency and inequality is thus the individual household member. The simulated average real income fluctuates around a more or less constant level until 1991, when it increases. This increase is smaller for households with children than for the entire population. The relative income of households with children decreases from about 90 per cent of average income at the beginning of the period to just above 80 per cent at the end. The tax reform might thus have favoured households without children, although the reform package included an increase in the child allowances to balance the tax changes.

The inequality in disposable income has increased in Sweden and, in particular, in the last two years of the simulation period (ie. 1991-1992). The model picks up these changes quite well. Figure 2 exhibits the simulated income distributions for 1984 and 1992. They show that the increase in inequality is the result of high income earners earning relatively more. To better understand why the simulated distributions of disposable income change, simulation results for labour force participation, annual hours of work and wage rates are also given in Appendix A.

Four different simulations have been run on MICROHUS:[7]

1. All modules with tax and transfer systems for each year 1984 to 1992. This simulation shows the full set of behavioural (or second and subsequent round) effects from the tax-transfer reform - including full labour supply effects.

2. As for Simulation 1, but in 1990-1992 no adjustment of annual work hours was permitted- that is, those who were employed in 1989 kept their 1989 hours until 1992. New entries to the labour market and exits caused by retirement or death were, however, permitted. This simulation shows a limited set of behavioural effects from the tax transfer reforms.

[7] All simulations were started with the same seed in order to minimise the influence of random variation when comparing simulations.

3. As for Simulation 1, but in 1990-1992 no demographic, housing, labour market or income changes were permitted - that is, the tax and transfer systems of 1990-1992 were applied to the income distribution etc. of 1989. In order not to allow inflationary price increases to distort comparisons across years, all wage rate and incomes were inflated by the CPI. This simulation shows the direct or first-round effects of the tax-transfer changes.

4. As for Simulation 1, but in 1990-1992 the 1989 tax system was used. This simulation thus applies the 1989 tax rules to the 1990-92 world (ignoring the transfer system).

The *direct* effect of the tax reform is obtained from simulation number 3, by comparing the 1991 and 1992 distributions with the distribution of 1989 from the same simulation. The comparison was based on 1989 rather than on 1990, because marginal tax rates had already started to decrease in 1990 - Table 3 gives the results. The direct effects of the tax reform are an increase in the average level of disposable income and increased income inequality. For instance, applying the 1992 tax and transfer system to the 1989 incomes and taking the difference in price levels into account results in a median real income and a Gini coefficient which are respectively 15 and 12 per cent higher than is the case when the 1989 tax and transfer system is

Figure 2: Simulated Distribution Functions of Disposable Income Per Equivalent Adult in 1984 and 1992

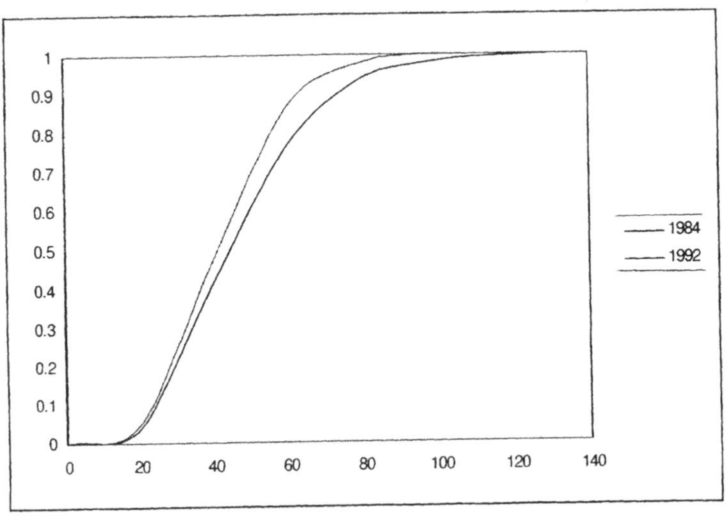

Table 3: Direct Effects of Tax Changes 1990-1992 (Simulation 3)

| | 1989 | 1990 | 1991 | 1992 |
|---|---|---|---|---|
| **All households** | | | | |
| Median disposable income per equivalent adult (1984 SEK) | 41577 | 42852 | 48680 | 47760 |
| Index | 100 | 103 | 117 | 115 |
| Coefficient of variation | 0.325 | 0.342 | 0.383 | 0.364 |
| Index | 100 | 105 | 118 | 112 |
| Gini coefficient | 0.179 | 0.188 | 0.209 | 0.200 |
| Index | 100 | 105 | 117 | 112 |
| **Households with children** | | | | |
| Median disposable income per equivalent adult (1984 SEK) | 36438 | 37299 | 41721 | 41390 |
| Index | 100 | 102 | 114 | 114 |
| Coefficient of variation | 0.247 | 0.272 | 0.325 | 0.311 |
| Index | 100 | 110 | 132 | 126 |
| Gini coefficient | 0.131 | 0.143 | 0.167 | 0.160 |
| Index | 100 | 109 | 127 | 122 |

used (Table 3). The increase in inequality is even higher (22 to 32 per cent) among households with children than it is for the whole population. These estimates are higher than those obtained before the tax reform was implemented. A partial explanation is that MICROHUS does not include the benefits which were not liable for taxation before the tax reform but became taxable after the reform. (In addition, it should be noted that in all of these results any differential effects of the increase in indirect taxes are not taken into account.)

The *behavioural* adjustment effects can be estimated by comparing simulations 1 and 2 with simulation 3. Simulation 2 does not permit adjustments in hours of work among those who are in the labour force. A comparison of Simulations 2 and 3 will thus reveal the effects on the income distribution of other behavioural adjustments - for instance, those related to the housing market and to the size and composition of households. A comparison with simulation 1, on the other hand, gives the full behavioural adjustment effect, including all labour supply effects.[8]

[8] This is not altogether true because, in addition to the changes in the tax and transfer systems, the model adapts to exogenous changes in the CPI, the real rate of return on assets, and the average unemployment rate. Of these, the unemployment rate might influence the distribution of income

When the 1991 tax and transfer system is applied to 1989 incomes, the model simulates, for all households, a median disposable equivalent annual income of 48,680 kronor (Simulation 3 in Table 4). If behavioural adjustments other than in hours of market work are permitted in 1991, the income level decreases by three per cent (Simulation 2). If a full adjustment is allowed, it decreases by four per cent (Simulation 1). Parallel to this decrease in income level there is an increase in income inequality. When everything else but hours of market work can deviate from their 1989 levels, the Gini coefficient in 1991 increases by seven per cent (Simulation 2). However, when market hours are also allowed to adjust to the new tax system, there is only a five per cent increase in inequality (Simulation 1). A general conclusion is that the behavioural adjustments cause income levels to decrease and inequality to increase. For households with children, the decrease in level is marginally stronger - and the increase in inequality marginally weaker - than for the entire population. The behavioural adjustment effects on inequality for all households amount to about one-third of the direct effects while, for households with children, they are smaller.

Table 4: Median Disposable Income Per Equivalent Adult and Income Inequality in 1991 and 1992 - Comparison of Results for Simulations 1, 2 and 3

| | 1991 | | | 1992 | | |
|---|---|---|---|---|---|---|
| Measure | 3 | 2 | 1 | 3 | 2 | 1 |
| **All households** | | | | | | |
| Median disposable income per equivalent adult (1984 SEK) | 48,680 | 47,084 | 46,779 | 47,760 | 45,700 | 44,901 |
| Index | 100 | 97 | 96 | 100 | 96 | 94 |
| Coefficient of variation | 0.383 | 0.419 | 0.406 | 0.364 | 0.408 | 0.391 |
| Index | 100 | 109 | 106 | 100 | 112 | 107 |
| Gini coefficient | 0.209 | 0.224 | 0.220 | 0.200 | 0.215 | 0.207 |
| Index | 100 | 107 | 105 | 100 | 107 | 104 |
| **Households with children** | | | | | | |
| Median disposable income per equivalent adult (1984 SEK) | 41,721 | 40,202 | 39,436 | 41,390 | 37,458 | 36,911 |
| Index | 100 | 96 | 95 | 100 | 91 | 89 |
| Coefficient of variation | 0.325 | 0.344 | 0.344 | 0.310 | 0.316 | 0.310 |
| Index | 100 | 106 | 106 | 100 | 102 | 100 |
| Gini coefficient | 0.167 | 0.175 | 0.174 | 0.160 | 0.167 | 0.161 |
| Index | 100 | 105 | 104 | 100 | 104 | 101 |

most, but none of these changes are believed to be of great importance for the inequality of disposable income.

What type of behavioural changes occurred in response to the tax-transfer changes? In simulation 3 all non-monetary variables are kept at their 1989 values after 1989. Table 5 compares these with the result of Simulation 1, which permits the model to adjust fully to the tax and transfer changes. Panel A shows that the behavioural responses to the tax changes include a small decrease in employment and a corresponding increase in unemployment and retirement. The numbers in the Table are proportions of unemployment, employment etc. in each year, relative to those in 1989 and expressed in index form. As the proportion of unemployed is very small, small changes in this proportion will result in large relative changes.

Table 5: Behavioural Adjustment Effects 1989 - 1992 (Simulation 1)

| | 1989 | 1990 | 1991 | 1992 |
|---|---|---|---|---|
| **A. LABOR FORCE PARTICIPATION ETC.** | | | | |
| Male adults | | | | |
| Unemployed (≥ 12 months) | 100 | 159 | 218 | 232 |
| Employed | 100 | 100 | 101 | 99 |
| Student | 100 | 98 | 92 | 93 |
| Retired | 100 | 100 | 98 | 106 |
| Other not in labor force | 100 | 96 | 95 | 85 |
| Female adults | | | | |
| Unemployed (≥ 12 months) | 100 | 125 | 187 | 225 |
| Employed | 100 | 100 | 98 | 99 |
| Student | 100 | 99 | 107 | 103 |
| Retired | 100 | 102 | 105 | 103 |
| Other not in labor force | 100 | 96 | 92 | 89 |
| **B. ANNUAL HOURS OF WORK** | | | | |
| Employed males | | | | |
| Mean | 100 | 99 | 101 | 100 |
| Coefficient of variation | 100 | 104 | 107 | 151 |
| Employed females | | | | |
| Mean | 100 | 98 | 99 | 95 |
| Coefficient of variation | 100 | 108 | 128 | 157 |
| **C. HOURLY WAGE RATES** | | | | |
| Employed males | | | | |
| Mean | 100 | 101 | 116 | 107 |
| Coefficient of variation | 100 | 113 | 115 | 116 |
| Employed females | | | | |
| Mean | 100 | 100 | 115 | 108 |
| Coefficient of variation | 100 | 144 | 112 | 132 |

Panel B shows that females have decreased their average hours of work after 1989 by a few per cent. This average, however, hides differences in behaviour among women. Those who worked short hours shortened their work hours even more and those who worked long hours increased their hours. Panel C shows that (nominal) wages increased for both genders, in particular in 1991 and 1992. Wage dispersion also increased (more for women than for men). The increased inequality in income due to behavioural changes can thus, at least partly, be traced to increased wage dispersion, increased dispersion in work hours among women and, to some extent, to increased unemployment and retirement. Demographic changes and changes in the housing market have a relatively small effect on inequality.[9]

The behavioural adjustment effect on income inequality should, however, perhaps not be attributed only to the tax reform. The model might have inherent dynamics which could have produced increased inequality independently of the tax reform. This is likely in particular for the sub-model which updates wages. A fourth approach to assess the effects of the income tax reform is to compare the income distributions which would have resulted if the 1989 income tax system had also applied in 1990-1992 (Simulation 4). Thus while the earlier simulations examined the impact of the 1991 tax-transfer system upon either the 1989 or the early 1990s worlds, this simulation tests the impact of the 1989 system upon the early 1990s world. Table 6 shows what changes in work hours, wage rates and income the model would produce if the 1989 tax system had also applied in 1990-92. Panel B indicates that the inequality of *wage rates* among males, as measured by the coefficient of variation, increases by 15 per cent from 1989 to 1991 and by 16 per cent from 1989 to 1992. These increases *exactly equal* the increase shown in Table 5, Panel C. Thus, the conclusion is that the increase in wage rate inequality for males cannot be the result of the tax reform. The same conclusion also holds for women. The rise in wage rate inequality during the early 1990s is thus due to labour market and other changes, rather than to the tax reforms in 1991 and 1992.

Panel A of Table 6 shows that work hours would increase by a few per cent if the 1989 tax system had been kept, and panel C shows that the income level would decrease a little and the inequality of disposable income would increase by 2 to 8 per cent.

---

9 For a simulation study of demographic effects on the income distribution see Klevmarken (1994).

Table 6: Changes in Hours of Work and Disposable Income 1989-1992 if the 1989 Tax System had Applied in 1990-1992 (Simulation 4)

| | 1989 | 1990 | 1991 | 1992 |
|---|---|---|---|---|
| **A. ANNUAL HOURS OF WORK** | | | | |
| **Employed males** | | | | |
| First quartile | 100 | 101 | 103 | 101 |
| Median | 100 | 101 | 102 | 100 |
| Third quartile | 100 | 102 | 103 | 102 |
| Mean | 100 | 102 | 103 | 101 |
| **Employed females** | | | | |
| First quartile | 100 | 102 | 104 | 100 |
| Median | 100 | 100 | 104 | 100 |
| Third quartile | 100 | 102 | 104 | 104 |
| Mean | 100 | 102 | 104 | 101 |
| **B. HOURLY WAGE RATES** | | | | |
| **Employed males** | | | | |
| First quartile | 100 | 88 | 105 | 95 |
| Median | 100 | 97 | 112 | 103 |
| Third quartile | 100 | 101 | 116 | 111 |
| Mean | 100 | 101 | 116 | 107 |
| Coefficent of variation | 100 | 113 | 115 | 116 |
| Gini | 100 | 111 | 115 | 118 |
| **Employed females** | | | | |
| First quartile | 100 | 94 | 114 | 100 |
| Median | 100 | 96 | 115 | 104 |
| Third quartile | 100 | 100 | 118 | 114 |
| Mean | 100 | 100 | 115 | 108 |
| Coefficent of variation | 100 | 143 | 112 | 132 |
| Gini | 100 | 114 | 112 | 120 |
| **C. DISPOSABLE INCOME** | | | | |
| **All households** | | | | |
| First quartile | 100 | 95 | 98 | 96 |
| Median | 100 | 95 | 99 | 97 |
| Third quartile | 100 | 96 | 103 | 97 |
| Mean | 100 | 96 | 101 | 98 |
| Coefficent of variation | 100 | 107 | 108 | 104 |
| Gini | 100 | 103 | 107 | 102 |
| **Households with children** | | | | |
| First quartile | 100 | 96 | 97 | 97 |
| Median | 100 | 97 | 98 | 96 |
| Third quartile | 100 | 97 | 102 | 98 |
| Mean | 100 | 97 | 100 | 97 |
| Coefficent of variation | 100 | 130 | 111 | 101 |
| Gini | 100 | 104 | 111 | 102 |

Table 7 shows the difference between two 'full behavioural response' simulations, one using the current tax and transfer systems (Simulation 1) and one imposing the 1989 tax system after 1989 (Simulation 4). For example, the index value of 97 for median employed males hours of work in 1990 means that with the 1990 tax system the median was 3 per cent less than with the 1989 tax system. Panel B indicates that the 1991-92 income taxes increased income inequality by 14 to 15 per cent for the whole population and by 20 to 25 per cent for households with children.[10] The same panel also shows that the 1991-92 tax system increased the average disposable income level of the whole population by 13 to 16 per cent. Panel A reveals that part of the increased income inequality comes from an increase in the dispersion of work hours for both males and females. In addition, the gender differences in average annual hours of work increased. There were almost no induced changes in labour force participation and there were no additional changes in wage dispersion compared with Table 6.[11]

## 5. Conclusions

The simulations with MICROHUS show that the 1990-91 Swedish tax reform increased the inequality of the distribution of disposable income per equivalent adult. The direct effects have been estimated at about 17 per cent in 1991 and 12 per cent in 1992 for the whole population of households. Our initial estimates of the effects of behavioural adjustments (Simulations 1 and 2) suggested about a 5 to 10 per cent additional increase in inequality once behavioural changes were also taken into account. However, most of this effect cannot be attributed to the tax reform but to the dynamic specification of the model which captured, for instance, the rising inequality of wage rates in Sweden in the early 1990s. A second approach to evaluate the effects of the tax reform involved comparing the income distributions obtained with the 1991-1992 tax systems (Simulation 1) with those which would have resulted if the 1989 tax system had still applied in 1991 and 1992 (Simulation 4). According to this comparison, the new tax system generated an increase in inequality of about 15 per cent. When this estimate is compared with the 17 per cent estimate of the direct effects, the indirect behavioural adjustment effect would thus seem to be very small.

---

[10] Thus the index number 114 for the Gini coefficient in 1992 means that the Gini coefficient from simulation 1 in 1992 exceeded the corresponding Gini coefficient from simulation 4 by 14 per cent.

[11] These results are not shown in Table 7.

Table 7: An Alternative Assessment of the Total Effects of the 1990-1992 Income Tax Changes - Current Tax System (Simulation 1) Compared to the 1989 Tax System (Simulation 4)

| | 1989 | 1990 | 1991 | 1992 |
|---|---|---|---|---|
| **A. ANNUAL HOURS OF WORK** | | | | |
| Employed males | | | | |
| First quartile | 100 | 97 | 97 | 100 |
| Median | 100 | 97 | 98 | 100 |
| Third quartile | 100 | 98 | 99 | 102 |
| Mean | 100 | 97 | 98 | 99 |
| Employed females | | | | |
| First quartile | 100 | 94 | 89 | 84 |
| Median | 100 | 96 | 94 | 95 |
| Third quartile | 100 | 98 | 99 | 101 |
| Mean | 100 | 96 | 95 | 94 |
| **B. DISPOSABLE INCOME PER ADULT EQUIVALENT** | | | | |
| All households | | | | |
| First quartile | 100 | 101 | 107 | 107 |
| Median | 100 | 101 | 114 | 111 |
| Third quartile | 100 | 101 | 116 | 115 |
| Mean | 100 | 101 | 116 | 113 |
| Coefficent of variation | 100 | 105 | 116 | 115 |
| Gini | 100 | 102 | 115 | 114 |
| **C. HOUSEHOLDS WITH CHILDREN** | | | | |
| First quartile | 100 | 100 | 104 | 102 |
| Median | 100 | 100 | 110 | 106 |
| Third quartile | 100 | 100 | 113 | 112 |
| Mean | 100 | 100 | 111 | 109 |
| Coefficent of variation | 100 | 111 | 125 | 124 |
| Gini | 100 | 102 | 120 | 121 |

The size of the indirect effect will, of course, depend very much on the model specification. In MICROHUS the labour supply adjustments are small, partly because the decision to join the labour force is independent of incentives from the tax system (by assumption) and partly because the estimated wage rate and income elasticities for those already employed are relatively small. The labour supply literature has shown that the participation decision depends more on wage rates and income incentives than does changes in hours of work. A re-specification which introduces more economic incentives into the module and which simulates entrances to the labour market might thus increase the adjustment effect. The application of supply functions for hours of work of the Hausman type could also be criticised. There is certainly scope for improvements and alternatives. For instance, a good labour supply model should give a better representation of the joint decisions of two spouses.

Although this work has not been able to demonstrate any major indirect, behavioural adjustment effects on the income distribution resulting from the Swedish tax reform of 1991 this might be a consequence of the model specification. It might also take longer than a year or two before these effects become manifest. Additional efforts to design flexible models which predict behaviour well are needed before we can be certain that estimates of direct effects alone provide good guides to the total effects.

## APPENDIX A

### Simulation 1: Reference Case With Full Model

**A. Age distribution (%) and sample size**

| | 1984 | 1985 | 1986 | 1987 | 1988 | 1989 | 1990 | 1991 | 1992 |
|---|---|---|---|---|---|---|---|---|---|
| Age | | | | | | | | | |
| 0 - 6 | 7.7 | 7.2 | 6.3 | 5.3 | 4.8 | 4.4 | 4.6 | 4.7 | 4.8 |
| 7 - 17 | 14.7 | 14.7 | 15.2 | 15.3 | 15.3 | 14.8 | 14.1 | 13.5 | 12.9 |
| 18 - 29 | 18.9 | 18.8 | 19.9 | 19.6 | 19.3 | 19.9 | 19.9 | 20.0 | 20.2 |
| 30 - 39 | 15.0 | 15.6 | 15.3 | 14.6 | 15.0 | 14.7 | 14.7 | 14.8 | 15.1 |
| 40 - 49 | 13.8 | 14.1 | 14.6 | 15.3 | 15.8 | 16.4 | 16.8 | 16.9 | 16.9 |
| 50 - 64 | 18.4 | 18.2 | 18.3 | 18.4 | 19.0 | 18.8 | 18.8 | 19.5 | 19.6 |
| 66 - 74 | 11.1 | 11.1 | 11.0 | 11.2 | 10.6 | 10.6 | 10.7 | 10.3 | 10.3 |
| 75 * | 0.2 | 0.3 | 0.4 | 0.3 | 0.2 | 0.3 | 0.4 | 0.4 | 0.4 |
| Sample size | 5250 | 5201 | 5170 | 5124 | 5050 | 5011 | 4968 | 4927 | 4866 |

* When a household head reaches the age of 75 the whole household is excluded from the sample. Only household members 75 years and older who live in a household with a head younger than 75 remain in the sample.

**B. Marital status (%); 18 years and older**

| | 1984 | 1985 | 1986 | 1987 | 1988 | 1989 | 1990 | 1991 | 1992 |
|---|---|---|---|---|---|---|---|---|---|
| Married | 53.1 | 52.6 | 52.2 | 52.1 | 51.5 | 51.3 | 50.7 | 50.9 | 50.3 |
| Cohabiting | 14.3 | 14.4 | 14.7 | 14.7 | 14.7 | 14.8 | 14.8 | 14.9 | 15.2 |
| Single | 32.6 | 33.0 | 33.1 | 33.2 | 33.8 | 34.0 | 34.5 | 34.2 | 34.5 |

**C. Labor force participation (%); 18 years and older**

| | 1984 | 1985 | 1986 | 1987 | 1988 | 1989 | 1990 | 1991 | 1992 |
|---|---|---|---|---|---|---|---|---|---|
| **Male adults** | | | | | | | | | |
| Unemployed (≥ 12 months) | 1.1 | 0.7 | 0.5 | 0.2 | 0.3 | 0.3 | 0.5 | 0.7 | 0.8 |
| Employed | 71.2 | 72.2 | 72.9 | 72.0 | 71.5 | 68.7 | 68.8 | 69.5 | 68.1 |
| Student | 5.3 | 5.3 | 5.3 | 7.2 | 8.0 | 9.0 | 8.8 | 8.3 | 8.4 |
| Retired | 18.5 | 17.8 | 17.9 | 17.4 | 17.1 | 19.0 | 19.0 | 18.6 | 20.0 |
| Other not in L.F. | 3.9 | 3.8 | 3.4 | 3.2 | 3.1 | 3.0 | 2.9 | 2.9 | 2.6 |
| **Female adults** | | | | | | | | | |
| Unemployed (≥ 12 months) | 1.3 | 1.8 | 1.2 | 0.5 | 0.2 | 0.4 | 0.5 | 0.8 | 0.9 |
| Employed | 64.1 | 64.5 | 65.4 | 63.6 | 64.1 | 63.3 | 63.3 | 62.1 | 63.0 |
| Student | 5.5 | 5.5 | 5.4 | 7.6 | 8.3 | 8.3 | 8.2 | 8.9 | 8.6 |
| Retired | 18.9 | 18.1 | 18.3 | 19.3 | 18.9 | 19.7 | 20.1 | 20.7 | 20.3 |
| Other not in L.F. | 10.2 | 10.1 | 9.7 | 9.1 | 8.5 | 8.2 | 7.9 | 7.5 | 7.3 |

**D. Annual hours of market work; employed**

| | 1984 | 1985 | 1986 | 1987 | 1988 | 1989 | 1990 | 1991 | 1992 |
|---|---|---|---|---|---|---|---|---|---|
| **Males** | | | | | | | | | |
| First quartile | 1750 | 1800 | 1764 | 1820 | 1820 | 1800 | 1768 | 1800 | 1813 |
| Median | 1924 | 1950 | 1924 | 1976 | 1976 | 1976 | 1932 | 1960 | 1976 |
| Third quartile | 2080 | 2080 | 2040 | 2132 | 2132 | 2091 | 2080 | 2132 | 2184 |
| **Females** | | | | | | | | | |
| First quartile | 1260 | 1300 | 1274 | 1300 | 1288 | 1300 | 1248 | 1200 | 1092 |
| Median | 1404 | 1440 | 1404 | 1428 | 1421 | 1456 | 1404 | 1428 | 1377 |
| Third quartile | 1551 | 1598 | 1560 | 1581 | 1600 | 1612 | 1612 | 1666 | 1700 |

**E. Hourly wage rates (current SEK); employed**

| | 1984 | 1985 | 1986 | 1987 | 1988 | 1989 | 1990 | 1991 | 1992 |
|---|---|---|---|---|---|---|---|---|---|
| **Males** | | | | | | | | | |
| First quartile | 38 | 42 | 39 | 41 | 37 | 42 | 37 | 44 | 40 |
| Median | 45 | 52 | 50 | 54 | 51 | 58 | 56 | 65 | 60 |
| Third quartile | 55 | 61 | 62 | 68 | 67 | 75 | 76 | 87 | 83 |
| Mean | 49 | 55 | 54 | 58 | 56 | 61 | 62 | 71 | 65 |
| Coefficent of variation | 0.439 | 0.443 | 0.508 | 0.510 | 0.585 | 0.548 | 0.619 | 0.629 | 0.634 |
| **Females** | | | | | | | | | |
| First quartile | 32 | 37 | 32 | 35 | 33 | 36 | 34 | 41 | 36 |
| Median | 38 | 44 | 41 | 46 | 42 | 48 | 46 | 55 | 50 |
| Third quartile | 46 | 53 | 50 | 57 | 54 | 62 | 62 | 73 | 69 |
| Mean | 40 | 46 | 43 | 48 | 46 | 52 | 52 | 60 | 56 |
| Coefficent of variation | 0.434 | 0.455 | 0.523 | 0.460 | 0.602 | 0.494 | 0.706 | 0.554 | 0.652 |

**F. Household disposable income per equivalent adult (constant 1984 SEK)**

| | 1984 | 1985 | 1986 | 1987 | 1988 | 1989 | 1990 | 1991 | 1992 |
|---|---|---|---|---|---|---|---|---|---|
| **All households** | | | | | | | | | |
| First quartile | 33792 | 35195 | 33645 | 33840 | 33135 | 33954 | 32602 | 35862 | 34936 |
| Median | 41691 | 44464 | 40475 | 42389 | 39870 | 41577 | 39826 | 46779 | 44901 |
| Third quartile | 51321 | 55475 | 50155 | 52983 | 50272 | 52325 | 50283 | 62945 | 58588 |
| Mean | 43682 | 46681 | 42817 | 44772 | 42645 | 44225 | 42925 | 51649 | 48721 |
| Coefficent of variation | 0.311 | 0.326 | 0.311 | 0.325 | 0.319 | 0.325 | 0.362 | 0.406 | 0.391 |
| **Households with children** | | | | | | | | | |
| First quartile | 32378 | 33803 | 32108 | 32474 | 31676 | 32040 | 30788 | 32545 | 31714 |
| Median | 37903 | 39860 | 36591 | 38173 | 35465 | 36438 | 35139 | 39436 | 36911 |
| Third quartile | 43991 | 47530 | 42879 | 45237 | 42123 | 42909 | 41525 | 49232 | 46780 |
| Mean | 39355 | 41653 | 38308 | 39693 | 37872 | 38190 | 37334 | 42569 | 40448 |
| Coefficent of variation | 0.263 | 0.277 | 0.264 | 0.262 | 0.288 | 0.247 | 0.354 | 0.344 | 0.310 |

## References


Aronsson, T., J.J. Heckman and J.R. Walker (1994), ' The effect of Sweden's tax and transfer system on labor supply', forthcoming in: R. Freeman, B. Swedenborg and R. Topel (eds.), *The Swedish Welfare State in Transition*, Chicago University Press.

Eriksen, T. (1973), *En prognosmodell för den allmänna tilläggspensioneringen*, Riksförsäkringsverket, Stockholm.

Gustafsson, B. and N.A. Klevmarken (1993), 'Taxes and transfers in Sweden: Incentive effects on labour supply', in: A.B. Atkinson and G. Viby-Mogensen (eds.), *Welfare and Work Incentives: A North European Perspective*, Clarendon Press, Oxford.

Klevmarken, N.A. (1973), 'En ny modell för ATP-systemet', *Statistisk Tidskrift*, (Statistical Review), Vol. 11, No. 5.

Klevmarken, N.A. (1994), 'The impact of demographic changes on the income distribution: Experiments in microsimulation', Paper presented to the 8th Annual Conference of the European Society for Population Economics, Tilburg, Holland, June.

Klevmarken, N.A., I. Andersson, P. Brose, L. Flood, P. Olovsson and A. Tasiran (1992), 'MICROHUS: A microsimulation model for the Swedish household sector - a progress report', Paper presented to the International Symposium on Economic Modelling, Gothenburg, Sweden, August.

Klevmarken, N.A. and P. Olovsson (1989), 'Hushållens ekonomiska levnadsförhållanden (HUS). Teknisk beskrivning och kodbok', Department of Economics, Gothenburg University.

Klevmarken, N.A. and P. Olovsson (1993), *Household Market and Nonmarket Activities: Procedures and Codes 1984-1991*, Almqvist and Wicksell International, Stockholm.

Klevmarken, N.A. and P. Olovsson (1994), 'Direct and behavioural effects of income tax changes - simulations with the Swedish model MICROHUS', Working Paper 1994:20, Department of Economics, Uppsala University, Uppsala.

# PART THREE

# Lifetime and Retirement Incomes

Jane Falkingham

London School of Economics
United Kingdom

Ann Harding

University of Canberra
Australia

# Chapter 11: Poverty Alleviation vs Social Insurance Systems: A Comparison of Lifetime Redistribution

## Abstract

*Among other objectives, modern social security and taxation systems seek to redistribute income from those with high to those with low incomes and to reallocate income across the lifecycle of individuals. While there are numerous studies of the extent to which these aims are achieved during a short time period, such as a week or a year, there are almost no studies of the lifetime redistributive impact of tax/transfer systems. In this chapter, two broadly comparable dynamic cohort microsimulation models are used to assess the lifetime redistributive impact of the British and Australian social security and direct taxation systems. The analysis suggests that the Australian system, with its emphasis on poverty alleviation, results in greater lifetime inter-personal redistribution than the British social insurance-based system.*

## 1. Introduction

All modern social security systems serve two linked, but distinct, aims. On the one hand, they are an updated 'Robin Hood' mechanism, for transferring resources raised through taxation from the better-off to those with low incomes - termed *inter-personal redistribution.* On the other hand, they are also intended to transfer resources from parts of the lifetime, when incomes are relatively high, to other times when the incomes of the same individual are low - *intra-personal redistribution.*

To date, most distributional analysis has focused on the first of these aims, largely consisting of studies of the annual net fiscal incidence of taxes and cash transfers (CSO, 1992; Harding, 1984; Saunders, 1984). These studies thus examine the redistributional impact of government during a short time period, such as a week or a year. However, a very different distributional picture emerges when the effects of social security and taxation are examined over the entire lifetime (Falkingham and Hills 1995; Harding, 1992a, 1992b, 1993a; Nelissen, 1993).

* The authors would like to thank Steve Gribble for helpful comments on an earlier version of this paper.

Social security systems can be broadly classified under two heads:

- social insurance-based systems, where an individual secures an entitlement to a range of income support payments because of his or her identifiable social insurance contributions whilst in the labour market; and

- social assistance systems, where entitlement to a benefit is based on citizenship rather than contribution, but where entitlement is subject to the imposition of income and assets tests and, frequently, for those of working age, requirements to engage in job search and training activities.

In reality, few social security systems are purely insurance-based, most having some elements of social assistance to cover individuals who fail to meet the contributory requirements.

One might predict that a primarily social insurance-based system would lead to a higher degree of intra-personal redistribution of resources over the lifetime - and that a primarily social assistance-based system, with its emphasis on poverty alleviation, would result in a greater degree of inter-personal redistribution. However, some degree of convergence in the redistributive impact of different systems when looked at in a lifetime perspective might also be expected. The 'insurance' principle leads, after the event, to *inter-personal redistribution*, as some will suffer the contingencies insured, like sickness and unemployment, while others will not. Similarly, periods of retirement will vary because of different life spans. The 'social' insurance principle that contributions should not reflect risk, that 'none should claim to pay less because he is healthier or in more regular employment', also leads to inter-personal redistribution (Beveridge, 1942, p. 13).

Likewise, recipients of income support payments under a poverty alleviation system may, when viewed across the life-cycle, simply be receiving back taxes they have paid at another age. At any one point in time, a large proportion of those with low income are retirees who might have enjoyed high incomes in the past, or students who will probably earn much higher incomes in the future. Social assistance schemes therefore implicitly reallocate income over the life-cycle, introducing an element of *intra-personal distribution*.

To examine whether a poverty alleviation system results in greater lifetime *inter-personal redistribution* requires data for complete lifetimes, which are exceedingly rare. Given this lack of data, an alternative is to examine what Summers referred to as the 'latent lifetime income distribution' (1956) - that is, the lifetime distribution which would result if current conditions continued indefinitely. Economists and econometricians have in the past used a number of techniques to simulate lifetime income profiles, but these have not allowed for the continual changes in the circumstances of individuals which occur in real life. The relatively recent technique of dynamic microsimulation, in contrast, allows the characteristics of individuals within the model to change constantly so that, for example, individuals may enter or leave the labour force, get married or divorced, bear children and so on (see Harding, 1993a). One of the most striking findings from longitudinal data is how much diversity people do experience over their lifetimes. Thus, dynamic microsimulation models are potentially a significant improvement on the static models used by econometricians in the past.

This paper investigates the hypothesis using data from two dynamic cohort microsimulation models developed in parallel for Australia (HARDING) and Great Britain (LIFEMOD). The examples of Australia and Britain are particularly germane to the question at hand. In the 1980s Australia undertook a radical reform of its social security system which left it with perhaps the 'purest' social assistance system in the industrialised world. The British system, in turn, still has its foundations in the Beveridge Report and remains the epitome of a social insurance-based system.

Section 2 discusses the construction of the two dynamic cohort models, focusing on their similarities and differences. Although developed in parallel and sharing a similar modular structure, the two models are not identical, perhaps more akin to 'kissing cousins' than sisters. Therefore, it is important to highlight where differences may be due to methodology rather than the operation of the respective social security systems. Section 3 summarises the social security systems for Australia and Britain, as modelled in HARDING and LIFEMOD.

Sections 4 and 5 contrast how these two examples of social insurance and poverty alleviation systems perform in the lifetime redistributive stakes. The Sections seek to provide answers to such questions as how redistributive overall are the systems; what balance between intra-personal and inter-personal redistribution do they produce; how far do the two systems converge in outcomes when compared over the lifetime; and, finally, is a poverty alleviation system really just that or does it, over the lifetime, imitate a social insurance system? Finally, Section 6 summarises the main findings.

# 2. The Models

The construction of both of the dynamic cohort models employed in this paper has been usefully described elsewhere (for HARDING see Harding, 1993a; for LIFEMOD see Falkingham and Lessof, 1991, 1992). Nevertheless, it is instructive to briefly describe their development, pointing out the differences and similarities.

## 2.1 Similarities

Both models simulate complete life histories for a pseudo-cohort of 2000 males and 2000 females. These individuals are dynamically aged from birth to death and experience major life events such as schooling, marriage, childbirth, children leaving home, employment and retirement. At this stage, there is no immigration or emigration into or out of the cohorts, and the only way in which the cohorts change size is from attrition due to mortality.

It is important to point out that the micro unit which both HARDING and LIFEMOD simulates is the individual, rather than the family or the household - and therefore it is only the characteristics of the cohort individuals *themselves* which are modelled. For example there is no comprehensive information on the children of each cohort member, but simply their age, parity, and whether or not they are participating in full-time education (that is, whether or not they are classified as dependent children). HARDING and LIFEMOD do not contain detailed information about the wider household composition of cohort members, unlike other micro-simulation models such as that developed at Tilburg University (Nelissen, 1993). However, since cohort members are selected to marry each other, the characteristics of an individual's spouse are also available at each stage during their union. This is of particular importance in the labour market, income tax and social security modules, where these characteristics interact and the characteristics of both partners in the preceding period determining the characteristics of the present period. Specification of income also depends on spouse's income.

The simulation process requires transition probabilities or behavioural equations as input for every variable generated. The various probabilities of demographic and other events occurring to people are estimated in the two models from a variety of sources (most notably published official statistics and secondary analysis of large sample social surveys such as the Australian Income Distribution Survey and the UK General Household and Family Expenditure Surveys).

Given the uncertainty surrounding future changes in marriage and birth rates, labour force participation rates and so on, both models assume a steady state world and, as far as possible, all input data are calculated for a specific year. Therefore, the HARDING cohort is 'born' into, and subsequently lives in, a world that looks like Australia in 1986. Unfortunately, due to data availability, the two models were not constructed on the same base year and the LIFEMOD cohort live their complete lives in a world which looks like Britain in 1985. Although the steady state assumption results in a highly stylised 'population', it has been found to provide a useful benchmark against which current government policies, and changes to those policies, can be evaluated. Both the Canadian DEMOGEN and German SFB3 dynamic cohort models also assume a steady-state world when evaluating the impact of existing and potential government policies (Wolfson, 1988, p. 233; Hain and Helberger, 1986, p. 63).

## 2.2 Differences

Aside from the actual transition probabilities to which each cohort member is exposed, and the institutional structure of taxes and benefits of Australia and Britain's social security systems, there are several key programming differences which should be highlighted. The two models used common core code for the earlier modules, particularly those concerned with family formation and dissolution. However, the models diverged in the approach taken towards the simulation of labour force participation and earnings. HARDING models the number of hours spent in paid employment, self-employment and/or unemployment at a given age, whereas LIFEMOD models the number of weeks spent in any given state during the year. HARDING also uses techniques to generate the inequality apparent within the labour market, such as the existence of concentrated long term unemployment. LIFEMOD, like the DEMOGEN and SFB3 models, uses econometric techniques to simulate the decision to enter or leave the workforce, whilst the HARDING model follows the DYNASIM approach in using simple matrices of probabilities of participation. However, the HARDING matrices include most of the dependent variables in the LIFEMOD equations and both studies found educational attainment and participation in the previous year to be amongst the main explanatory variables.

Because of data limitations, female labour force participation was modelled independently of spouse's employment patterns in HARDING. In LIFEMOD, however, the employment and earnings of married women were dependent upon their husband's employment status and income. This might have the effect of creating more 'double-unemployed' couples in Britain than in Australia (given the well known disincentive effect of husband's unemployment on wife's labour supply). However, when the lifetime distribution of earnings is presented in Table 2 there is little evidence of such an effect.

Another difference was in the treatment of maintenance income. Both HARDING and LIFEMOD modelled receipt of maintenance income by sole mothers following divorce. However, in LIFEMOD payment of maintenance income by the ex-husband was also modelled. This is taken to be 'negative' income and thus depresses men's original income. This may account for the somewhat flatter distribution of original income in Table 2 for Britain versus Australia. However, this paper compares the *change in the distribution between original and gross income (ie.* the role of cash benefits) and between gross income and net income (ie. the role of taxes), rather than examining the shape of the original income distribution. Consequently, differences in modelling methodology are not expected to greatly influence the overall results.

## 2.3 Caveats and qualifications

The use of such models in distributional analysis of the kind presented in this paper involves a number of difficult issues whose significance should be fully appreciated when interpreting the subsequent results. The problems apply equally both to the HARDING model and LIFEMOD and a common approach has been taken to their resolution. Thus, if biases exist, those biases should apply equally to the two models.

(1) *Indirect and local government taxes:* Both models currently ignore indirect taxes and are limited to the major cash transfers and income taxes administered by the central governments. Neither HARDING or LIFEMOD include taxes levied by other levels of government[1], the inclusion of which could (along with expenditure taxes) make a substantial difference to the findings. In the UK in 1985, income taxes and National Insurance Contributions amounted to 52 per cent of total central government revenue. In Australia in 1986, income taxes amounted to 51 per cent of all Federal government revenue. During the 1980s the British government pursued a policy of shifting the tax burden from one of 'tax as you earn' to 'tax as you spend', and this has obvious implications for redistribution now.

Benefits in kind from other government funded services also affect lifetime redistribution. Le Grand (1982) has identified the education system as being particularly regressive over the life-course, with the middle classes benefiting disproportionately from educational expenditures (particularly higher education) -

[1] In Britain in 1985 this was local authority rates, which were replaced by the infamous 'poll tax' in 1989-90 and succeeded by the new council tax in 1993-94. In Australia, all three levels of government levy taxes or rates which are not included within the scope of this analysis.

although this conclusion is disputed by Harding using Australian data (1993a, p, 72). Both models do allow for the allocation of expenditures on education (and, in the case of LIFEMOD, on the National Health Service); however, these are not included in this particular analysis (for a discussion of education transfers see Harding, 1995 and Barr and Falkingham, 1993).

(ii) *The counterfactual*: A second issue is that in assessing the impact of the social security system upon the distribution of income, the distribution after intervention necessarily has to be compared with the distribution before intervention. This immediately raises the question of what the most appropriate 'before' benchmark - or counterfactual - is. This study follows convention and measures the redistributive effect of the tax/transfer system against the original distribution of pre-tax and pre-transfer income. While it is clearly invalid to assume that the pattern of factor incomes would remain the same if there was no government, such an assumption is implicit in this study, as there are no data available indicating how the lifetime distribution of factor income in Australia or Britain would change if their governments miraculously disappeared.

(iii) *Assumed incidence of cash transfers and income taxes:* The benefit of cash transfers is assumed to be fully incident upon those to whom they are paid. This assumption is not uncontroversial. It is arguable that the benefit of family allowance is incident equally upon a husband and wife, or indeed upon the children, rather than solely upon the mother who is the formal recipient. Yet alternative assumptions are also not straightforward. Both governments have explicitly made the mother the recipient of child benefits (family allowances) because of doubts as to who received the benefit when it was paid to the husband. Similarly it is not clear that the benefits are fully incident upon children (Barro, 1974).

Likewise, the burden of taxation is assumed to be fully incident upon those legally liable to pay the taxes. Those with such liabilities are also assumed to pay them in full and no account is taken of possible tax evasion or of the underground economy. However, full compliance is not mirrored in the case of benefit receipt, and both models incorporate assumptions as to take-up rates. In LIFEMOD take-up of means tested benefits was related to the size of entitlement (the higher the entitlement the greater the likelihood of the individual claiming it). In HARDING, take-up of Family Income Supplement was affected by self-employment status, number of children and size of entitlement.

(iv) *Real economic growth*: A fourth issue is whether and how to allow for real economic growth. The 'unit of account' in both models is based on current earnings levels. One could allow for real earnings growth as well as career progression in earnings as people become older, but one would then also have to discount for the lower value of later receipts. Implicitly the approach adopted assumes that the effects of overall real economic growth and a real discount rate cancel each other out. In LIFEMOD, this means that, once in payment, benefits which are indexed to prices rather than earnings (such as public sector pensions or SERPS rights in retirement), are assumed to slip back each year at the rate of real earnings growth.

# 3. The Systems

## 3.1 Great Britain - a social insurance system

In LIFEMOD, the individuals within the model live their entire lives in the demographic and economic environment of Britain as it was in 1985. Therefore, the social security benefits they receive and their direct tax liabilities are calculated under the rules of the tax and benefit system as it was in April 1985. The simulation of the system is described in detail in Hills and Lessof (1993).

Entitlement to the majority of benefits within the British social security system depends upon an individual's National Insurance Contribution (NIC) record. NICs in any year depend upon earnings levels and the number of weeks of employment or self-employment. In 1985, employee contributions had a 'slab structure', with reduced rates payable in certain earnings bands. Earnings above the Upper Earnings Limit were exempt. The social security module of LIFEMOD follows this banded structure in modelling NIC liabilities.

In addition to NICs, contributions to the State Earnings Related Pension Scheme (SERPS) and, where applicable, contributions to occupational pension were also modelled. In 1985 it was possible for a (legally) married couple to choose between joint assessment of husbands and wives income or individual assessment. In LIFEMOD the tax liabilities under both these options are calculated. It is then assumed that couples will choose the option giving the lowest total liability: tax calculated jointly, or tax on the parts of wives' incomes which are deemed to be part of husband's income for tax purposes, is then apportioned between the husband and wife.

Old age pensions constitute by far the largest cash transfer programme on an annual basis and are also the most significant cash benefit when viewed over the life-course. Income support in old age in Britain takes the form of a two-tier pension. The first tier is provided by a universal flat rate basic state pension, for which an individual must have sufficient qualifying years - ie. years in which a given amount of NICs are paid. A person who has home care responsibilities may be credited with contribution years towards the basic state pension even if in that year she has paid insufficient NICs, either because of low or no earnings. This is known as Home Responsibility Protection. Receipt of HRP does not affect income in the year received but may affect future income streams.

The second tier is furnished by earnings-related pensions, which can be provided under SERPS or by an employer under an occupational pension scheme. In 1985 the number of private personal pensions were low, both in payment and in accumulation, and so these were not included in the modelling process. The majority of occupational pension schemes in Britain are defined benefit schemes with the pension linked to the level of final salary, whereas personal pensions pay a defined contribution. The latter are much more difficult to calculate in a steady state world. In LIFEMOD, pension rights are simulated along with SERPS and other pension entitlements as the cohort individuals age. The pension rights accumulated may be transferred to an individual's spouse in the event of them dying before pensionable age. Widows can also inherit occupational pensions.

Aside from old age support, there are a range of contingent benefits in Britain which are not subject to a means test: child benefit (which depends simply on the number of children and which was assumed in LIFEMOD to be incident on the mother except in the case of widowers); one parent benefit paid to lone parents; unemployment benefit paid for a maximum of 52 weeks in any two years to those who are unemployed and who have a good enough recent employment history; and invalidity benefit, severe disablement allowance, mobility allowance, attendance allowance and invalid care allowance, all of which depend on degree of disability and age of original receipt. Income from all of these, and tax liabilities, are modelled prior to calculating any eligibility for means-tested support.

Means-tested benefits include Supplementary Benefit (now Income Support) or Family Income Supplement (now Family Credit). The latter applies only to those with children and a period in the year of some full-time work. The amount received depends on the number and ages of children and on weekly income in relation to particular thresholds. Supplementary Benefit (SB) is available for individuals or couples with periods without full-time work during a year. The size of entitlement depends on family circumstances, age, previous receipt of SB, income and capital. Family Income Supplement is assumed to be received by women, except in the case of widowed fathers. Where a couple are entitled to SB, the amount received is taken to be part of the man's income. Additions to SB for housing costs, along with other housing-related cash transfers such as Housing Benefit and Rent and Rate Rebates, were not modelled in this version of LIFEMOD. Similarly subsidies to owner occupiers through mortgage interest tax relief were also not included.

Means-tested benefits are assessed on a weekly rather than an annual basis. Where income varies over the year (for instance, because of an episode of unemployment) it will generally underestimate benefit entitlement if receipt is calculated on the basis of average income over the whole year. Thus, annual income in LIFEMOD is divided between two periods - the period in which someone is employed full-time and the rest. Assessment for means-tested benefits is on a joint basis for couples (including those cohabiting but not married).

Student grants were simulated, but were treated as part of original income, rather than as a social security type cash benefit. This is different to the approach adopted in the Australian model.

## 3.2 Australia - a social assistance model

The Australian social security system principally consists of a range of income-tested payments which are available to those with particular characteristics (the simulation of the system is described in detail in Harding, 1993a). The social security cash transfers simulated in the model, with 1986 payment rates and income tests, were age pension, invalid pension, wife's pension, carer's pension, widow's pension, supporting parent's benefit, unemployment benefit, sickness benefit and special benefit. All of these transfers were income-tested upon the income of the individual (or husband and wife for married or de facto couples), and bore no relation to previous periods in the labour force or previous earnings. In addition to the basic rates, all of those receiving the above payments could receive a number of additional allowances, of which additional pension and benefit (paid for dependent children) and mother's/ guardian's allowance (paid to those receiving widow's pension or supporting parent's benefit) were included in the model. Rent assistance, which was paid to pensioners and beneficiaries who were private renters, was not included in the model, because of the lack of housing data on the 1986 Income Distribution Survey.

Family income supplement (FIS) was payable to low income families with dependent children who were not receiving any other form of income support from the Commonwealth - ie. effectively those who were in the labour force. Only two payments in 1986 were not income-tested and included within the model. One was family allowance, which was payable monthly to people with dependent children aged less than 16, full-time dependent students aged 16 and 17 not receiving education transfers, or similar students aged 18 to 24 in very low income families. Multiple birth payment to the parents of triplets or quads aged under 6 was also modelled.

In couples, all benefits and supplements were paid by the Department of Social Security to the husband, while pensions were split equally between partners but any additional payments for the children of pensioners were paid to the wife. Family allowance, multiple birth payment and FIS were all expressly paid to the mother in married couples. All these provisions were fully incorporated within the simulation.

The simulation of social security transfers was very comprehensive. A small number of payments were not modelled but, overall, 97 per cent of outlays on pensions, 99 per cent of outlays on benefits and 98 per cent of outlays on child transfers were captured within the simulation. Apart from waiting periods and a few special provisions, eligibility for social security cash transfers was essentially based upon weekly income. Consequently, eligibility for most payments was based on the income and circumstances of each family during each week of each year in the model.

In 1986 all pensions and supporting parent's benefit were both income and asset-tested. It was not possible to model the assets test adequately, because of the lack of data about assets in Australia. A number of techniques were therefore used to produce an appropriate degree of take-up, for both age pension, sickness and special benefit and FIS. (Age pension was the principal source of income for most retirees in Australia, with some 79 per cent of the population of age pension age receiving age pension in 1986. Others received occupational pensions from their employers, which were generally linked to final average salary and length of service.)

The Department of Education also provided two major transfers in 1986, income-tested upon the income of both parents and students - the Secondary Allowances Scheme (SAS) and the Tertiary Education Assistance Scheme (TEAS). Both of these were simulated in the model, as was the Postgraduate Awards Scheme. These payments were counted as part of the cash transfer system when assessing the redistributive impact of government (in contrast to the British methodology, which counted them as original income).

Moving to the income tax system, the tax unit in Australia is the individual, although those with particular family circumstances can claim a number of special deductions or rebates. The 1985-86 Australian income tax scales were modelled, as were the dependent spouse rebate, the sole parent rebate, and the pensioner and beneficiary rebates for social security recipients. The Medicare levy, which was a special tax supplement set at one per cent of taxable income, was also modelled. The Australian and British income tax scales are compared below in Table 1, along with some cash transfer payment rates.

Table 1: Comparison of the Australian and British Direct Tax Scales and Selected Cash Transfers

| Taxable Income | Marginal Tax Rate (for single taxpayer) |
|---|---|
| *Britain 1984-85* | |
| $0-$3,645 | 9%[1] |
| $3,645-$4,546 | 9% (above lower earnings limit for NIC) |
| $4,546-$26,804 | 39% (above single person allowance, until UEL) |
| $26,804-$39,587 | 40% |
| $39,587-$50,309 | 45% |
| $50,309-$66,598 | 50% |
| $66,598-$82,887 | 55% |
| $82,887 + | 60% |
| *Australia 1985-86* | |
| $0-$4,595 | 0% |
| $4,595-$12,500 | 25% |
| $12,500-$19,500 | 30% |
| $19,500-$28,000 | 46% |
| $28,000-$35,000 | 48% |
| $35,000 + | 60% |
| *Base pension in Australia at June 1986 for:* | |
| Single person | $102.10 per week |
| Married couple | $170.30 per week |
| *Supplementary benefit in Britain at June 1985:* | |
| Single person | $57.84 per week |
| Married couple | $93.90 per week |

[1] If earnings are less than $3,645 in a year then no national insurance is paid. Once earnings exceed $3,645, then nine per cent NIC is levied on the entire amount.

## 4. Lifetime Redistribution: Social Insurance vs Poverty Alleviation Systems

This section examines the overall lifetime redistributive impact of the British social security system as it was in 1985 compared to the Australian system in 1986. Individuals are ranked into decile groups by a measure of lifetime living standards. Rather than use total lifetime income to gauge these, the ranking uses each individual's *annualised lifetime family equivalent net income*. Four distinct decisions are being made here about the ranking measure.

First, lifetime income is *annualised* by averaging total lifetime net income over each year of life from age 16. This has the advantage of avoiding locating people in the upper part of the income distribution just because they have a long life, and so have longest to accumulate total income. Conversely, if annualised income were not used, those who died in their 20s and 30s would appear to have a low lifetime income and thus standard of living, whereas their incomes while they were alive might have been very high. Cohort individuals who die before the age of 17 are excluded from the analysis, as the majority of this group would not have entered the labour force and thus would have zero annualised income. Annualised income measures can be viewed as the average amount of income received during each adult year of life.

Second, *net* or disposable 'after-tax' income has been used in preference to gross income as it provides a more accurate measure of the living standards attained by individuals and families. Third, income has been *equivalised* to allow for different family circumstances between cohort individuals. If equivalent income were not used then, for example, a never married man with an income of half a million dollars would be regarded as having experienced that same lifetime standard of living as another male with the same lifetime income who supported a spouse and three children.

Fourth, although it is the individual who is the unit of analysis tracked, when a cohort member is living with other adults the equivalent income of the family is attributed to them. This assumes that each person in a family experiences the same standard of living ie. that income is shared equally within the family. Research by Edwards (1981) and Pahl (1990) suggests that this is not always the case. In other work the sensitivity of varying this assumption of equal sharing has been explored (Falkingham and Hills, 1995; Harding, 1993a), but here analysis is limited to this central case.

There are numerous difficulties with equivalence scales and it is not clear which is the most appropriate - especially for international comparative analysis (see Coulter et al., 1992 for a full discussion of the problems). The equivalence scale used in both the Australian and British results is that implicit in the British social security system in 1985 - the so called 'Mc.Clements scale' - but it has been rescaled so that a single adult is given the value of 1.00 (ie. to give family income per equivalent adult). (Additional adults are given the value of 0.64, children aged under 2 0.15, aged 5-7 years 0.34, aged 8-10 0.38, aged 11-12 0.41, aged 13-15 0.44 and aged 16-17 0.59.)

Table 2 presents the composition of lifetime income for the two model cohorts, ranked into decile groups by the measure of lifetime living standards (ie. equivalent net family income averaged over each year of life from age 16). One thing that is immediately clear from the table is that lifetime incomes, however measured, appear higher in Australia than in Britain. This may be due to differential length of life or other demographic/social factors, or it may be due to differences in the modelling. For example, the British world of 1985 has effectively been replicated up to 96 times (as all of the cohort are terminated at age 95), while one would expect Australian incomes a year later in 1986 to be somewhat higher (with the Australian consumer price index, for example, increasing by 8.4 per cent between June 1985 and June 1986). The difference may also be due to the exchange rate used to convert sterling to dollars (which was the average dollars to pounds rate during the Australian financial year 1985-86 - the best that could be achieved in an Australia in 1986 to Britain in 1985 comparison). The absolute amounts are very sensitive to the exact rate and so direct comparisons of absolute amounts between the two countries should be treated with some caution. Of more interest is the *pattern of income composition within deciles and its distribution.*

In contrast to traditional cross-sectional distribution studies, the distribution of lifetime income appears more equitable. Most notably, even the lowest deciles have significant amounts of income from earnings - constituting over half total lifetime gross income. Higher lifetime *original incomes* are the product of higher earnings and greater investment income and increased access to occupational superannuation (pensions). Although not shown in the table, the distribution of private pension income is highly skewed towards those in the top three deciles of lifetime income.

Table 3 shows the share of various income and tax measures received by each decile. It simply re-expresses the dollar values shown in Table 2 as decile shares, to aid interpretation. Table 3 suggests that, despite the higher dollar values for earnings and original factor income in Australia, the share of earnings received by each decile is very similar in the two countries. Nonetheless, at the top end of the earnings distribution there is some divergence, with the top 10 per cent of the population in Australia receiving 20.8 per cent of total gross lifetime earnings but the top decile in Britain receiving only 19.1 per cent.

Other components of original income appear to be more unequally distributed in Australia than in Britain (ie. private occupational pensions and investment income). Consequently, the top 10 per cent of Australians receive 23.2 per cent of total gross original income according to the HARDING model, while the top 10 per cent of Britons receive 19.9 per cent of British gross original income, according to LIFEMOD (Table 3). Some of these discrepancies may be due to different methods of modelling earnings, investment income and superannuation in the two models. However, it should be noted that this finding is not out of step with comparisons of the inequality of the *annual* (rather than lifetime) distribution of income in the two countries. For example, Mitchell found that the distribution of original income in Australia was more unequal than the distribution of original income in the UK (1991, p.123).

Cash transfers are much more important in Britain than in Australia for every decile. For example, as Table 2 shows, in Australia those in the 10 per cent of the population with the lowest lifetime incomes received on average $89,000 in cash transfers during their lives. This compares with the $134,000 received by the poorest 10 per cent of Britons. In Australia, cash transfers generally decline in absolute value as one moves up the income distribution. In contrast, in Britain the absolute value of lifetime total cash benefits is relatively flat across the whole income distribution (with a slight rise in the middle, followed by a slight dip). Thus, while the top decile of Australians received on average only $13,000 of cash transfers during their lifetimes, the top decile of Britons averaged $120,000 of benefits. The bottom decile in Australia thus received about 700 per cent more in cash transfers than the top decile while, in sharp contrast, the bottom decile in the UK received only 110 per cent more than the top decile.

Nonetheless, cash transfers in Britain were still progressive, amounting to a higher proportion of the income of the lifetime poor than of the lifetime rich. Figure 1 shows the average amount of cash transfers received by each decile, expressed as a percentage of their average gross lifetime income. In Britain, the lifetime cash transfers received by the bottom decile amounted to just over 35 per cent of their gross income, while the comparable figure for Australia was about 23 per cent. For

Table 2: Composition of Lifetime Income by Deciles of Annualised Lifetime Equivalent Disposable Income (in Australian dollars ('000s): $1 = £0.485 pence)

| | Decile of Annualised Lifetime Equivalent Disposable Family Income | | | | | | | | | | |
|---|---|---|---|---|---|---|---|---|---|---|---|
| | 1 | 2 | 3 | 4 | 5 | 6 | 7 | 8 | 9 | 10 | ALL |
| AUSTRALIA | | | | | | | | | | | |
| Earnings | 279 | 400 | 461 | 533 | 609 | 696 | 784 | 852 | 1007 | 1478 | 710 |
| Original Income | 297 | 430 | 498 | 591 | 672 | 785 | 903 | 1028 | 1257 | 1946 | 841 |
| Cash Transfers | 89 | 107 | 94 | 87 | 75 | 64 | 55 | 40 | 29 | 13 | 65 |
| Gross Income | 386 | 537 | 592 | 678 | 747 | 849 | 957 | 1068 | 1286 | 1959 | 906 |
| Income Tax | 45 | 77 | 97 | 128 | 154 | 193 | 240 | 290 | 395 | 745 | 237 |
| Net Income | 341 | 460 | 495 | 551 | 593 | 656 | 717 | 778 | 891 | 1214 | 670 |
| Equiv. Net Income | 359 | 477 | 518 | 577 | 619 | 676 | 724 | 800 | 887 | 1166 | 680 |
| Annualised Eq. Net | 5864 | 7540 | 8468 | 9395 | 10265 | 11184 | 12220 | 13530 | 15450 | 20157 | 11 |
| BRITAIN | | | | | | | | | | | |
| Earnings | 222 | 330 | 402 | 437 | 530 | 614 | 680 | 682 | 792 | 1070 | 558 |
| Original Income | 241 | 357 | 431 | 474 | 538 | 579 | 674 | 752 | 876 | 1221 | 612 |
| Cash Transfers | 134 | 132 | 134 | 140 | 136 | 146 | 136 | 146 | 123 | 120 | 136 |
| Gross Income | 375 | 487 | 563 | 614 | 674 | 726 | 810 | 899 | 1000 | 1342 | 745 |
| (NIC) | 38 | 57 | 74 | 87 | 103 | 115 | 140 | 167 | 200 | 315 | 39 |
| (ITAX) | 16 | 25 | 29 | 31 | 37 | 39 | 43 | 47 | 54 | 62 | 130 |
| Total Direct Taxes | 52 | 82 | 105 | 118 | 140 | 154 | 183 | 214 | 254 | 377 | 169 |
| Net Income | 323 | 404 | 458 | 497 | 534 | 571 | 627 | 682 | 746 | 965 | 581 |
| Equiv. Net Income | 328 | 421 | 476 | 525 | 555 | 604 | 643 | 715 | 769 | 964 | 599 |
| Annualised Eq. Net | 5670 | 7320 | 8124 | 8783 | 9402 | 10041 | 10845 | 11856 | 13260 | 16620 | 10 |

Table 3: Percentage Shares of Lifetime Income Accruing to Deciles of Annualised Lifetime Equivalent Disposable Income

| | Decile of Annualised Lifetime Equivalent Family Disposable Income | | | | | | | | | | |
|---|---|---|---|---|---|---|---|---|---|---|---|
| | 1 | 2 | 3 | 4 | 5 | 6 | 7 | 8 | 9 | Top | ALL |
| AUSTRALIA | | | | | | | | | | | |
| Earnings | 3.9 | 5.6 | 6.5 | 7.5 | 8.6 | 9.8 | 11.1 | 12.0 | 14.2 | 20.8 | 100 |
| Original Income | 3.5 | 5.1 | 5.9 | 7.0 | 8.0 | 9.3 | 10.7 | 12.2 | 14.9 | 23.2 | 100 |
| Cash Transfers | 13.6 | 16.4 | 14.4 | 13.4 | 11.4 | 9.8 | 8.4 | 6.1 | 4.4 | 2.0 | 100 |
| Gross Income | 4.3 | 5.9 | 6.5 | 7.5 | 8.2 | 9.4 | 10.6 | 11.8 | 14.2 | 21.7 | 100 |
| Income Tax | 1.9 | 3.2 | 4.1 | 5.4 | 6.5 | 8.2 | 10.2 | 12.2 | 16.7 | 31.5 | 100 |
| Net Income | 5.0 | 6.8 | 7.4 | 8.2 | 8.8 | 9.7 | 10.7 | 11.6 | 13.2 | 18.2 | 100 |
| Equiv. Net Income | 5.2 | 7.0 | 7.6 | 8.5 | 9.1 | 9.9 | 10.7 | 11.8 | 13.0 | 17.1 | 100 |
| Annualised Eq Net | 5.1 | 6.6 | 7.4 | 8.2 | 9.0 | 9.8 | 10.7 | 11.8 | 13.5 | 17.7 | 100 |
| BRITAIN | | | | | | | | | | | |
| Earnings | 4.0 | 5.9 | 7.2 | 7.8 | 8.9 | 9.5 | 11.0 | 12.2 | 14.2 | 19.1 | 100 |
| Original Income | 3.9 | 5.7 | 7.0 | 7.7 | 8.8 | 9.4 | 10.9 | 12.2 | 14.2 | 19.9 | 100 |
| Cash Transfers | 10.0 | 9.7 | 9.9 | 10.4 | 10.0 | 10.8 | 10.1 | 10.8 | 9.2 | 8.9 | 100 |
| Gross Income | 5.0 | 6.5 | 7.5 | 8.2 | 9.0 | 9.7 | 10.8 | 12.0 | 13.3 | 17.9 | 100 |
| (NIC) | 4.1 | 6.2 | 7.8 | 8.3 | 9.4 | 10.0 | 11.3 | 12.6 | 14.0 | 16.1 | 100 |
| (ITAX) | 2.3 | 4.5 | 5.8 | 6.7 | 8.0 | 8.9 | 10.8 | 12.9 | 15.4 | 24.2 | 100 |
| Total Direct Taxes | 3.0 | 4.9 | 6.2 | 7.0 | 8.3 | 9.2 | 10.9 | 12.9 | 15.1 | 22.4 | 100 |
| Net Income | 5.6 | 7.0 | 7.9 | 8.5 | 9.2 | 9.8 | 10.8 | 11.8 | 12.8 | 16.6 | 100 |
| Equiv. Net Income | 5.5 | 7.0 | 8.0 | 8.7 | 9.2 | 10.1 | 10.7 | 11.9 | 12.8 | 16.1 | 100 |
| Annualised Eq. Net | 5.7 | 7.1 | 7.9 | 8.4 | 9.2 | 9.7 | 10.7 | 11.5 | 13.0 | 16.8 | 100 |

the top 10 per cent of income recipients, cash transfers in Britain amounted to 9 per cent of lifetime gross income while, in Australia, the comparable figure was less than one per cent.

What about direct taxes? In Australia, Federal income tax is included within the simulation while, for Britain, the relevant comparison is with income taxes plus national insurance contributions. Direct taxes amounted to a higher proportion of gross lifetime income in Australia than in Britain - 26 per cent on average in Australia compared to 23 per cent in Britain (Table 2). For the poorest 10 per cent of the Australian cohort, total lifetime direct taxes amounted to $45,000 on average, compared to $52,000 for the British cohort. For the top decile in Australia, direct taxes averaged just under $750,000 while for the top decile in Britain total lifetime direct taxes summed to only $377,000.

Figure 1: Total Lifetime Cash Transfers Received as Percentage of Total Lifetime Gross Income

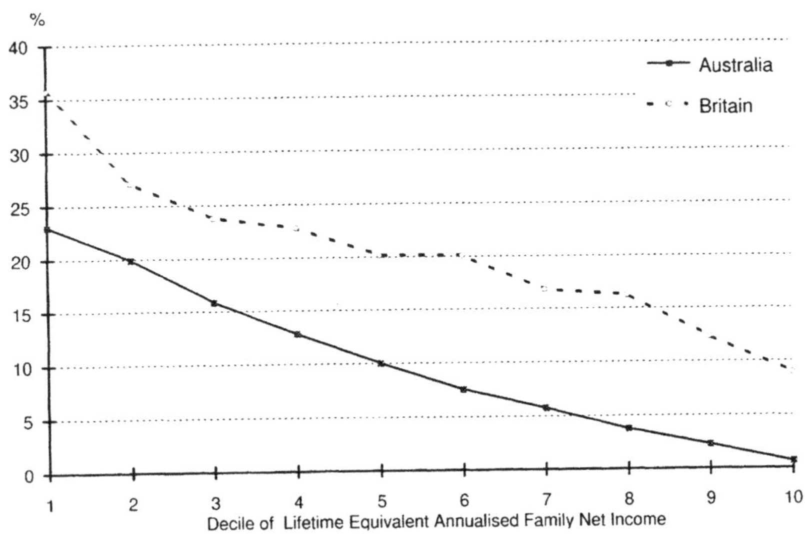


Table 3 shows the share of total lifetime direct taxes paid by each decile within the two countries. The poorest 10 per cent of the Australian cohort contributed 1.9 per cent of total lifetime direct taxes, compared to 3 per cent for the bottom British decile. In Australia, the top decile shouldered a larger slice of the lifetime direct tax burden than in Britain - 31.5 per cent versus 22.4 per cent for Britain. (This is in part a product of the interaction of higher incomes for the top decile in Australia, but is also due to a much more progressive income tax scale.)

Figure 2 shows direct tax burdens as a percentage of gross lifetime income for each decile. For the bottom decile, lifetime direct taxes amounted to 12 per cent of gross lifetime income in Australia and 14 per cent in Britain. The two tax incidence lines crossed at the fifth decile, with direct taxes amounting to 38 per cent of the gross income of the top decile in Australia and only 28 per cent in Britain. On the face of it, Figures 1 and 2 suggest that both the cash transfer and direct tax systems were more progressive over the lifetime in Australia than in Britain. Does this necessarily mean that the net effect of the tax/transfer system was more progressive in Australia?

Figure 2: Total Lifetime Direct Taxes Paid as Percentage of Total Lifetime Gross Income

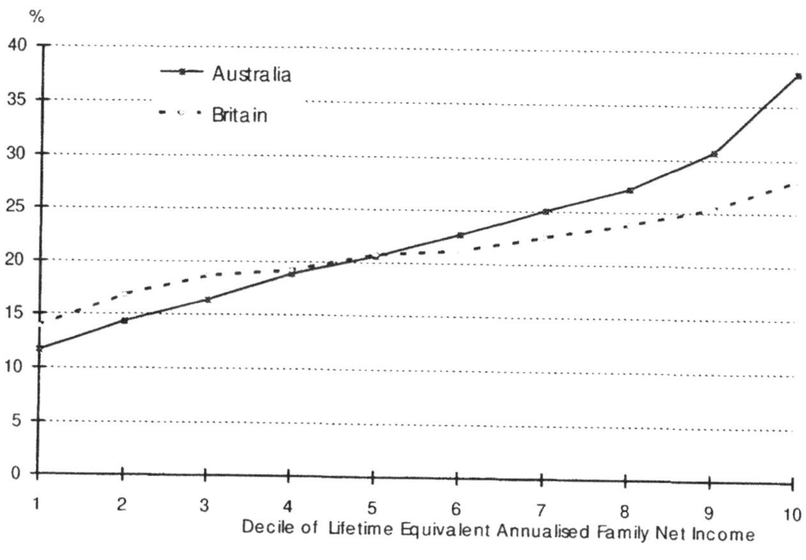


Direct comparison of the redistributive impact of the tax and benefit systems is clouded by the fact that the two countries start from different underlying distributions. What is of central concern to the hypothesis, however, is not that lifetime incomes appear to be more unequally distributed in Australia than in Britain, but rather the impact of the tax and benefit system on that distribution. The effect of the social security system is shown by the difference between the distribution of original income and that of gross income, while the impact of the direct tax system is reflected in the difference between the distribution of gross income and that of net income (also termed disposable income). The redistributive effect of the other main instrument of redistribution, the family, can be illustrated by comparing net income with net equivalent income.

Figure 3 shows the difference made by the tax/transfer system to the lifetime shares of income accruing to each decile within the two countries. The figure shows that the cash transfer system increased the decile share of the lowest lifetime income decile in Britain by more than in Australia. That is, while the lowest British decile received 3.9 per cent of lifetime original income, this share was increased by cash transfers to 5.0 per cent of gross income - the 1.1 per cent increase illustrated in Figure 3. In contrast, the share of income accruing to the lowest decile in Australia was only increased by 0.6 per cent as a result of the cash transfer system. This thus suggests that, although the cash transfer system appeared less progressive in Britain, it nonetheless had a larger redistributive impact over the entire lifetime for those at the very bottom of the lifetime income distribution than the Australian system - no doubt due to the higher absolute level of transfers. However, as Figure 3 shows, it was only for the bottom decile that this effect was apparent; for lifetime deciles two to five, the impact of the cash transfer system on decile shares was very similar in both countries.

Interestingly, despite the far higher magnitude of cash transfers being paid to upper income families in Britain than in Australia, for the top 20 per cent of the cohort the British cash transfer system nonetheless produced a larger decline in income share than the Australian transfer system. Thus, the share of income taken by the top decile in Britain declined by two per cent (from 19.9 to 17.9 per cent) as the income measure was shifted from original to gross income. In contrast, the share of income taken by the top decile in Australia was only reduced by 1.5 per cent by cash transfers. Moving to the tax system, with the single exception of the lowest income decile, the Australian tax system appears to change net income shares to a greater extent than the British system. Consequently, for deciles two to five, the net effect of the Australian tax and transfer system is to increase the income share accruing to the lower half of the income distribution by more than the British tax/transfer system. Similarly, the more progressive Australian direct tax structure resulted in the income share of the top 30 per cent of the cohort being reduced by more as a result of the tax/transfer system than in Britain. It is notable that in both countries the cross-over point between net gain and net loss occurs at the seventh decile.

Figure 3: Changes in Lifetime Income Shares Accruing to Each Decile, As a Result of the Transfer and Direct Taxation Systems

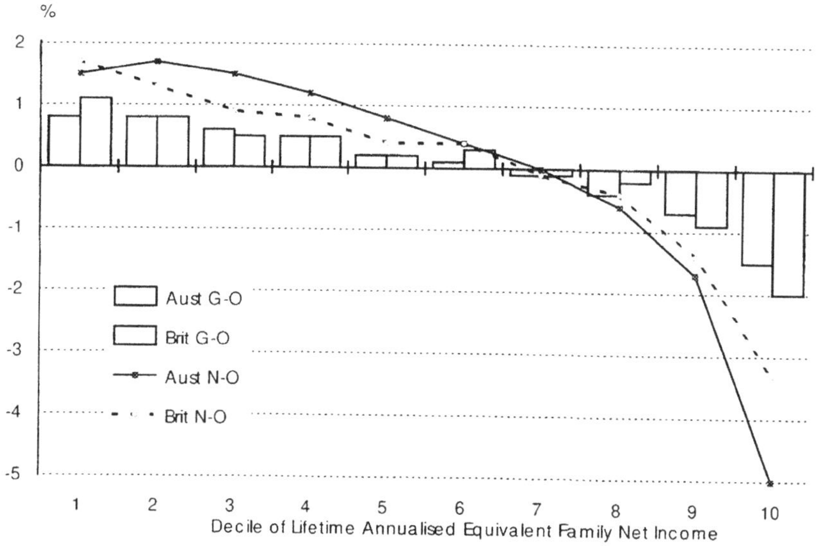


Note: Aust G-O measures the impact of the cash transfer system, and is the percentage share of gross income accruing to a particular decile minus the percentage share of original income accruing to the same decile. A positive result thus means that the cash transfer system increases the income share accruing to a decile. Aust N-O measures the combined impact of the tax/transfer system, and shows the percentage share of net income accruing to a particular decile minus the percentage share of original income accruing to that decile.

A useful summary measure of inequality is provided by Gini coefficients and Table 4 below shows these for various lifetime income measures for the HARDING and LIFEMOD cohorts. The Gini co-efficient has a value of one when one person receives all of the income in a society - and a value of zero if income is equally distributed among all persons. The higher the Gini co-efficient, the more unequal the distribution of income being measured.

The cash transfer system in Australia reduces the Gini coefficient (from original income to gross income) by 0.038. In Great Britain, although the absolute amounts of cash benefits in Table 2 were much flatter across the deciles, the Gini coefficient is reduced by 0.055. Thus, the cash transfer system has a greater equalising effect in Britain than in Australia.

Table 4: Gini Coefficients for Lifetime Income Measures

| Income Measure | Australia | Britain |
|---|---|---|
| Original income | 0.370 | 0.327 |
| Gross income | 0.332 | 0.272 |
| Disposable (net) income | 0.273 | 0.245 |
| Equivalent disposable family income | 0.220 | 0.204 |
| *Difference between*: | | |
| - gross income minus original income | -0.038 | -0.055 |
| - disposable income minus gross income | -0.059 | -0.027 |
| - disposable income minus original income | -0.097 | -0.082 |
| - equivalent disposable income minus disposable income | -0.053 | -0.041 |

The impact of the direct taxation system is to further reduce the Gini coefficients. The additional reduction in Britain is relatively small - only 0.027, even though the absolute amounts of direct taxes paid exceed cash transfers received. In Australia, however, the magnitude of the reduction is more prominent - 0.059. Overall, the net effect of the tax/transfer system in Britain, reflected in the difference between the disposable and original income distributions, is to reduce the Gini co-efficient by 0.082. In Australia, the net effect is greater, at 0.097. Nonetheless, the lifetime distribution of disposable income - at least as measured by the Gini co-efficient - remains more equal in Britain than in Australia.

All of the preceding analysis simply shows the income personally received by cohort individuals within the two models. In many cases this would not provide an accurate indicator of the standard of living achieved by individuals. This is because those with a low original income of their own might have lived in a family where they shared in the earnings of their spouse. To examine the redistributive impact of the family, it is possible to look at the difference between the distribution of the disposable income of individuals and the disposable income of the families in which those individuals lived. Table 4 suggests that the role played by the family in redistributing lifetime income is as great as that of the tax system in Australia. Thus, while the Australian tax system reduced the Gini by 0.059 the impact of the family on disposable income reduces the Gini by 0.053. In Britain the family plays less of a role, but still remains very influential, reducing the Gini by 0.041. One reason for this is that the LIFEMOD cohort experience a higher incidence of marital dissolution than the HARDING cohort and thus spend a greater proportion of their lifetimes in single adult households, where there is no opportunity to benefit from income sharing.

# 5. Intra-Personal vs Inter-Personal Redistribution

From the models, it is possible to identify receipts of cash benefits which are 'paid for' at another stage in an individual's life (intra-personal redistribution) and those which represent net transfers from others (inter-personal redistribution). In order to separate out the net transfers which are taking place within the model populations, it is necessary to make an assumption as to the taxes allocated to pay for them. This can be done in two alternative ways:

(a) assuming only *direct taxes are used to finance benefits;* thus, in the British case it is assumed that cash benefits are financed by all of the National Insurance Contributions and 58 per cent of the income tax paid by each group; in Australia, 27.6 per cent of all income taxes paid by the cohort would finance all cash transfers received by the cohort.

(b) assuming that a proportion of *both direct and indirect taxes is used.* Indirect taxes are not modelled in the original versions of HARDING or LIFEMOD (although see Harding, 1993b) but, in recent years, the combined effect of direct and indirect taxes in Britain has become close to being proportional to gross income[2]. The combination of direct and indirect taxes is therefore proxied by assuming financing from the percentage of each individual's gross income required to pay for benefits in aggregate. As a percentage of gross income, in HARDING 7.2 per cent of the lifetime gross income received by all cohort members would finance all cash transfers received by the cohort. In Britain the equivalent figure is 16.3 per cent.

Two possible financing scenarios are thus tested. By adopting such an approach, the counterfactual assumptions used are that, in the absence of cash transfers from the welfare state, individuals would have to pay correspondingly less tax :

- in proportion to their direct tax liabilities; or
- in proportion to their gross income.

Thus it is hoped that systemic differences in taxation structures between the two countries are controlled for.

---

[2] For instance, for Britain the CSO (1993, Appendix 4, Table 2) shows the Gini coefficients of equivalised gross income and equivalised post tax income (ie. after both direct and indirect taxes) as equal, at 0.32 in 1985.

## 5.1 Financing from direct taxes

The first line of Table 5 shows for each decile - ranked by the measure of living standards, which is annualised lifetime disposable family income - total lifetime cash benefit receipts. The following line shows what each group would contribute in direct taxes to finance the cash benefits of the entire cohort. The calculations in the remainder of the table show net receipts and payments using these 'allocated' tax payments.

During each year of life individuals may pay tax, receive cash benefits, or (for perfectly good reasons) both. In the last case, some or all of the benefits they receive in a year will be paid for by their own direct tax in the same year - that is, will represent intra-personal redistribution *within the same year* (sometimes known as 'churning'). Examples of this would be people receiving:

- taxable benefits (such as the state pension);
- non-taxable benefits but who pay tax on other income (for instance, women with earnings receiving Child Benefit); or
- people receiving benefits because they are out of work for part of a year, but who have earnings and pay tax in the rest of it.

People may also pay for their own benefits in other years of their lives. Over their whole lives, individuals are either net lifetime taxpayers or net lifetime benefit recipients (or, conceivably, neither, for those who break-even).[3]

---

[3] Algebraically the various measures can be expressed as follows. Let Bi be the gross cash benefits received by individual i in year t, and Ti be the gross (allocated) tax paid by them in that year. Then the aggregate gross cash benefits for the model population over its entire lifetime is given by:

$$G = \sum_i Bi = \sum_i Ti$$

Self-financed benefits (SFB) are given by:

$$SFB_i = Ti \text{ (if } Bi > Ti)$$
$$= B_i \text{ (if } Bi < Ti)$$

An individual's net lifetime gain is:

$$Ni = B_i - T_i$$

Total *inter-personal* redistribution (IR) is the sum of positive net lifetime gains and is given by:

$$IR = \sum_i (Ni\ Ni > 0)$$

Total *intra-personal* redistribution (IA) is the sum of self-financed benefits:

$$IA = \sum_i SFBi \qquad \text{(and by construction, G=IR+IA)}$$

The table illustrates the scale of such 'self-financing' of benefits (or intra-personal redistribution, together with the net lifetime taxes or benefits of individuals in excess of self-financed benefits (inter-personal redistribution). Taking, for example, the bottom Australian decile first, this decile received on average a total of $89,000 in cash transfers during their entire lifetimes. About 28 per cent of the total income tax which they paid during their lifetimes was put into the pot which financed all cash transfers, so that their allocated direct tax burden was about 28 per cent of the $45,000 of lifetime direct tax shown in Table 2 (ie. was about $12,000). In other words, out of all the cash transfers paid to the entire cohort during their lifetimes, the bottom decile contributed on average $12,000 of the direct taxes which financed those cash transfers.

Of this $12,000 allocated tax burden, $11,000 was paid back to these same individuals in the form of cash transfers (line five of Table 5). The lowest decile thus financed $11,000 of the $89,000 of cash transfers which were paid to them. Most individuals in the bottom Australian decile were net lifetime beneficiaries after taking account of cash transfers received and allocated taxes paid. This was shown in the average lifetime net benefit for the entire decile of $78,000. However, a few individuals within this decile were net lifetime taxpayers. In other words, a few individuals in the bottom decile paid out more in allocated taxes during their lifetimes than they received in cash transfers. For the decile as a whole, the lifetime allocated tax which they paid out and which financed the benefits of others averaged about $1,000. (Consequently, the average lifetime gain for this group was $77,000). In summary, therefore, the 'self-financed benefits' rows in Table 5 show the scale of intra-personal redistribution. The 'average lifetime gain' rows show the scale of inter-personal redistribution.

What about the bottom decile in Britain? This decile received much more in cash transfers than the bottom Australian decile, but they also paid much more allocated direct tax. The poorest decile in Britain thus paid for about 28 per cent of the cash transfers which they received (in contrast to only 12 per cent for the lowest Australian decile). However, the bottom decile in Britain still received on average $93,000 of cash transfers financed by others - much higher than the net $77,000 received by the bottom Australian decile. For deciles two to five, however, the Australian tax/transfer system delivered more inter-personal redistribution. Consequently, the average lifetime gain for those in the bottom half of the lifetime income distribution averaged $55,000 for Britain and $63,000 for Australia.

The proportion of gross benefits received by each decile which represented intra- rather than inter-personal redistribution grew rapidly as one moved up the income distribution. In Britain, by the third decile the majority of the group's gross receipts were self-financed; the same was true by the sixth decile in Australia. For the top group, $108,000 out of the $120,000 received on average in Britain was self-financed and the net taxpayers within the group paid out $189,000 (averaged over the whole group). The pattern for the entire cohort is shown in the final column in Table 5. On average, HARDING model individuals received $65,000 in cash transfers over their lifetimes, with $25,000 or 38 per cent of this representing benefits which individuals effectively paid for themselves. In Britain, individuals averaged a much higher $136,000 in lifetime cash transfers, $84,000 of which - or 62 per cent - they financed from their own taxes.

This pattern is summarised in Figure 5. The top line for each country shows average gross benefits going to each decile group, and the bottom line the average gross (allocated) taxes which they pay. Deducting the totals of annual and lifetime clawback leaves the two unshaded areas in the top and bottom of the diagram - the net lifetime benefits of lifetime gainers and the net lifetime taxes of lifetime losers. It is these unshaded areas which represent the inter-personal redistribution which is going on. Judged by Robin Hood criteria, this redistribution is clearly progressive (although somewhat approximate, with some net lifetime taxpayers at the bottom and a rather larger number of net lifetime gainers at the top), and is much greater in Australia than in Britain.

Meanwhile, the shaded areas (of equal size for each group at top and bottom of the diagram) show the intra-personal redistribution going on. In the Australian system, on average, $25,000 ie. 38 per cent of lifetime benefits are paid for by the same individual at another stage in their lives whilst 62 per cent are paid for by others. By a remarkable coincidence, the proportion are exactly reversed for Britain with 62 per cent of benefits constituting intra-personal redistribution and 38 per cent inter-personal redistribution. So far, the null hypothesis seems unequivocally unrejected!

## 5.2 Financing from all taxes

The above results are based on the assumption that cash benefits are financed by progressive direct taxes. Table 6 shows the results of making the alternative assumption that financing comes from a share of all tax revenues. The assumption is that tax is proportional rather than progressive, with allocated tax amounting to a

Table 5: Lifetime Totals of Net Taxes and Benefits by Decile of Lifetime Income Distribution Funded by 'Allocated' Direct Taxes (in Australian dollars ('000s): $1 = £0.485 pence)

| | Decile of Annualised Lifetime Equivalent Family Disposable Income | | | | | | | | | | |
|---|---|---|---|---|---|---|---|---|---|---|---|
| | 1 | 2 | 3 | 4 | 5 | 6 | 7 | 8 | 9 | Top | ALL |
| AUSTRALIA | | | | | | | | | | | |
| Cash Benefits | 89 | 107 | 94 | 87 | 75 | 64 | 55 | 40 | 29 | 13 | 65 |
| Allocated Direct Tax | 12 | 21 | 27 | 29 | 43 | 54 | 66 | 80 | 109 | 206 | 65 |
| Lifetime Net Benefit | 78 | 88 | 71 | 58 | 42 | 27 | 15 | 6 | 2 | 0 | 40 |
| Lifetime Net Tax | 1 | 2 | 4 | 6 | 10 | 17 | 26 | 46 | 82 | 193 | 40 |
| Self Financed Benefits | 11 | 19 | 23 | 23 | 33 | 37 | 40 | 34 | 27 | 13 | 25 |
| Average Lifetime Gain | + 77 | + 86 | + 67 | + 52 | + 32 | + 10 | -11 | -40 | -80 | -193 | |
| BRITAIN | | | | | | | | | | | |
| Cash Benefits | 134 | 132 | 134 | 140 | 136 | 146 | 136 | 146 | 123 | 120 | 136 |
| Allocated Direct Tax | 41 | 68 | 84 | 97 | 113 | 126 | 148 | 173 | 204 | 297 | 136 |
| Lifetime Net Benefit | 97 | 74 | 64 | 64 | 52 | 54 | 39 | 35 | 21 | 12 | 52 |
| Lifetime Net Tax | 4 | 10 | 14 | 21 | 29 | 34 | 51 | 62 | 102 | 189 | 52 |
| Self-Financed Benefits | 37 | 58 | 70 | 76 | 84 | 92 | 97 | 111 | 102 | 108 | 84 |
| Average Lifetime Gain | + 93 | + 64 | + 50 | + 43 | + 23 | + 20 | -12 | -27 | -81 | -177 | |

Figure 4: Lifetime Taxes and Benefits by Decile of Annualised Lifetime Equivalent Family Disposable Income

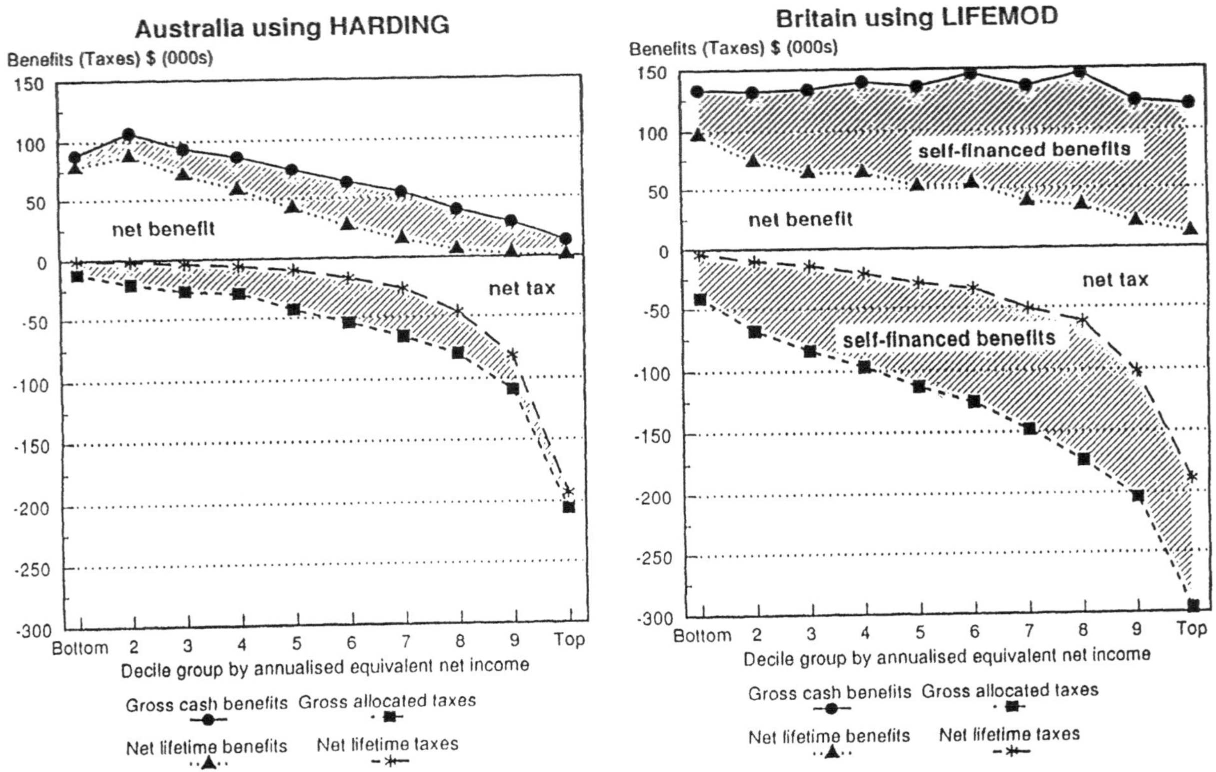

constant proportion of gross income in both countries. Four features stand out in comparison to the earlier results which assumed that progressive taxes financed cash transfers:

- the proportion of intra-personal redistribution rises from 38 per cent to 52 per cent in Australia and from 62 per cent to 71 per cent in Britain;

- the average net gains to the bottom deciles are reduced and, symmetrically, the net loss of the top groups are decreased. Thus, there is less inter-personal redistribution;

- the inter-personal redistribution involved, whilst of a smaller scale than when direct taxes are assumed to be the source of finance, is none the less progressive;

- even controlling for the finance mechanism, the British social security system results in more intra-personal redistribution than in Australia and less inter-personal redistribution.

## 6. Conclusion

All social security and income tax systems are intended to generate both inter-personal redistribution - from those with high to those with low incomes - and intra-personal redistribution - from one period of an individual's lifecycle to another. Until now, most analysis of the redistributional impact of the tax/transfer system has been limited to analysis of the inter-personal redistribution generated at a single point in time. The lack of lifetime data has limited the capacity to answer questions about the intra- and inter-personal redistributive impact of the modern welfare state over a longer time period. This paper represents a first attempt to undertake international comparative work about the lifetime redistributive effect of social systems which emphasise social insurance goals compared to those which emphasise poverty alleviation goals.

The work was based on two dynamic cohort microsimulation models, which put 4,000 simulated Britons and 4,000 simulated Australians through their entire lifecycles, from birth to death. Although some aspects of life have been modelled differently, the models are in many respects unusually comparable, having being developed with a high degree of co-operation.

Table 6: Lifetime Totals of Net Taxes and Benefits by Decile of Lifetime Income Distribution Funded by 'Allocated' Gross Income (in Australian dollars ('000s): $1= £0.485 pence)

| | Decile of Annualised Lifetime Equivalent Family Disposable Income | | | | | | | | | | |
|---|---|---|---|---|---|---|---|---|---|---|---|
| | 1 | 2 | 3 | 4 | 5 | 6 | 7 | 8 | 9 | Top | ALL |
| AUSTRALIA | | | | | | | | | | | |
| Cash Benefits | 89 | 107 | 94 | 87 | 75 | 64 | 55 | 40 | 29 | 13 | 65 |
| Allocated Direct Tax | 28 | 38 | 42 | 49 | 54 | 61 | 70 | 77 | 93 | 141 | 65 |
| Lifetime Net Benefit | 64 | 73 | 58 | 46 | 33 | 21 | 11 | 5 | 2 | 0 | 31 |
| Lifetime Net Tax | 3 | 4 | 6 | 8 | 12 | 18 | 26 | 42 | 66 | 128 | 31 |
| Self-Financed. Benefits | 25 | 34 | 36 | 41 | 42 | 43 | 44 | 35 | 27 | 13 | 34 |
| Average Lifetime gain | + 61 | + 69 | + 52 | + 38 | + 21 | + 3 | -15 | -37 | -64 | -141 | |
| BRITAIN | | | | | | | | | | | |
| Cash Benefits | 134 | 132 | 134 | 140 | 136 | 146 | 136 | 146 | 123 | 120 | 136 |
| Allocated Direct Tax | 77 | 88 | 103 | 110 | 122 | 130 | 145 | 162 | 178 | 242 | 136 |
| Lifetime Net Benefit. | 72 | 55 | 47 | 49 | 41 | 44 | 32 | 33 | 19 | 10 | 40 |
| Lifetime Net Tax | 5 | 11 | 16 | 19 | 27 | 28 | 41 | 49 | 74 | 132 | 40 |
| Self Financed. Benefits | 72 | 77 | 87 | 91 | 95 | 102 | 104 | 113 | 104 | 110 | 96 |
| Average Lifetime gain | + 67 | + 44 | + 31 | + 30 | + 14 | + 16 | -9 | -16 | -55 | -122 | |

The Australian lifetime income distribution appeared more unequal than the British distribution, and average gross lifetime incomes in Australia, at $906,000, were substantially higher than in Britain ($745,000). This appeared due to a multitude of factors, including an earlier base year for Britain (1985 as opposed to 1986 for Australia), higher unemployment and lower education levels in Britain, and socio-demographic differences, including higher mortality, divorce and sole-parenthood rates in Britain. It is important to keep these differences in the original income distribution firmly in mind when assessing the following results. For example, because the top 10 per cent of the Australian cohort had an original income which was 40 per cent higher than the top 10 per cent of the British cohort, one would expect their income tax contributions to be substantially higher (because of the progressive tax system). Thus, these underlying differences complicate the comparison and only broad conclusions can be drawn about how the two welfare states appear to be redistributing lifetime income.

Some broad conclusions did, however, emerge. Total lifetime cash transfers seemed to be much greater in Britain than in Australia, amounting to $136,000 on average for Britain but only $65,000 for Australia. The absolute value of lifetime cash transfers received declined swiftly as one moved up the Australian income distribution. In contrast, the absolute value remained relatively stable across the whole income distribution in Britain.

Cash transfers amounted to a higher proportion of gross income in Britain than in Australia for every decile. For example, cash transfers amounted to about 35 per cent of the gross income of the bottom decile in Britain and 23 per cent in Australia. For the top decile, the comparable figures were nine per cent for Britain and less than one per cent for Australia. According to Gini co-efficients, the British cash transfer system reduced the inequality of the original income distribution by more than the Australian.

On the direct tax side, average lifetime income taxes in Australia amounted to $237,000, compared to an average $169,000 for the British cohort. Interestingly, in both countries income taxes (plus national insurance contributions in Britain) amounted to half of total tax revenue. The Australian direct tax system looked more progressive, with average tax rates being lower for the bottom Australian decile than for the British - but higher for the top decile. However, to some extent this apparent difference was due to discrepancies in the original income distribution against which progressivity was assessed. The direct tax system reduced the Australian Gini coefficient by 0.059, more than double the 0.027 reduction achieved by the British direct tax system.

It was possible to look at the extent to which receipts of cash benefits by an individual were paid for by that person at another stage in their life (lifetime intra-personal redistribution) and the extent of net transfers from other people (inter-personal redistribution). Assuming that all cash benefits were financed from direct taxes, on average out of the $65,000 each Australian received in cash transfers over the lifetime, 38 per cent of these were effectively paid for by the individuals themselves, whereas 68 per cent were paid for by others. These proportions were reversed for Britain.

The proportion of cash transfers that represented intra-personal redistribution increased as one moved up the income distribution. In Britain, by the third decile, the majority of the group's gross receipts were self-financed over the lifetime; the same was true by the sixth decile in Australia. Alternatively assuming that cash transfers were financed by a proportional rather than a progressive tax resulted in the proportion of transfers that were self-financed increasing from 38 per cent to 52 per cent in Australia and from 62 per cent to 71 per cent in Britain.

Thus, controlling for the greater redistributive effect of the more progressive Australian tax system does not alter the finding that a primarily social assistance-based system (like Australia), with its emphasis on poverty alleviation, results in a greater degree of inter-personal redistribution. Conversely a primarily social insurance-based system (like Britain), with its emphasis on the link between contributions and benefits, results in a greater degree of intra-personal redistribution.

Finally, it must be acknowledged that this study examines the redistributive impact of the UK and Australian systems by comparing the post-tax/transfer income distribution with the original income distribution prevailing within *each* country. An alternative method would be to simulate the rules of the tax/transfer systems of both countries against the *same* original income distribution (ie. run the rules for both tax/transfer systems against either the Australian or the UK original income distributions). Such an exercise would provide a very interesting complement to the study presented here. However, it would not necessarily provide more definitive answers about the lifetime redistributive impact of each country's systems, as this would depend upon the extent to which the prevailing tax/transfer rules had already affected behaviour - thus the 'original' income distribution - within each country.

## References

Barr, N. and J. Falkingham, 'Paying for learning', Welfare State Programme, Discussion Paper WSP/94. London: London School of Economics.

Barro, R. (1974), 'Are government bonds net wealth?', *Journal of Political Economy*, Vol. 82.

Beveridge, W. (1942) *Social Insurance and Allied Services*, Cmd. 6404, HMSO, London.

Central Statistical Office (UK) (1992), 'The effects of taxes and benefits on household income 1989', *Economic Trends*, No. 459, January.

Coulter, F., F.A. Cowell and S.P. Jenkins (1992), 'Differences in needs and assessment of income distributions', *Bulletin of Economic Research*, Vol. 44, No. 2.

Edwards, M. (1981) *Financial Arrangements Within Families*, National Women's Advisory Council, Canberra.

Falkingham, J. and C. Lessof (1991), 'LIFEMOD - The formative years', Welfare State Programme Research Note WSP/RN/24, London School of Economics, London.

Falkingham, J. and C. Lessof (1992), 'Playing god: The construction of LIFEMOD, a dynamic cohort microsimulation model', in: R. Hancock and H. Sutherland (eds.), *Microsimulation Models for Public Policy Analysis: New Frontiers*, STICERD Occasional Paper No.17, London School of Economics, London.

Falkingham, J. and J. Hills eds. (1995) *The Dynamic of Welfare*, Harvester-Wheatsheaf, London.

Hain, W. and Helberger, C. (1986), 'Longitudinal microsimulation of lifetime income' in: G. Orcutt, J. Merz and H. Quinke (eds.), *Microanalytic Simulation Models to Support Social and Financial Policy*, North-Holland, New York.

Harding, A (1984), *Who Benefits: The Welfare State and Redistribution*, Social Welfare Research Centre Reports and Proceedings No. 45, University of New South Wales, Sydney.

Harding, A. (1990a), 'Dynamic microsimulation models: Problems and prospects', Welfare State Programme Discussion Paper WSP/48, London School of Economics, London.

Harding, A. (1993a), *Lifetime Income Distribution and Redistribution: Applications of a Dynamic Cohort Microsimulation Model*, North Holland, Amsterdam.

Harding, A (1993b), 'The lifetime incidence of indirect taxes: Wholesale sales tax and excises', in: J. Head (ed.), *Fightback: An Economic Analysis*, Australian Tax Research Foundation.

Harding A. (1995) 'Financing higher education: An assessment of income-contingent loan options and repayment patterns over the life cycle', *Education Economics*, Vol. 3, No. 2.

Hills, J. and Lessof, C. (1993), 'Modelling direct tax and social security over the lifetime', Welfare State Programme Research Note WSP/RN/25, London School of Economics, London.

Le Grand, J. (1982), *The Strategy of Equality: Redistribution and the Social Services*, Allen and Unwin, London.

Mitchell, D. (1991), *Income Transfers in Ten Welfare States*, Avebury, Aldershot.

Nelissen, J.H.M (1993), *The Redistributive Impact of Social Security Schemes on Lifetime Labour Income*, Report No. 22, Tilburg University.

Pahl, J. (1990), 'Household spending, personal spending and the control of money in marriage', *Sociology*, Vol. 24, No. 1.

Saunders, P. (1984), 'Evidence on income redistribution by governments', OECD Economics and Statistics Department Working Paper No. 11, OECD, Paris.

Summers, R. (1956), 'An econometric investigation of the size distribution of lifetime average income', Department of Economics Technical Report No. 31, Stanford University, Standord, USA

Winter, D. (1994), 'Modelling lifetime earnings', Mimeo paper, London School of Economics, London.

Wolfson, M. (1988), 'Homemaker pensions and lifetime redistribution', *Review of Income and Wealth*, Vol. 34, No. 3.

J.H.M. Nelissen

Tilburg University
The Netherlands

# Chapter 12: Social Security and Lifetime Income Redistribution: A Microsimulation Approach

## Abstract

*This chapter analyses the vertical redistributive impact of social security schemes on lifetime labour income for the cohorts born in 1930, 1940, 1950 and 1960 in the Netherlands. The analysis has been carried out using the dynamic microsimulation model NEDYMAS. The reduction in income inequality achieved by the social security system on a lifetime basis appears to be rather limited. This does not, however, imply that some groups do not receive major benefits from the system. It also appears that Atlantic-type flat-rated schemes do not necessarily produce greater redistribution than Continental-type wage-related schemes.*

## 1. Introduction

A social security system generally contains both vertical and horizontal income redistribution objectives. Vertical redistribution involves a redistribution of income between households in different income groups, whereas horizontal redistribution refers to the redistribution of income between households with different characteristics. To what extent the social security system results in redistribution is, however, determined by many factors, which generally differ between generations. Research about the redistributive impact of social security schemes on *lifetime* income is very scarce (see Davies, et al., 1984; Harding 1993; and Falkingham et al.,1993).

The greatest problem in analysing the impact of social security on lifetime income is the almost complete absence of data, which implies that we have to 'construct' a database. Because of the complex conditions for receiving a social security benefit, many characteristics at the micro level have to be known - for example, age, marital status, sex, household size and composition, income, (former) economic activity,

* The author would like to acknowledge gratefully the funding provided by the Research Program for Population Studies of the Dutch Scientific Organisation NWO.

labour supply, income of other household members, and so on. A very suitable method for this purpose - and probably the only one - is microsimulation, which offers the possibility of creating a fictive panel. As a consequence, individual socioeconomic life histories can be simulated and the method provides the "... capability of examining the entire distribution of effects, not just an aggregate or a mean" (Haveman and Hollenbeck, 1980).

This paper examines lifetime vertical redistribution in the Netherlands. Section 2 provides a description of the social security system in the Netherlands. In Section 3, the microsimulation model is described and some problems of the definition of income, the income unit and the standardisation of income are treated. Section 4 contains the results. Section 5 presents the conclusions.

## 2. The Social Security System in the Netherlands

In the Netherlands, there are two types of *Social Insurance* - general insurance and employee insurance. Both types of insurance are based on the Pay-As-You-Go system. The general insurances are Atlantic-oriented, flat-rated benefits, covering all residents. Premiums have to be paid by all residents and are proportional to income up to a fixed maximum income. The solidarity principle is the starting-point, not the insurance principle. Current general insurances comprise old-age pensions (AOW), widows and orphans pensions (AWW), long-lasting disability pensions (AAW), family allowances (AKW) and exceptional medical expenses (AWBZ).

The employee insurances only apply to *employees in the private sector*. The insurance or equivalence principle is the point of departure and they are compulsory. Premiums are a fixed proportion of income (which can be seen as the insured risk), whereas benefits are usually related to (former) income. Both are limited by a ceiling. Current employee insurances at work are: sickness benefits (ZW); long-lasting disability pension (WAO); unemployment benefits (WW) and health insurance (ZFW). The paper also looks at the impact of some former types of family allowances (KW, KWR and KWL).

The Netherlands also has some *social provisions*, mainly intended to support those who cannot provide for themselves, principally including social assistance (ABW). Benefits are flat-rated, means-tested and financed by public funds. Other schemes are a former unemployment provision which was wage-related (WWV) and family allowances for self-employed persons (KKZ).

A large number of *supplementary pensions* are available for both civil servants and private employees. This study takes account of all premium payments, but limits the benefit side to the civil servants pensions fund (ABP), which includes supplementary old age pensions (ABP-OP), widows and orphans pensions (ABP-WP) and disability pensions (ABP-IP). These contributions and benefits are wage-related without ceiling. There exists a franchise due to the state pension and the scheme is capital funded, although some elements of the Pay-As-You-Go system are included.

Finally, a number of state arrangements exist for civil servants and employees in public firms (called *direct payments*). They are comparable to the employee insurances but differ because they are financed via public funds. The benefits are wage-related and do not have a ceiling. The most important ones are the civil servants unemployment benefits (WRO), the sickness payments for civil servants (DSO) and the former family allowances for civil servants (KTO).

Currently, about one-third of Dutch net national income is redistributed in the form of social security benefits - including fund-raising by pension funds, the percentage exceeds 40 per cent of national income. The main element is social insurances, with general insurances constituting 40 per cent of total benefits and employee insurances 28 per cent. Both social provisions and supplementary pensions account for about 15 per cent, whereas direct payments by the employer constitute no more than a few per cent of the total social security benefits.

The extent to which the various social security schemes are treated in this paper is shown in Table 1. Only 47 per cent of total benefits in 1950 and about 70 per cent from 1970 onwards are covered (due to lack of data about supplementary pensions and compensation for medical expenses). Less than half the supplementary pensions are treated, as it is not possible to simulate private pension benefits. Only pensions under the civil servants pension fund, ABP, are included.

With the exception of the 1940s and 1950s, about 70 per cent of the social insurance benefits have been included in the model. The key excluded benefits are those for health insurance and exceptional medical costs, because of lack of data. These health cost insurances accounted for 28 per cent of social insurances in 1950, 21 per cent in 1960 and 29 per cent in 1988.

On the financing side, only social insurances and supplementary pensions are included. The other social provisions and benefits paid directly by the employer -

Table 1: Extent of Coverage of the Various Social Security Schemes in the Simulation

| Type | 1950 (%) | 1960 (%) | 1970 (%) | 1980 (%) | 1988 (%) |
|---|---|---|---|---|---|
| *Benefit side* | | | | | |
| - social insurances | 40 | 69 | 72 | 72 | 70 |
| - supplementary pensions | 34 | 33 | 39 | 40 | 42 |
| - social provisions | 83 | 71 | 67 | 69 | 74 |
| - direct payments | 75 | 94 | 98 | 96 | 97 |
| TOTAL | 47 | 64 | 69 | 69 | 67 |
| *Financing side* | | | | | |
| - social insurances | 56 | 82 | 91 | 92 | 99 |
| - supplementary pensions | 100 | 100 | 100 | 100 | 100 |
| TOTAL | 72 | 87 | 92 | 93 | 99 |

Sources: Ministerie van Sociale Zaken en Werkgelegenheid (1984 and 1989) and own calculations.

that is, the state - are both financed by public funds. From about 1965 onwards coverage of the financing side is more than 90 per cent (99 per cent in 1988).

Contribution rates for the general insurances and the maximum income to which these premiums have to be paid (the annual ceiling) are shown for a number of years in Table 2. The premium percentages are the sum of both employer and employee premium percentages. Contribution rates for the old age state pension more than doubled between 1960 and 1985, mainly due to increases in the benefit in the 1970s and the change in the age composition of the Dutch population. A further increase between 1985 and 1990 was caused by a change in the annual ceiling, as a consequence of the Oort tax reform.

The rate for family allowances has increased due to the abolition of the specific family allowances schemes for employees, civil servants and self-employed persons and their incorporation into the AKW, and the changes in the extent to which the contributions are financed by public funds (see also Table 4). The contribution rate for exceptional medical costs has also grown continuously over the period examined, due to extended coverage of medical and surgical treatments and increasing health care costs. Overall, in 1990, non-elderly individuals paid 22.1 per cent of their income in general contributions. This is partly done by their employers. No contributions are levied on income below what is called the tax exemption (generally

Dutch Florins - Dfl. 4,568) and the maximum contribution rate is levied on Dfl. 42,123 (the ceiling).

Table 2: Contribution Rates* and Annual Ceilings for the General Insurances

| | AOW old age pension (%) | AWW widows pension (%) | AKW family allowances (%) | AAW disability pension (%) | AWBZ health insurance (%) | ceiling (Dfl.) |
|---|---|---|---|---|---|---|
| 1960 | 5.5 | 1.25 | - | - | - | 7.450 |
| 1965 | 8.7 | 1.5 | 1.8 | - | - | 12.000 |
| 1970 | 9.5 | 1.5 | 2.075 | - | 1.245 | 17.450 |
| 1975 | 10.4 | 1.5 | 1.2 | - | 2.7 | 31.750 |
| 1980 | 10.25 | 1.35 | 2.0 | 4.2 | 3.2 | 46.400 |
| 1985 | 11.7 | 1.4 | 3.95 | 6.05 | 4.25 | 63.200 |
| 1990 | 14.3 | 1.25 | 0 | 1.15 | 5.4 | 42.123 |

* paid by both employers and employees

Table 3: Contribution Rates* and Ceilings for Employee Insurances

| | KB family allowance (%) | ZW sickness benefit (%) | WW unemp-loyment benefit (%) | WAO disability benefit (%) | ZFW health costs (%) | ceiling ZFW (Dfl) | ceiling others (Dfl) | exemption WAO (Dfl) |
|---|---|---|---|---|---|---|---|---|
| 1950 | 5.0 | 4.0 | - | - | 3.6 | 3,750 | 3.432 | - |
| 1955 | 5.0 | 4.0 | 1.4 | - | 4.0 | 6,000 | 4,992 | - |
| 1960 | 5.3 | 4.2 | 1.44 | - | 4.8 | 7,450 | 6,240 | - |
| 1965 | 3.0 | 5.1 | 1.16 | - | 5.4 | 10,000 | 8,424 | - |
| 1970 | 3.3 | 6.3 | 1.0 | 5.3 | 7.5 | 12,480 | 23,140 | - |
| 1975 | 3.5 | 7.85 | 1.16 | 9.2 | 9.2 | 23,400 | 42,640 | - |
| 1980 | - | 8.05 | 1.27 | 14.1 | 8.1 | 34,320 | 61,100 | 14,560 |
| 1985 | - | 6.57 | 3.55 | 16.15 | 9.2 | 40,820 | 68,120 | 23,660 |
| 1990 | - | 6.35 | 3.18 | 12.15 | 7.9 | 43,500 | 68,900 | 23,920 |

* paid by both employers and employees

The premium percentages for the employee insurances (Table 3) reflect the changes in risk and changes in the contributions from public funds. These contributions have varied considerably over time (see Table 4). The rate for family allowances in 1965 was considerably lower than in earlier years, due to the introduction of the AKW in 1963. Contributions to sickness insurance have grown due to changes in sick-leave.

The large increase in the contribution rate for unemployment benefits in the 1980s was caused by the withdrawal of contributions from public funds, in combination with an increase in unemployment. A rise in the number of disabled persons increased the rate for disability insurance. However, in the second half of the 1980s this increase ceased, due to reforms in 1987 (which diminished the benefit level from 80 per cent to 70 per cent of the former wage) and failure to fully index the exemption ceiling to inflation.

Table 4: Contribution from Public Funds to Social Security Programs

| | Total | AOW old age pension | AWW widows pension | AKW family allowances | AAW disability pension | AWBZ health insurance | KB family allowance | ZW sickness benefit | WW unemp-loyment benefit | WAO disability benefit | ZFW health insurance |
|---|---|---|---|---|---|---|---|---|---|---|---|
| | (%) | (%) | (%) | (%) | (%) | (%) | (%) | (%) | (%) | (%) | (%) |
| 1955 | 3.4 | - | - | - | - | - | 4.6 | 2.9 | 31.9 | - | 4.5 |
| 1960 | 3.6 | 1.7 | 1.4 | - | - | - | 2.5 | 0 | 24.2 | - | 0 |
| 1965 | 5.7 | 6.2 | 1.1 | 1.1 | - | - | 0.3 | 0 | 23.9 | - | 0 |
| 1970 | 5.6 | 3.7 | .4 | 0.5 | - | 40.9 | 5.1 | 0 | 19.5 | 0 | 0 |
| 1975 | 8.5 | 6.7 | .4 | 27.7 | - | 24.9 | 1.4 | 0 | 43.7 | 0 | 0 |
| 1980 | 15.5 | 10.4 | 7.1 | 52.4 | 7.7 | 24.4 | - | 0 | 23.8 | 8.7 | 5.4 |
| 1985 | 3.0 | 0.5 | 0.5 | 0.3 | 2.9 | 3.3 | - | 0 | 0.1 | 0 | 0 |
| 1989 | 9.2 | 0.3 | 0.0 | 100 | 0.3 | 12.1 | - | 0 | 0 | 0 | 13.0 |

* Total refers to the programs listed here.

Since the early 1970s the levels of benefits under the general insurance state pensions have been linked to the net minimum wage. For example, the net old age pension for a married couple equals the net minimum wage, while for a single person it amounts to 70 per cent. The benefit rate for a married couple, defined as the gross pension benefit divided by gross average income, increased from 0.30 in 1957, to 0.32 in 1961, 0.46 in 1966, 0.51 in 1971 and about 0.60 from 1976 on (Verbon, 1988).

The benefit rate for the AKW for the first child remains almost continually at somewhat more than 3 per cent of the net minimum wage. For subsequent children, the benefit rate varies. For example, the benefit rate for the second child amounted to almost 0.05 in 1963, to 0.05 in 1975 and to about 0 .04 in 1990. For the sixth child the benefit rates were 0.05, 0.08 and 0.05, respectively. The level of the benefit is determined not only by income trends but also by other - political - reasons.

The wage-related benefits paid under the sickness, unemployment and disability benefit schemes amounted to 80 per cent of the (former) wage up to 1987, and 70 per cent thereafter. From 1969 a minimum level was in place (the net minimum wage). The maximum benefit equals the contribution ceiling.

## 3. The Microsimulation Model NEDYMAS

The NEDYMAS Model is a dynamic cross-sectional model (Nelissen, 1994). Demographic processes are explicitly simulated, which means that the size of the micro-database changes during the simulation period. A first useable sample can be derived from the 1947 Census data. As a consequence, the first birth generation which can be followed over time is the generation born in 1930. Once the initial population has been determined, the characteristics of the individuals (and thus the households) during the period under investigation are updated. The modules used in the current version of NEDYMAS and the sequence of treatment are shown in Table 5. Like all microsimulation models, NEDYMAS is a recursive model. First, all demographic transitions are made, followed by education, changes in economic activity, labour income, income transfers and taxes.

The model is not able to simulate capital income, because it does not contain a module for private consumption. As a result, savings cannot be determined and, therefore, neither can wealth nor income from wealth. Therefore, analysis is limited to the redistributive impact of the social security system on lifetime *labour* income.

The various transition rates are based on observations, if available. However, especially for the period 1947-1965, additional assumptions had to be made. The future demographic transition rates are based on the forecasts of the Netherlands Central Bureau of Statistics. The transition probabilities with respect to the education submodules are held constant at the 1988 level, while future developments in labour force participation and unemployment are based on the forecasts of the Dutch Department of Social Affairs. National income is assumed to grow annually by two per cent. From 1991 onwards, the social security premiums are determined endogenously on the basis of the simulated benefits and income.

An equivalence scale has been applied to the income, benefits and contributions of households, using scales derived from the Subjective Poverty Line method, in which households were asked what income they needed to reach a minimum level of

Table 5: Program Module Sequencing for Each Individual in NEDYMAS

| PROGRAM MODULE | CONDITIONAL UPON |
|---|---|
| **A. DEMOGRAPHIC MODULE** | |
| Immigration | year of birth, age, marital status, year of previous emigration if applicable, prob of immigration without family, probability of re-migration like immigration |
| Emigration | like migration |
| Old people's home | year of birth, age, sex, marital status |
| Death | year of birth, age, sex, marital status |
| Marriage | year of birth, age, sex, marital status, level of education |
| Divorce | year of birth, age, sex, marital status, age difference between partners, number of children |
| Child custody | year, sex |
| Dehabitation | like divorce |
| Cohabitation selection | like marriage |
| Fertility | year of birth, age, sex, marital status, living together or not, parity |
| Splitting-off children | year of birth, age, sex, nuptiality rate, probability of cohabitation, participation in higher education |
| **B. LABOUR AND INCOME MODULE** (first part) | |
| Education | year, sex, availability certificates |
| Scholarship | year, age, type of education, parents' income |
| Income percentile | year, level of education, sex |
| Labour supply | year, age, sex, pre-tax household income, shadow wage |
| Transitions from school | year, age, sex, labour supply |
| Transitions from disablement | year, age, sex, labour supply |
| Transitions from military service | year, age, labour supply |
| Transitions from being employed | year, age, sex, type of employment, level of education |
| Transitions from being unemployed | year, age, sex, marital status, level of education, labour supply, duration of unemployment |
| Transitions from the state houseman/housewife | year, age, sex, labour supply |
| Retirement | age |
| Labour income | year, age, hours of work, unemployment rate, growth in national income, labour force, level of education |
| **C. SOCIAL SECURITY MODULE** | |
| Private pension premiums | contribution rate, wage income, amount of income exempted for contribution |
| Pension premiums for civil servants | contribution rate, wage income, amount of income exempted for contribution |
| Deduction civil servants | contribution rate of various schemes, income |

| PROGRAM MODULE | CONDITIONAL UPON |
|---|---|
| Old age state pension benefits | year, age, sex, marital status |
| Widowers state pension benefits | year, age, sex, marital status, presence of minor children, age at widowhood |
| Widow, widower and orphan pensions for civil servants | year, former income, term of service |
| Family allowances | year, sex, marital status, number of children, age of the children |
| Disability state pension benefits | year, former income, degree of disability |
| Sickness insurance benefits | year, income, marital status, minimum wage |
| Disability insurance benefits | year, income, marital status, minimum wage, degree of disability |
| Disablement pensions civil servants | year, former income, term of service, degree of disability |
| Old age pensions for civil servants | year, former income, term of service |
| Unemployment insurance benefits | year, former income, sex, marital status, minimum wage, weeks worked during last 5 years, age of children |
| Unemployment provision benefits | year, age, sex, marital status, minimum wage, numbers of days worked in the last year, duration of employment |
| Supplementary benefits | kind of benefits(s), marital status, partner's income |
| Provision older and partly disabled employees | age at first day of unemployment, weeks worked during last 5 years, degree of disability, marital status, partner's income |
| Social assistance benefits | year, age, marital status, wealth, income, partner's income, labour supply |
| Sickness insurance contributions | contribution rate, income |
| Health insurance contributions | contribution rate, income, number of children, partner's income |
| Disability insurance contributions | contribution rate, income, amount of income exempted for premium payment |
| Unemployment insurance contributions | contribution rate, income |
| Old age state pension contributions | idem |
| Widowers state pension contributions | contribution rate, income, age, sex, marital status |
| Disability state pension contributions | contribution rate, income |
| Family allowance contributions | idem |
| Exceptional medical expenses contributions | contribution rate, income, number of children, partner's income |
| Contribution to civil servants pension fund | contribution rate, income, exemption, age |
| **B. LABOUR AND INCOME MODULE** (second part) | |
| Taxes | year, age, sex, marital status, income, presence of children, type of employment, degree of disability |

welfare (Diederen, 1983). Single adults have an equivalence scale value of 0.70 and subsequent adults a value of 0.30, with the value for children depending upon their age and the number of children in the household (ranging from 0.11 on average for children aged less than six to 0.20 for 12-17 year olds). This equivalent income is attributed to each individual in the household unit. This implies that the income measure takes full account of the variance in household circumstances. The resulting amounts have been deflated and discounted to 1990, using a discount rate of two per cent, which is about the real interest rate in the Netherlands in this century.

# 4. Simulation Results

The redistributional effects of the social security system are assessed in this section, using the average lifetime wage and the average benefits from and contributions to the various social security schemes; the effect of the various premiums and benefits on the Theil coefficient; the contribution of the various social security benefits and premiums towards income per decile; and the benefit-contribution ratio per decile. These four elements are given for the cohorts born in the years 1930-1935 (cohort 1930), 1936-1945 (cohort 1940), 1946-1955 (cohort 1950) and 1956-1965 (cohort 1960). Persons who migrated have been excluded. The calculations are based on ten runs, each with a different set of random numbers, all starting with a micro-database of 10,000 persons in the year 1947. The simulations cover the period 1947-2060. The redistributive impact has been measured via comparison with gross wages, because there are no data available to simulate a world in which government is absent. Further, the burden of benefits (contributions) is assumed to be fully incident upon those who receive (pay) the amount.

Because the model does not contain a module for capital income, the taxes are only imposed on wages and social security income. As a result, only a part of all tax transfers are considered. Therefore, the analysis is limited to the redistributive impact of social security benefits and of social security contributions and does not consider contributions from public funds. This means that in 1989, for example, about 9 per cent of the financing side of the social insurances is excluded (Table 4).

## 4.1 Average equivalent lifetime income

As Table 6 shows, lifetime gross wages are on average Dfl. 3,148,500 for cohort 1930, compared with Dfl. 3,719,300 for cohort 1960 (in 1990 Dfl.). The net benefit is the difference between received benefits and paid contributions. This holds for all schemes, with the exception of the health care schemes (ZFW and AWBZ), for which only premium payments are taken into account. The lifetime calculations have been developed using a real discount rate of two per cent. Since this rate looks reasonable for the Netherlands, the net benefit can be considered as a real gain from the system. In the terminology of Burkhauser and Warlick (1981) and Wolff (1991), the net benefit can be considered as the transfer component of the scheme under consideration. This net benefit results from the use of equivalence factors, tax finance and intergenerational transfers (see Nelissen, 1995 for the effects of the AOW).

The total net benefit from the social security system amounts to Dfl. 430,300 (13.7 per cent of gross wages) for cohort 1930; Dfl. 418,400 (12.1 per cent) for cohort 1940; Dfl. 328,000 (8.8 per cent) for 1950; and Dfl. 297,700 (8.0 per cent) for 1960. The gain from the social security system is thus declining rapidly, as a percentage of gross wages. The decline can be related almost completely to the old age state pension AOW. The oldest generation gains Dfl. 200,000 from the AOW, but the gain for the youngest cohort is only Dfl. 18,000. Family allowances lead to a large net gain for all cohorts, but the gain is largest for the youngest cohort. This is the result of shifting to the public funding of family allowance premiums in 1988.

The net gains from the disability schemes (AAW and WAO) are initially also considerable. The net benefits are Dfl. 82,000 for cohort 1930, but only Dfl. 52,000 for cohort 1960. The gain from the disability state pension (AAW) is positive for all cohorts, but cohorts 1950 and 1960 lose on the disablement insurance (WAO): their premium contributions are greater than the received benefits. The net gain from unemployment insurance (WW) is positive for all cohorts and, with the exception of the oldest cohort, fairly constant (about Dfl. 25,000). For sickness benefits (ZW), each cohort pays more in premiums than is received in the form of benefits. The net effect from the widows state pension (AWW) is very limited, moving between a net loss of Dfl. 7,400 for the youngest cohort to a net gain of Dfl. 6,200 for the oldest cohort. The trend is comparable with that for the AOW: the net gain declines, the younger the cohort.

The pension scheme for civil servants (ABP), however, shows the reverse trend, as the net gain is larger the younger the cohort. The net gain for cohort 1930 amounts to Dfl. 54,000, compared to Dfl. 103,000 for cohort 1960 (due to the reduction in the ABP premiums, for political reasons, in the 1980s). The social assistance ABW also shows increasing benefits over time, reaching Dfl. 142,700 for the youngest cohort.

Table 6: Average Equivalent Lifetime Wages and Social Security Benefits and Premiums for the Four Cohorts[1]

| COHORT | 1930 | | 1940 | | 1950 | | 1960 | |
|---|---|---|---|---|---|---|---|---|
| | (Dfl '000) | s.d. | (Dfl. '000) | s.d. | (Dfl '000) | s.d. | (Dfl '000) | s.d. |
| Gross wages | 2903.6 | 89.2 | 3206.1 | 54.4 | 3445.5 | 44.8 | 3460.3 | 70.2 |
| Wages (including employer contributions) | 3148.5 | 93.3 | 3466.7 | 56.0 | 3721.7 | 49.9 | 3719.3 | 76.8 |
| **Employers' contributions** | 244.9 | 5.0 | 260.6 | 4.0 | 276.2 | 7.8 | 259.0 | 12.9 |
| ABP - civil servants | 61.3 | 4.5 | 62.6 | 1.9 | 73.0 | 5.9 | 69.9 | 1.8 |
| AKW - family allowance | 16.2 | 0.4 | 16.8 | 0.4 | 15.9 | 0.3 | 13.9 | 0.3 |
| KW/KWL/KWBL - old FA | 17.1 | 0.3 | 14.9 | 0.3 | 11.5 | 0.5 | 3.9 | 0.0 |
| AAW - disability pension | 23.2 | 0.3 | 23.1 | 0,5 | 23.2 | 0.4 | 19.6 | 0.5 |
| WAO - disability benefit | 22.4 | 0.5 | 23.5 | 0.7 | 20.3 | 0.5 | 17.1 | 0.5 |
| ZW - sickness benefit | 48.8 | 0.9 | 56.8 | 1.0 | 61.9 | 1.2 | 64.8 | 1.6 |
| WW - unemp. benefit | 7.8 | 0.1 | 10.7 | 0.5 | 12.7 | 0.5 | 14.1 | 0.5 |
| ZFW - health cost | 19.2 | 0.6 | 23.6 | 0.5 | 29.6 | 0.8 | 31.8 | 0.4 |
| AWBZ - health insurance | 29.2 | 0.6 | 28.7 | 0.5 | 28.2 | 0.4 | 23.9 | 0.3 |
| **Employees' contributions** | 337.1 | 7.7 | 446.5 | 10.5 | 573.6 | 5.0 | 705.7 | 26.4 |
| AOW - old age pension | 166.5 | 3.3 | 209.2 | 3.1 | 261.0 | 3.3 | 326.1 | 4.3 |
| AWW - widows pension | 24.0 | 1.1 | 31.0 | 1.1 | 38.2 | 1.9 | 45.6 | 2.0 |
| AKW - family allowance | 3.7 | 0.2 | 3.3 | 0.2 | 2.9 | 0.1 | 3.0 | 0.1 |
| AAW - disability pension | 10.4 | 0.4 | 19.0 | 0.5 | 30.2 | 0.7 | 44.2 | 0.9 |
| WAO - disability benefit | 35.7 | 1.1 | 53.7 | 1.5 | 72.5 | 1.3 | 86.5 | 2.2 |
| ZW - sickness benefit | 10.0 | 0.0 | 12.6 | 0.5 | 15.0 | 0.0 | 17.3 | 0.5 |
| WW - unemp. benefit | 11.6 | 0.5 | 16.0 | 0.4 | 19.7 | 0.5 | 21.0 | 0.6 |
| ZFW - health costs | 17.9 | 0.9 | 20.4 | 1.1 | 23.4 | 1.7 | 22.1 | 2.9 |
| AWBZ - health insurance | 24.7 | 0.6 | 47.0 | 1.7 | 73.3 | 2.1 | 102.5 | 2.4 |
| ABP - civil servants | 33.3 | 3.4 | 34.3 | 0.9 | 37.4 | 1.5 | 37.4 | 1.3 |
| **Benefits** | 1012.3 | 23.7 | 1125.5 | 18.6 | 1177.8 | 16.3 | 1262.4 | 19.5 |
| AOW - old age pension | 369.0 | 9.9 | 365.0 | 6.8 | 341.7 | 8.0 | 344.1 | 9.6 |
| AWW - widows pension | 29.9 | 3.9 | 33.2 | 3.3 | 39.2 | 2.4 | 38.1 | 2.9 |
| AKW - family allowance | 70.0 | 2.8 | 59.9 | 1.4 | 57.1 | 1.1 | 100.9 | 2.7 |
| AAW - disability pension | 94.9 | 3.8 | 110.8 | 5.3 | 119.4 | 2.1 | 133.5 | 5.0 |
| KW/K(B)WL/KKZ/KTO | 52.6 | 0.8 | 71.6 | 1.1 | 56.4 | 0.7 | 39.7 | 0.6 |
| WAO - disability benefit | 77.3 | 5.2 | 80.4 | 3.9 | 79.2 | 2.8 | 86.1 | 4.6 |
| ZW/DSO - sickness benefit | 62.3 | 1.5 | 74.9 | 1.6 | 85.4 | 1.5 | 92.2 | 1.2 |
| WW - unemp. benefit | 30.9 | 2.3 | 49.7 | 5.5 | 59.8 | 3.7 | 58.2 | 2.4 |
| WWV - unemp. provision | 8.8 | 1.1 | 10.1 | 0.5 | 10.7 | 0.6 | 9.5 | 0.5 |
| WRO - civil servants | 2.0 | 0.6 | 2.4 | 0.6 | 3.6 | 0.4 | 4.2 | 0.7 |
| ABP-OP - civil servants | 72.6 | 12.2 | 73.8 | 4.4 | 89.2 | 6.1 | 102.4 | 6.3 |
| ABP-WP - civil servants | 18.9 | 4.0 | 15.8 | 3.2 | 17.8 | 2.2 | 19.8 | 2.8 |
| ABP-IP - civil servants | 57.0 | 10.5 | 66.8 | 5.3 | 84.2 | 4.9 | 90.0 | 5.1 |
| ABW - social assistance | 65.8 | 11.1 | 107.4 | 7.6 | 134.8 | 8.3 | 142.7 | 8.8 |
| **Net Benefits** | 430.3 | 17.0 | 418.4 | 8.1 | 328.0 | 12.0 | 297.7 | 21.7 |
| **Gross Lifetime Income** | 3578.8 | 99.9 | 3885.1 | 52.5 | 4049.7 | 46.0 | 4017.0 | 70.7 |
| **Taxes** | 1270.3 | 53.6 | 1295.2 | 29.8 | 1296.9 | 22.0 | 1278.0 | 25.4 |
| **Net Lifetime Income** | 2308.6 | 52.8 | 2589.8 | 30.3 | 2752.7 | 30.9 | 2739.1 | 50.3 |

[1] All prices in 1990 Dfl.; s.d = standard deviation.

## 4.2 The redistributive impact of the social security system

The redistributive impact of the Dutch social security system, as measured by the Theil coefficient, is shown in Table 7. (The Theil coefficient ranges in value from 0 - when all individuals have the same income - to log n - when one individuals holds all income and n is the number of individuals.) The Theil coefficient for gross wages, including employer contributions, is 0.161 for cohort 1930, 0.133 for 1940, 0.125 for 1950 and 0.116 for 1960. This suggests that lifetime wage inequality within cohorts is decreasing rapidly.

With the exception of the pension scheme for civil servants (ABP), all *benefits* have an equalising effect. The old age state pension (AOW) has the largest effect, diminishing income inequality by 16.4 per cent for the oldest cohort and by 12.7 per cent for the youngest cohort. The disability state pension AAW also has a comparatively large equalising effect of 7 to 9 per cent. The disablement insurance (WAO) benefit has a somewhat less equalising effect, lowering the Theil coefficient from 3.4 per cent for cohort 1950 to 4.3 per cent for the youngest cohort. After a decline in the redistributive impact of general family allowances (AKW), there is an increase for the youngest cohort (probably due to the increases in the benefit per child in the 1980s). The effect of the widow state pension (AWW) benefit is very limited, which is, of course, related to the fact that the average benefit is very small (eg. Dfl. 29,900 for cohort 1930 - or less than one per cent of 'before tax' lifetime income).

The effect of unemployment insurance (WW) on the Theil coefficient is also very limited. Generally, the equalising effect amounts to less than one per cent and is not always significant. The effect of sickness benefit (ZW) is somewhat larger, but remains less than three per cent. Social assistance (ABW) has - as may be expected - a large redistributive impact. The Theil coefficient is reduced by 7.5 per cent for cohort 1930 and by almost 12 per cent for the youngest cohort.

In summary, for all benefits under the general insurances the Theil coefficient decreases by about 25 per cent while, for benefits under the employee insurances, it decreases about 10 per cent. Thus, over the lifetime, the Dutch insurance scheme benefits reduce inequality.

With the exception of the civil servants pension (ABP) premiums, all *contributions* increase income inequality. One reason for this is the existence of a maximum

Table 7: Redistributive Impact of Social Security Benefits and Contributions

| Cohort | 1930 | 1940 | 1950 | 1960 | 1930 | 1940 | 1950 | 1960 | 1930 | 1940 | 1950 | 1960 |
|---|---|---|---|---|---|---|---|---|---|---|---|---|
| THEIL COEFFICIENT | | | | | | | | | | | | |
| Employee's gross wages | .171 | .142 | .132 | .122 | | | | | | | | |
| - including employer contributions | .161 | .133 | .125 | .116 | | | | | | | | |
| **CHANGES IN THEIL COEFFICIENT (in per cent)** | | | | | | | | | | | | |
| | **Benefits** | | | | **Contributions** | | | | **Net Effect** | | | |
| Old age pension (AOW) | -16.4 | -15.0 | -12.4 | -12.7 | 5.2 | 6.0 | 7.0 | 9.3 | -13.0 | -10.8 | -7.4 | -6.1 |
| Widows pension (AWW) | -2.1 | -2.0 | -2.0 | -2.2 | 0.8 | 0.9 | 1.0 | 1.2 | -1.3 | -1.0 | -1.0 | -0.9 |
| Family allowances (AKW) | -4.2 | -3.4 | -2.6 | -5.0 | 0.6 | 0.6 | 0.5 | 0.6 | -3.7 | -2.9 | -2.1 | -4.5 |
| Disability pension (AAW) | -7.1 | -7.6 | -7.6 | -9.1 | 0.8 | 0.8 | 1.0 | 1.4 | -6.4 | -7.0 | -6.8 | -8.1 |
| Old schemes (KW(L)/KBWL/ KKZ/KTO) | -3.2 | -4.1 | -2.8 | -1.9 | 0.7 | 0.6 | 0.4 | 0.1 | -2.5 | -3.5 | -2.4 | -1.8 |
| Disability benefit (WAO) | -4.0 | -3.9 | -3.4 | -4.3 | 0.6 | 0.7 | 0.9 | 1.1 | -3.2 | -3.4 | -2.7 | -3.4 |
| Sickness benefit (ZW/DSO) | -2.3 | -2.4 | -2.3 | -2.6 | 1.7 | 1.7 | 1.6 | 1.8 | -0.8 | -0.9 | -0.9 | -1.0 |
| Unemployment benefit (WW) | -1.1 | -0.4* | -.09 | -1.2 | 0.5 | 0.5 | 0.6 | 0.7 | -0.6 | +0.1* | -0.4* | -0.5* |
| Old unemployment benefit WWV | -0.4 | -0.5 | -0.5 | -.05 | | | | | | | | |
| Civil servants pensions (ABP-OP) | 3.0 | 3.0 | 3.0 | 3.5 } | | | | | | | | |
| ABP-WP | 0.6 | 0.4* | 0.2* | 0.0* } | -1.8 | -1.4 | -1.5 | -1.7 | +1.5 | +1.8 | +0.7* | +0.4* |
| ABP-IP | 0.5 | 0.3* | -0.8 | -1.3 } | | | | | | | | |
| Social assistance (ABW) | -7.5 | -9.6 | -9.8 | -11.5 | | | | | | | | |
| Health insurance (AWBZ) | | | | | 1.7 | 2.0 | 2.6 | 3.1 | | | | |
| Health costs (ZFW) | | | | | 3.0 | 3.1 | 3.1 | 3.1 | | | | |
| General insurances | -25.8 | -24.3 | -21.3 | -24.7 | 7.7 | 9.1 | 10.4 | 14.7 | -21.9 | -19.8 | -15.5 | -18.2 |
| Employee insurances | -9.8 | -9.9 | -8.7 | -9.2 | 3.8 | 3.8 | 3.9 | 4.0 | -7.1 | -7.3 | -6.0 | -6.4 |
| All insurances | -35.3 | -35.5 | -32.5 | -36.6 | 12.3 | 14.2 | 16.1 | 21.8 | -26.2 | -26.7 | -22.1 | -24.8 |

* Not significant at 0.05 level.

premium income, for all schemes except the ABP. Excluding the old age AOW, all contributions affect the Theil coefficient by less than three per cent. There is an increase in the Theil coefficient caused by the AOW of 5.2 per cent for cohort 1930, 6.0 per cent for 1940, 7.0 per cent for 1950 and 9.3 per cent for 1960.

The net impact of benefits and contributions for the various social security schemes is given in the right panel of Table 7. Most noteworthy is that the disability state pension (AAW) reduces the Theil coefficient for cohort 1960 by more than the old age state pension (AOW). The general family allowances (AKW) and the disablement insurance (WAO) also lead to a considerable decline in the Theil coefficient. The effect of the former family allowances (KW and so on) is somewhat less, whereas the effects of the unemployment insurances (WW) and sickness insurance (ZW) are very limited. As one might expect, the civil servants pension (ABP) leads to an increase in income inequality. All general insurances together lead to a reduction in the Theil coefficient of 15 to 22 per cent, whereas employee insurances reduce income inequality by 5 to 7 per cent. Taken together, all these insurances lead to a reduction in income inequality of 26 per cent for cohort 1930, 27 per cent for 1940, 22 per cent for 1950 and 25 per cent for 1960.

The effect on the Theil coefficient, of course, depends on the amounts involved. One possible measure for this is the 'adjusted redistributive impact', which shows how much the scheme affects the lifetime income distribution, assuming that the scheme amounts to one per cent of gross lifetime wages (Nelissen, 1994). In this way, the measure offers the possibility of comparing the redistributive impact of the various schemes under the assumption that the various schemes do not differ with respect to their size. For example, the adjusted redistributive impact of the general insurances is only a bit larger than that of the employee insurances. The 'average' general insurance decreases income inequality by about 1.15, if its size amounts to one per cent of gross wages, whereas the reduction for the 'average' employee insurance amounts to 0.90 per cent on average (Nelissen, 1994). This implies that the wage-related employee insurances do not necessarily result in a smaller redistributive impact than the flat-rated general insurances. It depends, among other things, on the level of the maximum contribution limit.

## 4.3 Social security benefits and contributions by decile

Another way of looking at the redistibutive impact of the social security system is via contributions and premium payments by decile. All individuals have been sorted into 10 equally sized groups (deciles) on the basis of household lifetime equivalent gross wages. In interpreting Figure 1, it should be appreciated that both the premiums paid within a year and the benefits received within a year have been adjusted for household composition. For example, for pension schemes, this implies that the equivalent premiums are generally lower than the non-equivalent premiums, while the equivalent benefits are sometimes higher than the non-equivalent benefits. This means that the pension premiums have been paid during a period in which the cohort's average household size is proportionally large, whereas benefits have been received during a period in which the average cohort's household size is proportionally low. This partly explains why the (equivalent) lifetime benefits for civil servants pensions (ABP) are larger than the (equivalent) lifetime premium payments.

The key role of the old age pension schemes becomes immediately clear from Figure 1. Both contributions to and benefits from the old age state pension AOW - and, to a lesser extent, those of the civil servants pension ABP - rise well above the others. The AOW benefits are almost equally distributed over the deciles. Only the first decile - which contains proportionally many people who die at a young age - receives a somewhat smaller average benefit. On the other hand, contributions increase the higher the decile. This results in a smaller net benefit for the AOW the higher the decile. The continual increase in premium contributions by decile is also found for the other general insurances (AWW, AKW, AAW and AWBZ). The lowest and highest deciles have the smallest general family allowances (AKW) benefit, with benefits being very similar for the other deciles. Benefits from the disability pension (AAW) are strongly skewed towards the lower deciles, while the top decile pays more than is received via benefits.

The linkage of the benefit from employee insurances to (former) income results in less redistribution from those with higher to those with lower lifetime income. Nonetheless, the net loss made by higher income deciles has increased for recent cohorts. For example, for cohort 1930, decile eight was the first one to pay more than it received from the WAO scheme, whereas this holds for decile five in cohort 1960. For sickness benefits ZW, contributions exceed benefits for the top five deciles. The health insurance (ZFW) clearly shows the existence of an upper income limit on premiums, as payments increase up to decile five and decrease thereafter.

For the important civil servants pension fund, ABP, the higher the income, the higher both benefits and contributions. This is because the contribution rate and benefits received are both a percentage of income, with no upper limit for either. The net benefit is positive for all deciles, but very small for the two lowest deciles.

Social assistance ABW delivers the highest benefits, after the AOW and ABP old age pension schemes. The ABW benefit is, of course, most beneficial to lower income groups, but the higher deciles also receive a large amount from the scheme. However, their benefits are probably overstated, because no account is taken of wealth and other income - and the higher deciles, in particular, derive the most advantage from these.

## The benefit-contribution ratio

Another interesting indicator of lifetime income redistibution between generations is the benefit-contribution ratio (equivalent benefits divided by equivalent contributions). For cohort 1930, the benefit-contribution ratio for all social security schemes examined is 1.91 (Table 8). This means that cohort 1930 gets Dfl. 1.91 for each Dutch guilder invested in the social security system, so that the net gain amounts to 91 per cent. The net gain is only 27 per cent for cohort 1960. The largest reduction takes place between cohorts 1940 and 1950, with a fall of 17 per cent. Cohorts 1940 and 1960 are confronted with a fall of 13 and 8 per cent, respectively, compared with their preceding cohort.

The lower benefit-contribution ratio for cohort 1940 compared with cohort 1930 is due to the general insurances. This benefit-contribution ratio declines by 19 per cent, whereas that for the employee insurances is almost constant. For the 1950 cohort, which had the largest decrease in the total benefit-contribution ratio, the decrease resulting from the general insurances is about twice that of the employee insurances (20 and 11 per cent, respectively). The drop in the ratio for the youngest cohort is the result of a decline of 9 per cent in the general insurances and a decline of 8 per cent in the employee insurances.

Comparing the results for cohort 1960 with those of cohort 1930 shows that the benefit-contribution ratio for all insurances decreases by 34 per cent. For the general insurances, the decline amounts to 41 per cent and, for the employee insurances, 18 per cent.

This decline in net gain is reflected in the benefit-contribution ratios for the various schemes. The benefit-contribution ratio for the old age state pension (AOW) declines with the youth of the cohort and the decline is very steep - 20 per cent or even more between successive cohorts. In the case of cohort 1960, the benefit-contribution ratio for the AOW is 52 per cent lower than that for cohort 1930. The latter receives 2.19

Figure 1: Equivalent Social Security Benefits and Contributions (Dfl '000) by Decile

Cohort 1930

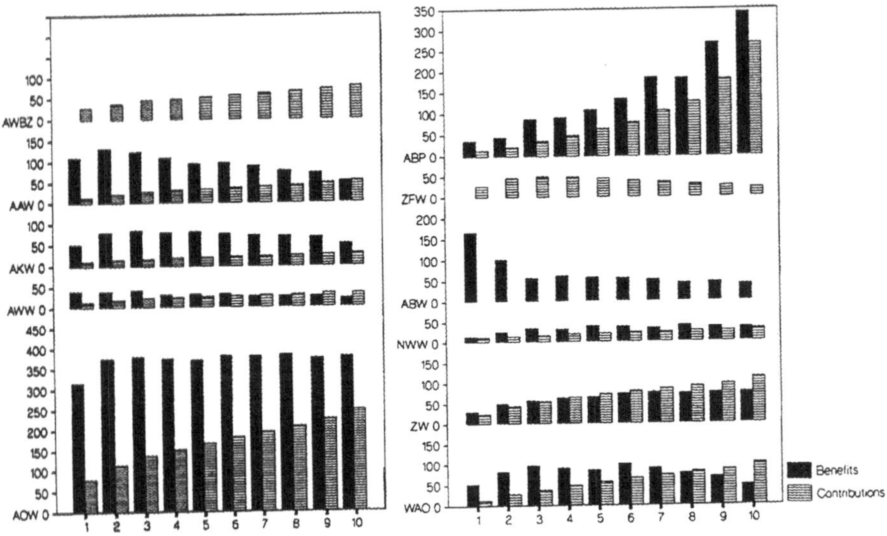


Cohort 1940

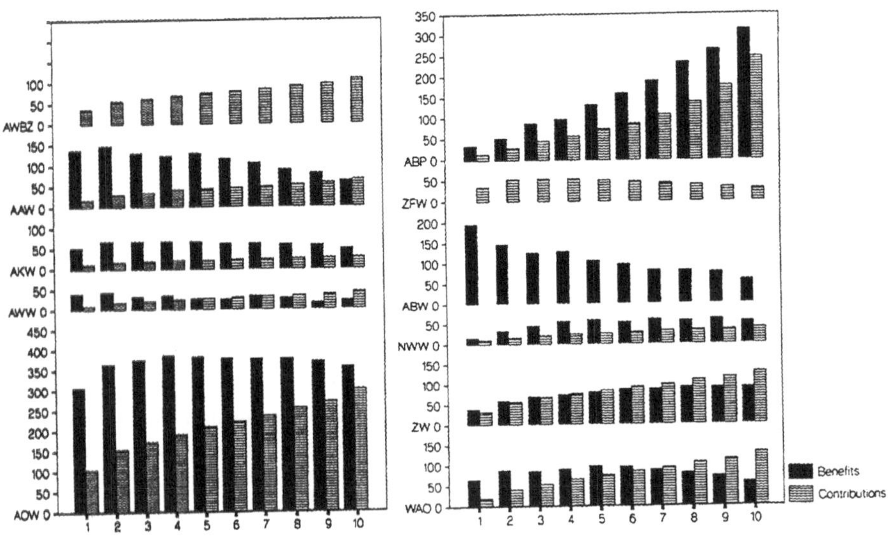

(ranked on the basis of equivalent gross lifetime income)

Cohort 1950

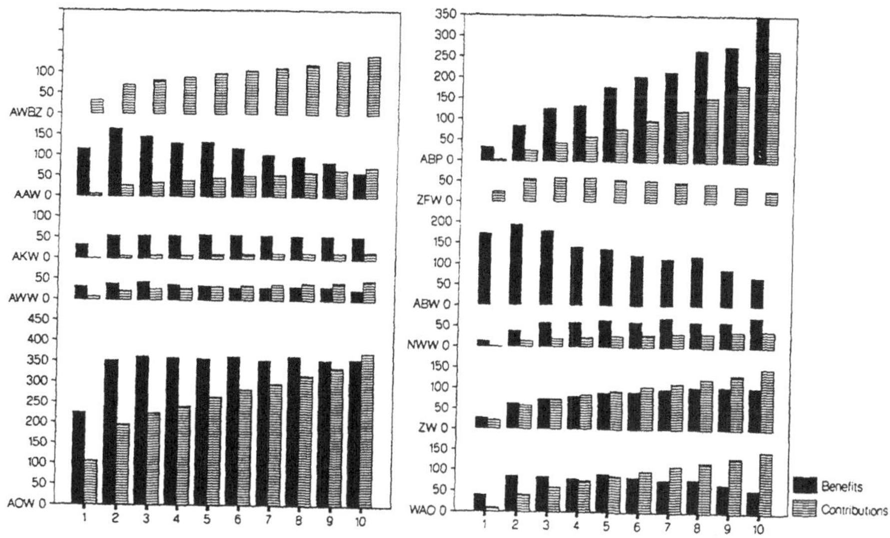


Cohort 1960

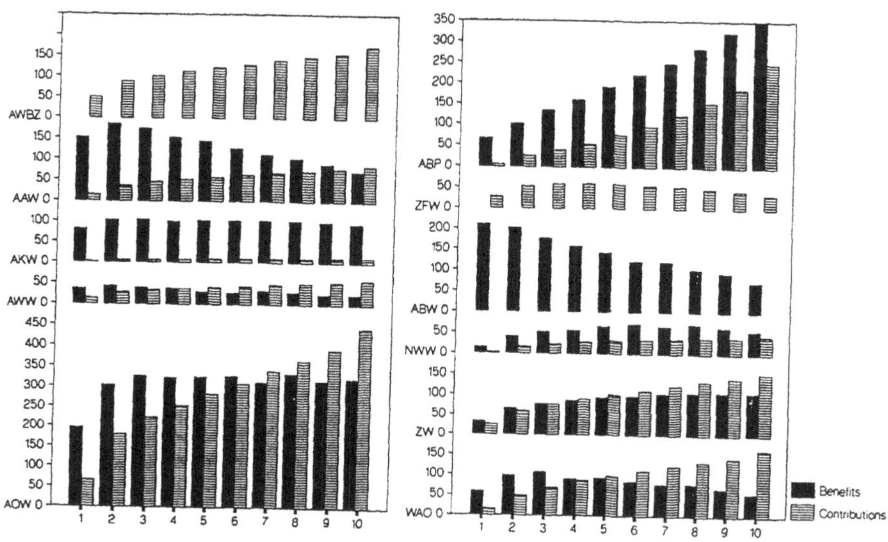

times the premiums paid, whereas the former receives only 6 per cent more than its contributions. The decline in the widow state pension (AWW) is particularly evident and the youngest cohort pays more for the AWW than will be received (the benefit-contribution ratio amounts to 0.84). In contrast, the youngest cohort has a higher benefit-contribution ratio for the general family allowances (AKW) than the oldest cohort. The ratio is 73 per cent higher, due to the change in the way the AKW has been financed since 1988 (ie. from public funds).

In contrast, for the individual employee insurances, only the benefit-contribution ratio for the disablement insurance WAO has declined. This ratio was 38 per cent lower for cohort 1960 (0.83) than it was for cohort 1930 (1.33). This is partly the result of the reform of the social security system in 1987, when the wage-related benefits were decreased by 10 per cent. During the economic crisis of the early 1980s a proportionally large number of persons in cohort 1930 - and to a lesser extent in cohort 1940 - entered the disablement schemes (Aarts and De Jong, 1990). While these two earlier cohorts make a net gain, the net benefit is negative for both cohort 1950 and 1960 (whose benefit-contribution ratios amount to 0.86 and 0.83, respectively)[2].

The benefit-contribution ratio for unemployment insurance WW is the largest for cohort 1940, at 1.88. This cohort made a proportionally large appeal to these schemes during the economic crisis of the early 1980s. Besides this, younger cohorts were affected by the cuts in the benefit level and changes in the duration of the benefit as a consequence of the adoption of the WW. These negative implications are larger the younger the cohort. Sickness insurance ZW shows a stable pattern, with each generation losing about 12 per cent of its contributions (due to proportionally high administration and transaction costs).

The civil servants pension (ABP) shows an equal benefit-contribution ratio for cohort 1930 and cohort 1940, whereas the ratio rises for the other two cohorts. This was caused by the decrease in the contribution rate in the 1980s.

---

[2] In sharp contrast, the youngest cohort receives greater benefits from the employee family allowances, with a benefit-contribution ratio almost twice that of cohort 1950. The abolition of this scheme plays an important role. The youngest cohort averaged 19 years of age at the time of this abolition. This implies that the benefits earmarked for them as a child have been received (but, of course, not the benefits earmarked for their children), whereas premium payments can only be imputed during, on average, those 19 years.

# 5. Conclusions

This paper has analysed the redistributive impact of social security schemes on lifetime income, using the microsimulation model NEDYMAS. Although the model has been applied to the Netherlands, it is still possible to generalise from the results. The social security system in the Netherlands is very well suited to such a purpose, because it includes both a needs-based welfare system (the general insurances and social provisions) and a wage-related social security system for employees and civil servants (the employee insurances and schemes for civil servants) - referred to below as the Atlantic system and Continental system, respectively. Other countries generally focus on one of the two. For example, Great Britain, Denmark, Norway and Sweden have opted for the Atlantic approach, whereas Austria, Belgium and Germany chose the Continental system.

Over time, elements of the Continental system have been added to the Atlantic system and *vice versa*. In the Atlantic system, additional benefit schemes are typically created for some categories - for example, for the elderly in the United Kingdom and Sweden. Similarly, basic allowances have been introduced in the Continental system - such as the guaranteed income for the elderly in Belgium. Until the late 1980s this trend toward convergence of the two systems persisted. In countries with Atlantic schemes there is now fast growth in supplementary occupational arrangements, whereas only very slight increases in such arrangements have occurred in countries with Continental schemes.

In the late 1980s and 1990s, cuts in government expenditure had an impact on the social security system. Cuts to social security schemes have led to discussion of the relevance of wage-related benefits above the social minimum level. This has led some to suggest the abolition of wage-related schemes and the limitation of the social security system to (sometimes even means-tested) flat-rate benefits. If a person wants a higher benefit, so the argument goes, (s)he has to insure against risk in the private insurance market. In other words, it is advocated that government limits itself to the Atlantic-type schemes. This development has occurred in the United Kingdom (where wage-related sickness and unemployment benefits were abolished in the 1980s) and in the Netherlands (in the proposals for the WAO and the discussion of the so-called '*ministelsel*'.) Another trend is toward the introduction (or strengthening) of the individual's employment history as a determinant of the level of benefit under the Continental system (shown in recent discussion on Old Age Insurance (OAI) in the United States and the reforms of unemployment insurance in the Netherlands).

Table 8: Benefit-Contribution Ratios of the Schemes for the Four Cohorts*

| Scheme | | Cohort 1930 | Cohort 1940 | Cohort 1950 | Cohort 1960 |
|---|---|---|---|---|---|
| Old age state pension | AOW | 2.19 | 1.75 | 1.31 | 1.06 |
| Widow state pension | AWW | 1.26 | 1.07 | 1.02 | 0.84 |
| General family allowances | AKW | 3.49 | 2.98 | 3.06 | 6.03 |
| Disability state pension | AAW | 2.86 | 2.62 | 2.23 | 2.08 |
| Family allowances for employees | KW/etc | 3.07 | 4.85 | 4.90 | 10.15 |
| Disability insurance | WAO | 1.33 | 1.04 | 0.86 | 0.83 |
| Sickness insurance | ZW | 0.88 | 0.89 | 0.89 | 0.88 |
| Unemployment insurance | WW | 1.59 | 1.88 | 1.85 | 1.67 |
| Civil servants pension | ABP | 1.57 | 1.61 | 1.68 | 1.89 |
| Family allowances | AKW + KW | 3.29 | 3.76 | 3.76 | 6.81 |
| Disability schemes | AAW + WAO | 1.90 | 1.61 | 1.37 | 1.30 |
| General insurances | 1 | 2.29 | 1.87 | 1.50 | 1.36 |
| Employee insurances | 2 | 1.35 | 1.36 | 1.21 | 1.11 |
| TOTAL | 3 | 1.91 | 1.67 | 1.38 | 1.27 |

1 AOW, AWW, AKW and AAW
2 WAO, ZW, WW and KW and related schemes
3 AOW, AWW, AKW, AAW, WAO, ZW, WW and KW and related schemes
* Equivalent benefits divided by equivalent contributions

What can we conclude from this study? First, the lifetime redistributive impact of the benefits differs only slightly between the Atlantic-type flat-rate benefits and the Continental-type wage-related benefits (after adjusting for the size of the schemes). This implies that earnings-related benefits do not necessarily have a much smaller redistributive impact than flat-rated ones. This, of course, depends on the level of the maximum wage-related benefit as, the larger the maximum benefit, the larger the difference between the two types of schemes. If the benefit has no limit, the income-equalising effect of the Continental-type scheme becomes very small.

Second, means-tested schemes have a relatively large redistributive impact. For example, the size of the social assistance ABW benefits is comparable with that of the disability state pension (AAW) but the redistributive impact of the former is about 50 per cent greater than that of the latter if we adjust for its size. Moreover, the redistributional impact of the ABW is underestimated, because NEDYMAS contains no capital income and wealth module.

Third, the existence of a ceiling on contributions results in an increase in income inequality and, the lower the ceiling, the larger the increase. Again, comparing the age pension (AOW) and sickness benefits (ZW), the adjusted redistributive impact of the ZW is about 77 per cent of that of the AOW for generation 1930 and about 57 per cent for generation 1960. The level of premium ceilings thus has a large effect on the redistributive impact. Because Atlantic-type schemes generally have a lower ceiling than Continental-type schemes, income inequality is usually enlarged more by the former type (via the contributions) than by the latter type.

Fourth, the existence of an exemption income level in the contribution schedule (below which contributions do not have to be made) greatly reduces the increase in income inequality. The disability insurance (WAO) has an exemption which amounts to about 35 per cent of the ceiling from 1982 onwards and about 25 per cent in the period before 1982. The premium ceiling for the unemployment insurance (WW), for example, equals that of the WAO, but the adjusted redistributive impact of the WAO contributions is only half that of the WW contributions, which has no exemption.

Fifth, if no contribution ceiling exists, the levying of contributions in combination with a contribution exemption results in a reduction in income inequality, but the decrease is small in comparison with the effect of flat-rate benefits (as shown by civil servants pensions).

Sixth, the existence of an exemption in the benefits, in combination with no upper bound for the benefit, results in an enlargement of income inequality - such as that exemplified in the old age pension for civil servants (ABP-OP) in the Netherlands.

Overall, the Dutch social security system shows that the redistributive impact of Atlantic-type schemes does not necessarily differ greatly from that of Continental-type schemes. The income-equalising effect of the flat-rate benefit in comparison with a wage-related benefit (in combination with a maximum level of the benefit) can be neutralised by a lower contribution ceiling. That is why there is only a relatively small difference in the redistributive impact of the general and employee insurances in the Netherlands.

The benefits and contributions structure of the Dutch social security system results in only a limited redistribution of lifetime earnings. The schemes under consideration lead to a reduction in the Theil coefficient that varies from 22.1 per cent for cohort 1950 to 26.7 per cent for cohort 1940. This reduction is even smaller than the reduction in lifetime wage income inequality between cohorts 1930 and 1960, which amounts to 28 per cent. Of course, the extension of the social security system has diminished income differences, but its impact is smaller than that of some other factors.

However, this does not imply that some groups cannot have a large benefit - in relation to their contribution - from the system. However, the relative benefits are less generous for lower deciles in later cohorts. Decile one in cohort 1930 receives about four times more per contributed guilder to the social insurances (discounted by two per cent) than does decile 10. For cohort 1960, this ratio of return is decreased to about two.

It is remarkable that the redistributive impact of the social security system hardly differs between generations. The notion that a comprehensive social security system will result in an extensive change in income distribution does not necessarily hold for lifetime income. This does not imply that, on an annual basis, the redistribution cannot be large. Indeed, in the Netherlands, it can be (see Muffels et al., 1986). The scheme that results in by far the largest redistribution on an annual basis - the old age state pension (AOW) - has a limited redistributive impact on a lifetime basis. After standardising for the amount of money spent on the programs, its redistributive impact is smaller than that of the disability state pension AAW or the combined family allowances schemes in the youngest cohort. The explanation is simple: the AOW is paid to all people who reach 65 years, whereas the AAW only benefits people who are disabled - and the probability of becoming disabled decreases as income increases.

Another striking point is that the supplementary pension scheme for civil servants (ABP) results in a significant enlargement of income inequality for cohorts born around 1930 and 1940. This is an unintended consequence of the low contribution rates in the 1980s and 1990s, attributable to cuts in government expenditure.

All generations have a benefit-contribution ratio that is larger than 1.00. This means that all generations "profit" from the social security system. However, this does not take account of financing via public funds, which amounts to 10 to 15 per cent for the generations considered here. Further, the use of a higher discount rate results in a lower benefit-contribution ratio. If a discount rate of three per cent is used *and* account is taken of public funding, then cohort 1960 only breaks even. This suggests that younger generations will probably lose from the social security system. Intergenerational redistribution results in a gain for the generations up to cohort 1960 but a loss to the generations born after about 1965. This type of intergenerational redistribution, from younger cohorts to older cohorts, is often justified by the increase in non-transfer income over time. However, this reasoning only holds when (non-transfer) income continues to increase. As Table 6 shows, income growth appears to have come to a halt. If the generational losses become larger and larger, there may be growing tension between younger and older generations in the future.

## References

Aarts, L.J.M. and P.R. De Jong (1990), *Economic aspects of disability behavior*, Ph.D Dissertation, Erasmus University, Rotterdam.

Burkhauser, R.V. and J.L. Warlick (1981), 'Disentangling the annuity from the redistributive aspects of social security in the United States', *Review of Income and Wealth*, Vol. 27, No. 4.

Davies, J., F. St-Hilaire and J. Whalley (1984), 'Some calculations of lifetime tax incidence', *American Economic Review*, Vol.74, No. 4.

Diederen, H.M.N. (1983), 'Gestandaardiseerde inkomensverdeling 1980', *Sociale Maandstatistiek*, Vol. 31, No. 11.

Falkingham, J., J. Hills and C. Lessof (1993), 'William Beveridge versus Robin Hood: Social security and redistribution over the lifecycle', Welfare State Programme, Discussion Paper No. 88, London School of Economics, London.

Harding, A. (1993), *Lifetime Income Distribution and Redistribution; Applications of a Dynamic Cohort Microsimulation Model*, North Holland, Amsterdam.

Haveman, R.H. and K. Hollenbeck (1980), 'Preface', in: R.H. Haveman and K Hollenbeck, (eds.), *Microeconomic Simulation Models for Public Policy Analysis, Volume 2*, Academic Press, New York.

Ministerie van Sociale Zaken en Werkgelegenheid (1984), *Financiële Nota Sociale Zekerheid 1985*, Tweede Kamer, vergaderjaar 1984-1985, 18612, Nos. 1-2.

Ministerie van Sociale Zaken en Werkgelegenheid (1989), *Inkomensbeleid 1990*, Tweede Kamer, vergaderjaar 1989-1990, 21307, No. 2.

Muffels, R.J.A., J.H.M. Nelissen and W.J. Nuyens (1986), 'De inkomensherverdelende werking van sociale zekerheidsregelingen', *Sociaal Maandblad Arbeid*, Vol. 41, No. 1.

Nelissen, J.H.M. (1994), *Income Redistribution and Social Security: An Application of Microsimulation* Chapman and Hall, London.

Nelissen, J.H.M. (1995), 'Lifetime income redistribution by the Old Age State Pension in the Netherlands', *Journal of Public Economics* (forthcoming).

Verbon, H. (1988), *The Evolution of Public Pension Schemes,* Springer Verlag, Berlin.

Wolff, E.N. (1991), 'The distributional implications of social security annuities and transfers on household wealth and income', Paper presented to the 1991 ESPE Meeting, Pisa, June.

Heinz P. Galler

Martin-Luther-University
Germany

# Chapter 13: Microsimulation of Pension Reform Proposals: Modelling the Earnings of Couples

## Abstract

*In Germany the reform of widow's pensions is back on the political agenda. An updated version of the Frankfurt dynamic microsimulation model is used to analyse the impact of alternative reform proposals. The chapter discusses the question of how stochastic dependencies in the earnings capacities of individuals forming a couple can be taken into account in microsimulation. In addition, using data from the German socioeconomic panel study, first estimates of the correlation structure are presented for an error components model for wage rates. These estimates indicate a moderate positive correlation across individuals.*

## 1. Introduction

Despite the fact that the German pension system underwent a reform in 1992, the problem of widows pensions is still on the political agenda. There is political pressure to introduce a minimum pension level in order to reduce old age poverty (which is mainly a problem of widows with low dependents benefits and no, or only a small, personal pension claim of their own). The discussion has gained additional momentum as, after German unification, a minimum pension was introduced in the 'new' eastern states of the Federal Republic, in order to avoid pensioners there falling below the poverty line in the transition process. However, this regulation will expire in 1997. The questions now being discussed are whether or not such a regulation should be continued beyond this date and whether or not it should be extended to other states of the federation.

The main reasons for female old age poverty are women's comparatively lower earnings and their interrupted work histories, caused by child raising and household obligations. As pension claims, in principle, are proportional to average earnings over the life cycle in the German pension system, interruptions and part-time work result in significantly lower pensions for women. In the last reform, attempts were made to reduce the problem by granting women, for each child born to them, up to three years of pension insurance without contributions and based on an average imputed income. However, this reform does not solve the problem of women who remained out of the labour force for longer periods.

Common features of the reform proposals are the introduction of a minimum pension and the payment to the surviving spouse of an individual pension claim, based on contributions, instead of the traditional dependent's benefit. The basic idea is to split the pension claims of a couple in a given year between the spouses, regardless of who actually earned the income. In order to attain a sufficient level of claims, persons not working in the labour market are supposed to pay contributions based on an imputed income that is defined as a percentage of the average earnings in the population covered by the system. The only exemptions to this are for people raising children and those taking care of a disabled person, for whom credits without contributions have been proposed. The costs resulting from these exemptions are to be paid from the general budget.

Because of the complex structure of the pension system, a microsimulation approach is the most appropriate way to analyse the effects of the reform on the distribution of personal income, both from a cross-sectional and a lifecycle perspective. Furthermore, the effects on revenue and expenditure can be estimated more accurately if microdata containing information on the relevant characteristics of the individuals are used. Consequently, an updated version of the dynamic Frankfurt microsimulation model is used for these analyses.[1]

The basic approach is, first, to simulate for a representative sample of micro-units both the contributions to the pension system and the pensions being received under each reform proposal. In a second step, the impact of the reform on both personal income distribution and aggregate revenue and expenditure can be inferred from the sample data without much difficulty.

If only a short period is to be considered, a cross-sectional sample and static microsimulation techniques are appropriate. However, if the redistribution of income over time is to be considered, lifetime earnings profiles are required for the

---

1 This model has been developed by the author and a research team at the University of Frankfurt with support from the Deutsche Forschungsgemeinschaft. It has been used extensively for the analysis of pension reform proposals in the early eighties- see Krupp et al (1981). Since then, updated versions have been applied to analyses in the field of population dynamics, social security and pensions.

micro-units. These lifetime earnings profiles can be used to calculate both the individual contributions to the pension system and the benefits received during the life cycle of the individuals.

Longitudinal data collected, for instance, by retrospective interviews are not fully suited for such an analysis. Even if all information required is available in such a dataset, it describes lifetime earnings in the past 60 years. As economic conditions have changed substantially in Germany during that period, it is difficult to infer the effects of the pension system in the future from lifetime earnings profiles of today's retired people. This is true even if a period of, for instance, only 20 years is considered. From this point of view, it is preferable to use a dynamic microsimulation model to generate likely future earnings profiles on the basis of current settings.

## 2. Modelling the Earnings of Couples

From a substantive point of view, dependencies between the earnings profiles of the individuals forming a couple are of major interest for the analysis of widows pensions. Such dependencies result from the joint decisions on labour supply within a family or couple. In addition, the income distribution among couples may be affected by selective matching of the partners: a positive correlation of earnings capacities will add to the dispersion of the income distribution, while a negative correlation will reduce income disparities. Thus, if the personal income distribution is of interest, a model is required that allows us to simulate individual income dynamics while taking into account both structural relations and stochastic dependencies in the behaviour of the partners within a couple.

Dependencies between the earnings of spouses have already been considered to some extent in the earlier versions of the Frankfurt microsimulation model. The model structure is based on a male chauvinist approach. In each period the behaviour of men and unmarried women is simulated on the basis of the state at the end of the previous year. Then, in a second round, the labour force participation and earnings of married women are simulated, given the results for their husbands. Thus, structural dependencies between spouses are included in the model in a recursive way. In addition, in order to allow for selective matching, the education of the partners is used as one of the characteristics in the distance function that governs the matching of partners in marriages or consensual unions.[2]

---

[2] In the model couples are formed by first selecting randomly the 'desired' characteristics of the partner from the conditional distribution of the characteristics of men, given the characteristics of the

However, education is only an imperfect measure of the earnings capacity of a person. This is also true for other variables such as experience.[3] As a consequence, the explanatory power of earnings equations is limited, and a substantial part of the variation of earnings is attributed to the latent error terms of the equations.[4] If the entire dispersion of the empirical income distribution is to be reproduced by a microsimulation model, the random errors must be included in the simulation. Technically, this poses no problems if a stochastic model is available for the joint distribution of the error components. Based on this distribution the random components of earnings can be generated in a microsimulation model by the same Monte-Carlo methods that are used to simulate other processes.

An error components structure provides a simple framework that allows the introduction of stochastic dependencies between individual earnings. In the most simple case, one error component represents the heterogeneity of the individual micro-units and a second component gives the effects of unobserved factors that are uncorrelated with the heterogeneity component. Usually, earnings models of this type are only considered for individual persons. But, in principle, an error components model with two equations can also be specified for couples. In such a model, the error components may be correlated across the individuals. A correlation of the individual-specific heterogeneity component may result from selective matching, while the residual components may be correlated because the same factors affect the earnings of both partners.

Basically, a standard human capital approach has been used to model individual earnings. The log wage $w_{it}^{g}$ rate of an individual $g \in \{m, f\}$ observed in couple $i$ in period $t$ is modelled as a function of variables such as schooling, work experience, and dummy variables representing job characteristics. An error components structure is assumed for the error term $u_{it}^{g}$ with an individual-specific error component $\mu_{i}^{g}$ that is constant over time, and an independent, identically distributed residual component $\varepsilon_{it}^{g}$ that is uncorrelated with the individual component. Period-specific effects are modelled explicitly by including dummy variables corresponding to the year of the observation in the vector of explanatory variables. For an individual $i$ of gender $g$ being observed for $T_i$ periods, the earnings function is given by:

---

woman - and then searching the population for a man that is available for matching and has characteristics most close to the desired values as measured by a distance function.

[3] For an overview, see. Willis (1986) and Weiss (1986).

[4] In empirical analyses usually less than half of the variance of earnings can be explained.

$$w_{it}^g = x_{it}^{g\prime}\alpha^g + u_{it}^g = x_{it}^{g\prime}\alpha^g + \mu_i^g + \varepsilon_{it}^g \quad ; \quad g = m, f \; ; \quad t = 1, \ldots, T_i$$
$$E(\mu_i^g) = E(\varepsilon_{it}^g) = E(\mu_i^g \varepsilon_{it}^g) = E(\varepsilon_{it}^g \varepsilon_{i\tau}^g) = 0 \quad \textit{for } t \neq \tau \tag{1}$$

Both the individual-specific and the residual error components may be correlated across the equations for two individuals $m$ and $f$ forming a couple. But they are assumed to be independent between different couples in the sample. If the error terms of the two equations for a couple are stacked into a vector $u_{it}' = (u_{it}^m, u_{it}^f)' = (\mu_i^m, \mu_i^f)' + (\varepsilon_{it}^m, \varepsilon_{it}^f)' = \mu_i' + \varepsilon_{it}'$, the covariances are given by:

$$
\begin{aligned}
E(\mu_i \mu_i') &= \Sigma_\mu \\
E(\mu_i \varepsilon_{it}') &= 0 \quad \forall\, t \\
E(\varepsilon_{it} \varepsilon_{i\tau}) &= \begin{cases} \Sigma_\varepsilon & \textit{for } t = \tau \\ 0 & \textit{else} \end{cases} \\
E(u_{it} u_{i\tau}) &= \begin{cases} \Sigma_\mu + \Sigma_\varepsilon & \textit{for } t = \tau \\ \Sigma_\mu & \textit{else} \end{cases}
\end{aligned} \tag{2}
$$

When estimating the parameters of such relations from empirical data, attention must be paid to the problem of sample selectivity. It occurs since wages and working hours are observed only for individuals participating in the labour market. If the residuals of the wage equations are stochastically dependent on the decision to work or not, estimates that are derived from the subsample of working people without taking into account these dependencies will be biased. In principle, unbiased estimates can be obtained by adding relations to the model explaining labour force participation and jointly estimating the augmented model. However, since different patterns of labour force participation are observed over time in a panel, accounting for sample selectivity of that type is more complicated in panel data as compared to conventional cross-sections.[5]

Additional sample selectivity may occur due to panel attrition, since a part of the sample is not observed over the full period. If attrition is related to the same factors that enter into the wage equations, sample selection would be informative with regard to wages and biased estimates would result. However, previous analyses with the German Socioeconomic Panel indicate that attrition does not result in significant selectivity bias (Rendtel and Buechel, 1994). Therefore, panel attrition is not taken into account explicitly in the empirical analysis.

---

[5] The problem of sample selectivity in the context of panel data has been discussed by Verbeek and Nijman (1992a and b).

A third source of eventual sample selectivity is the formation of couples. In the sample a significant portion of individuals at a given point of time live as single persons without a partner. Most of these individuals live with a partner for some time during their life cycle, but the characteristics of that partner are not observed in the data. On the other hand, the correlation between individual earnings can only be estimated from couples with both partners working. Thus, partial observability of couples should also be taken into account.

In principle, a model can be specified that describes the dependencies between family formation, labour force participation and earnings. However, such an approach is not feasible due its complexity, at least if no restrictive simplifying assumptions are being imposed.[6] In addition, restricting the sample to two-earner couples would imply at least a substantial loss of degrees of freedom, since a large number of observations of single earner couples or of working singles would be excluded from the sample. Besides the loss of precision of the estimates this might also lead to biased estimates since, for instance, younger persons with high education would be excluded systematically from the sample.

Therefore a simplified approach has been used that rests on the sequential estimation of conditional models. In a first step, earnings functions are estimated for individual persons, allowing for sample selectivity related to individual labour force participation. These estimates are based on the observations for all persons, regardless of whether they are singles or live with a partner. Then, in a second step, the correlation structure is estimated from the observations of couples with both partners working. In principle, this is similar to a two-step estimation procedure in a seemingly unrelated regression problem using the single equation parameter estimates of the first stage to estimate the correlation structure in the second stage.

---

6 For a discussion of some aspects of the problem see the survey by Killingsworth and Heckman (1986). In the microsimulation model conditional independence has been assumed between family formation on the one hand and labour force participation and earnings on the other. The basic assumption is that labour force participation and earnings are not correlated with the latent factors in the demographic sub-model if the basic demographic variables like gender, age, marital status and education are controlled for. This may appear to be rather restrictive but, otherwise, the model structure would have become too complex. An additional reason is that in the demographic sub-model parameter values are used that have been estimated from aggregated data since the survey data available do not contain enough observations in order to estimate, for instance, mortality or fertility rates with sufficient precision. Even in a sample of about 15000 individuals only a few cases of birth or death are observed in a given year. This makes the joint estimation of a comprehensive model of both demographic and socio-economic processes not feasible.

A basic assumption is that, given the explanatory variables, the earnings equations are the same regardless of the living arrangement of the individuals. If this is true, and if exogeneity is assumed for the explanatory variables, the parameters of the earnings functions can be estimated from data on individuals in the sample. This assumption may appear questionable. However, findings of other studies on earnings functions indicate that the single equation bias should not be large, at least with regard to variables like education, work experience and family characteristics.[7]

Since in the first step data on all individuals are used independent of whether they live with a partner or not, and panel attrition is assumed to have no effect, only sample selectivity due to labour force participation and unemployment must be considered. For this purpose, probit equations modelling the probability of observing the individual wage rate have been specified for both men and women. As the sample has been restricted to persons who have left the education system and who are not receiving a pension, practically all men are in the labour force. Basically, the wage rate is not observed only for unemployed men. Thus, the probit equation for men gives the probability of a man not being unemployed. For women, on the other hand, both the states of being out of the labour force and of being unemployed are important. But in order to simplify matters, both states have been collapsed into one state of non-working, with the probit equation giving the probability of a women being in the labour force and working.[8]

In analogy to the wage equations, an error components structure can also be assumed for the probit equations predicting employment with individual-specific error components $\gamma_i^g$ and residual components $\eta_{it}^g$. If dummy variables $d_{it}^g$ are used to indicate whether the individual is observed to work ($d_{it}^g = 1$), the random effects probit equations have the following form with explanatory variables $x_{it}^g$ and parameter vector $\beta^g$:

$$
\begin{aligned}
z_{it}^g &= x_{it}^g \beta^g + v_{it}^g = x_{it}^g \beta^g + \gamma_i^g + \eta_{it}^g \\
d_{it}^g &= \begin{cases} 1 & if\ z_{it}^g \geq 0 \\ 0 & else \end{cases} \qquad (3) \\
E(\gamma_i^g) &= E(\eta_{it}^g) = E(\gamma_i^g \eta_{it}^g) = E(\eta_{it}^g \eta_{i\tau}^g) = 0 \quad for\ t \neq \tau
\end{aligned}
$$

---

[7] Eg. Mincer and Polachek (1978); Nakamura et al (1979).

[8] An additional argument in favour of such an approach is that unemployment is to some extent a transitory state of women who actually leave the labour force but collect unemployment benefits as long as they are eligible. Thus in practice it is difficult to make a clear distinction for women between being voluntarily out of the labour force and involuntary unemployment.

It seems to be plausible that the error components of the probit equations and the equations for the wage rates may be correlated. First, unobserved individual characteristics may affect both the probability of being in work and the wage rate received. The same is true for short run random effects that have an effect only in a single period. On the other hand, if period-specific effects are controlled by dummy variables, the assumption of stochastic independence of different couples seems to be adequate. Collecting the error components of the probit equations in analogy to the wage equations into the vectors $\gamma_i$ and $\eta_{it}$, the following covariance structure is assumed for a couple, allowing for a correlation of the individual-specific and of the residual error components respectively:

$$\begin{aligned} &E(\mu_i \gamma_i') = \Sigma_{\mu\gamma} \\ &E(\mu_i \eta_{it}') = E(\gamma_i \varepsilon_{it}') = 0 \quad \forall\, t \\ &E(\varepsilon_{it} \eta_{i\tau}') = \begin{cases} \Sigma_{\varepsilon\eta} & if\ t = \tau \\ 0 & else \end{cases} \end{aligned} \tag{4}$$

The complete model defines a system of four equations for a couple, or two equations for each individual. However, if only a single person without a partner is observed, only the marginal information corresponding to two equations is available. However, a simple relation is obtained between the complete model and the two marginal models if a joint multivariate normal distribution is assumed for the error components. In this case, the system can be split into two sets of two marginal equations for each individual, with the covariance structure given by the corresponding part of the full covariance matrices. These marginal models can be estimated separately from data on individuals.

If empirical estimates are available for the parameters of the model, including the covariance structure of the error components, microsimulation of the earnings is straightforward. Basically, two steps of the simulation model are affected. First, in the matching procedure that is used to simulate the marriage market, the correlation between the individual-specific error components is taken into account by including the error component into the vector of characteristics that is used to match the individuals. Each individual in the simulation is assigned at random a value of the specific error component. Then, in the simulation of the marriage market, for women the 'desired' value of the error component of the partner is generated at random from the conditional normal distribution, besides the other characteristics. Finally a man meeting these requirements is selected from the candidates available in the sample.

Given the individual error components for a couple, the simulation of the individual wage rates in a given period poses no problems. First, the conditional expectation of the log wage rate is computed for each individual, based on the values of the explanatory variables and the parameter estimates. Then the individual specific error component is added. Finally, the value of the residual component is generated by a random drawing from the bivariate normal distribution of the residual error components. If only one person in a couple is working, the marginal distribution is used to generate the residual component.

## 3. First Parameter Estimates

The parameters of the model have been estimated using data from the first eight waves of the German Socioeconomic Panel Study. For the estimates being considered here, the sample has been restricted to (West) German nationals of age 20 to 60 who have finished education and who do not receive a pension. This might result in some additional sample selection bias if the propensity to stay in the education system or to retire is related to the error term in the wage equation. However, to avoid additional complications, this has not been taken into account. Self-employed persons have been included into the sample because the same wage equation is used in the model for self-employed and employed persons. The wage rates have been computed by dividing gross income received in the last month by the number of hours actually worked. No additions have been made for extra payments.

In principle, the parameters of such a model can be estimated by maximum likelihood methods. As an alternative, a generalised two step method of the Heckman type can be used to derive unbiased estimates (Verbeek and Nijman, 1992). However, for both approaches integrals of the multivariate normal density must be evaluated numerically. This proved to be impractical for the full model, given the size of the data set and the number of parameters to be estimated.[9] Therefore, a simplified approach resting on the estimation of marginal models was adopted. This implies a loss of efficiency. However, given the large number of observations this should not be too important.

---

[9] While the maximum likelihood estimator performed well for a small problem using simulated data, the estimation of the earnings model with a comparatively large number of parameters and a large number of observations was not feasible due to numerical problems. Since only a limited number of points can be used in the numerical integration procedure, the derivatives of the likelihood functions can only be derived with limited precision. Apparently due to this lack of precision the optimisation procedure did not converge. Eventually, this problem can be overcome by improved numerical methods.

The estimation problem can be simplified in different ways. A first possibility is to use a two step approach, considering in the first step only the marginal model for individuals and estimating the covariances of the residuals across partners in a second step. In addition to simplifying the estimation problem, this has the advantage that data on single persons can be included in the estimation without problems. A second approach is to ignore the longitudinal character of the data and to consider the marginal model for different periods. Given the model structure, the parameters can be estimated consistently from such cross-sections. This results in a loss of efficiency but, on the other hand, standard estimation procedures are available for cross-sections. Thus, a estimation strategy might be to first estimate the model from the different cross-sections and then to derive a combined estimate based on the estimates for all different periods using a minimum distance estimator.

In the estimation process, the different approaches have been combined. First, cross-sectional models allowing for sample selectivity have been estimated for men and women respectively for each year from 1984 to 1991, in order to test for eventual selectivity bias. While for men unemployment seems to imply no significant sample selection bias, significant effects have been obtained for women in a two-step Heckman procedure.

Given these results, the wage equation for men has been estimated by generalised least squares using the panel observations on men who have been in the sample for at least two periods. As a different number of periods was observed for different individuals, an estimator for an unbalanced panel design was applied.[10] The equation represents a standard human capital model with work experience disaggregated into the duration of the last spell of work and the duration of previous spells. Time specific effects are modelled explicitly by dummy variables for the different years. The results are presented in Table 1 (See Appendix A for variable definitions).

The parameter estimates do not vary very much between the simple pooled least squares and the feasible generalised least squares regression and are in line with earlier results from German data. Civil servants and public employees receive somewhat lower wage rates, in exchange for a lower unemployment risk. The wage rates of self-employed men are somewhat lower due to the higher number of working hours. The dummy variables for the different years capture wage increases over time.

---

[10] The GAUSS procedure TSCS has been used for estimation.

Table 1: Regression Estimates for the Wage Function of Men 1984-1991

| Variable | Pooled OLS regression Coefficient | Pooled OLS regression Std. Error. | GLS-panel-model Coefficient | GLS-panel-model Std. Error |
|---|---|---|---|---|
| Constant | 3.434158 | 0.025300 | 3.517092 | 0.144089 |
| M_educ | 0.075522 | 0.001576 | 0.067873 | 0.010752 |
| M_public | -0.037488 | 0.011353 | -0.049791 | 0.022536 |
| M_civil | -0.109721 | 0.011676 | -0.104627 | 0.035344 |
| M_self | -0.192667 | 0.015444 | 0.077207 | 0.029237 |
| M_exp_las | 0.028870 | 0.001304 | 0.023379 | 0.003274 |
| M_exp_las^2 | -0.000523 | 0.000032 | -0.000394 | 0.000061 |
| M_exp_prior | 0.013690 | 0.002138 | 0.017571 | 0.005570 |
| M_exp_prior^2 | -0.000314 | 0.000082 | -0.000345 | 0.000173 |
| M_un_tot | -0.024844 | 0.006698 | 0.018988 | 0.021209 |
| M_training | 0.002721 | 0.002138 | 0.003912 | 0.011703 |
| Year_84 | -0.207882 | 0.016356 | -0.212451 | 0.020055 |
| Year_85 | -0.192106 | 0.016144 | -0.199744 | 0.018153 |
| Year_86 | -0.166451 | 0.016231 | -0.173065 | 0.016472 |
| Year_87 | -0.153082 | 0.016221 | -0.155220 | 0.014940 |
| Year_88 | -0.092029 | 0.016374 | -0.099863 | 0.013710 |
| Year_89 | -0.075636 | 0.016449 | -0.080925 | 0.012716 |
| Year_90 | -0.022631 | 0.016655 | -0.024790 | 0.012104 |
| N. of obs. | 12767 | | 12767 | |
| N. of groups | 2451 | | 2451 | |
| $R^2$ | 0.228 | | | |
| R-bar$^2$ | 0.226 | | | |
| F-Value | 220.882 | | 1489.936 | |
| dgf.: f1,f2 | 18,12749 | | 18,12749 | |
| Lagrange Chi-2(1 dgf) | | | 6922.5366 | |
| Hausman Chi-2 (17 dgf) | | | 13.1445 | |

Data base: German Socioeconomic Panel waves 1-8. Variables are defined in Appendix A.

For women, a similar panel data model could not be applied directly since sample selectivity had to be taken into account. Attempts to estimate a panel data model with sample selectivity by maximum likelihood methods (Verbeek and Nijman 1992a) proved to be too difficult due to numerical problems. Therefore, a two-step Heckman type procedure was adopted. First a random effects probit model was estimated for labour force participation. Then, in a second step, an augmented wage equation was estimated by pooled OLS with two additional regressors accounting for the conditional expectation of the residual of the wage equation given that the woman works. This is a direct extension of the standard Heckman method for cross-sections to the case of panel data.

Basically, the Heckman method consists of including in the wage equation as an additional regressor the conditional expectation of the residual, given that the wage is actually observed. The residual of the augmented equation has zero expectation and is independent of the regressors. Thus, unbiased estimates are obtained.

However, the residual is heteroscedastic and, as a consequence, the estimates are not fully efficient. However, for a large sample, the loss of efficiency should not be very important.

For a panel data model with a multivariate normal error component structure, the conditional expectation of the error term in the wage equation can be derived from the random effects probit equation using standard results for the conditional moments of the truncated multivariate normal distribution by Johnson and Kotz (1972) and McGill (1992) respectively. After some algebraic manipulations, the following expression can be obtained for the conditional expectation of the combined error term $u_{it}^f$ in the wage equation:

$$\begin{aligned} E\left(u_{it}^f \middle| d_i\right) &= \sigma_{\varepsilon\eta}^f \lambda_{it} + \sigma_{\mu\gamma}^f \overline{\lambda}_i \\ with & \\ d_i &= (d_{i1}^f, \ldots, d_{iT}^f) \\ \overline{\lambda}_i &= \sum_t \lambda_{it} \\ \lambda_{it} &= (2d_{it}^f - 1)\frac{\partial}{\partial z_{it}} ln\Phi\left(diag(2d_{it}^f - 1)X_i^f \beta^f\right) \end{aligned} \tag{5}$$

Here $\lambda_{it}$ is the element of the hazard gradient of the multivariate normal distribution of the error terms of the random effects probit equation corresponding to period $t$ for observation $i$. This is a direct generalisation of the inverse Mill's ratio that is used as an additional regressor in the standard Heckman procedure for cross-sectional models with sample selectivity. However, for panel data models, in addition to the hazard rate for the current period, the sum of the hazards for all periods is also used as an additional regressor. Thus, the equation to be estimated on the second step has the following form:

$$\begin{aligned} w_{it}^f &= x_{it}^f \alpha^f + \sigma_{\varepsilon\eta}^f \lambda_{it} + \sigma_{\mu\gamma}^f \overline{\lambda}_i + \Delta\mu_i^f + \Delta\varepsilon_{it}^f \\ &= x_{it}^f \alpha^f + \sigma_{\varepsilon\eta}^f \lambda_{it} + \sigma_{\mu\gamma}^f \overline{\lambda}_i + \Delta u_{it}^f \end{aligned} \tag{6}$$

Basically, this is again a random effects panel data model. However, the error components are heteroscedastic and auto-correlated. Therefore, the model cannot be estimated by standard procedures for panel data models. A simple solution to this problem is to estimate the augmented model by ordinary least squares, using the pooled sample of all observations. This estimate can easily be obtained and is consistent. It is, however, inefficient but, given the size of the sample, inefficiency is not a major problem.

Parameter estimates for both the probit model and the wage equation are presented in Table 2. The specification of both equations is more or less standard. In the wage equation work experience is measured in detail by discriminating between full-time and part-time work and between work experience in the last spell and that in previous spells. In addition, the duration of full-time work in the last spell is measured separately for women with continuous work histories and women who experienced interruptions. As for men, job characteristics are also used as explanatory variables. The regressors in the probit equation include variables that define the earnings capacity of the women (ie. education and work experience), family characteristics that affect the marginal utility of time use and, for women with a partner, a set of variables that describe the earnings capacity of the man.

Again, the estimates are in line with earlier results. Labour force participation and the wage rate of women are both positively related to education and work experience. The estimated rate of return on education is approximately the same for women as for men. With regard to work experience, the wage profile of women shows a stronger curvature than that of men: the larger coefficients of the linear and the quadratic experience variables imply that the growth rate of wages is higher for women in the first years of work but the growth rate reduces more rapidly than for men. While the effect of being self-employed is approximately the same for men and women, working in the public sector has a positive effect on the female wage rate as compared to a negative effect for men. This can be explained by the fact that in the public sector, in principle, the same wage is paid for a given job independent of gender, while in the private sector women are discriminated against to some extent.

Marriage has a stronger negative effect on labour force participation than living in a consensual union. As expected, children, and especially children younger than three years, have a negative impact on labour force participation. In addition, the wage rate is negatively related to the number of children in the household. An explanation might be that women with (minor) children are more restricted in their choice of jobs. Besides this, unobserved individual specific factors play a major role, at least for female labour force participation. The (significant) estimate for Theta implies that 57 per cent of the variance of the error term of the probit equation can be attributed to individual specific effects. For the wage equation, such an estimate cannot be derived directly, since only a pooled regression has been performed for the subsample of working women. The estimates for $\lambda$ and $\overline{\lambda}$ imply a small positive correlation between the individual-specific error components of the probit model and the wage equation, while the correlation of the residual components is insignificant.

Table 2: Parameter Estimates for Women, 1984-1991

| | Random Effects Probit Equation | | Wage equation (Pooled OLS) | |
|---|---|---|---|---|
| Variable | Coefficient | Std. Error | Coefficient | Std. Error |
| Constant | 1.2976 | 0.1547 | 3.0817 | 0.0381 |
| F_age | -0.0607 | 0.0033 | | |
| F_educ | 0.1097 | 0.0107 | 0.0790 | 0.0027 |
| F_full_perm | 0.0941 | 0.0039 | 0.0401 | 0.0024 |
| F_full_perm^2 | | | -0.0008 | 0.0001 |
| F_full_inter | 0.0548 | 0.0056 | 0.0177 | 0.0054 |
| F_full_inter^2 | | | -0.0004 | 0.0002 |
| F_full_prior | 0.0516 | 0.0036 | 0.0168 | 0.0022 |
| F_full_prior^2 | | | -0.0005 | 0.0001 |
| F_full_5 | 0.0210 | 0.0025 | 0.0041 | 0.0014 |
| F_part_last | 0.1103 | 0.0046 | 0.0064 | 0.0013 |
| F_part_prior | 0.0507 | 0.0054 | -0.0003 | 0.0014 |
| F_un_last | -0.2870 | 0.0205 | -0.0245 | 0.0171 |
| F_un_before | 0.0055 | 0.0202 | | |
| F_no | -0.1594 | 0.0043 | -0.0205 | 0.0072 |
| F_no_before | 0.0439 | 0.0041 | | |
| F_training | 0.0295 | 0.0228 | 0.0164 | 0.0046 |
| F_public | | | 0.1177 | 0.0123 |
| F_civil | | | 0.0738 | 0.0255 |
| F_self | | | -0.2053 | 0.0266 |
| Marriage | -0.3410 | 0.0576 | | |
| Partner | -0.1494 | 0.0768 | | |
| Children | -0.0778 | 0.0213 | -0.0301 | 0.0073 |
| Child_03 | -1.2363 | 0.0489 | | |
| Child_06 | -0.4012 | 0.0571 | | |
| M_unemp | -0.6508 | 0.3079 | | |
| M_working | 0.2265 | 0.1497 | | |
| M_educ | -0.0185 | 0.0110 | | |
| M_public | 0.0322 | 0.0761 | | |
| M_civil | 0.2234 | 0.0756 | | |
| M_self | 0.3657 | 0.0792 | | |
| M_exp_last | 0.0028 | 0.0027 | | |
| M_exp_prior | -0.0011 | 0.0052 | | |
| M_un_tot | 0.0225 | 0.0400 | | |
| M_training | -0.0184 | 0.0129 | | |
| Year_84 | -0.1898 | 0.0574 | -0.2439 | 0.0210 |
| Year_85 | -0.1244 | 0.0545 | -0.1970 | 0.0209 |
| Year_86 | -0.1724 | 0.0537 | -0.1705 | 0.0210 |
| Year_87 | -0.1291 | 0.0554 | -0.1559 | 0.0209 |
| Year_88 | -0.1853 | 0.0541 | -0.1154 | 0.0213 |
| Year_89 | -0.1023 | 0.0549 | -0.0848 | 0.0213 |
| Year_90 | -0.0119 | 0.0582 | -0.0184 | 0.0212 |
| Theta | 1.6382 | 0.0558 | | |
| Lambda-Bar | | | 0.0803 | 0.0083 |
| Lamda | | | 0.0065 | 0.0184 |
| Observations | 15836 | | 8756 | |
| LogLik. | -3978.36 | | | |
| LogLik. pooled | -4449.89 | | | |
| Chi-2 (1) | 943.06 | | | |
| r2 | | | 0.23 | |
| r-bar2 | | | 0.22 | |
| F(26,8729) | | | 98.09 | |

Data base: German Socioeconomic Panel Study, waves 1-8. Variables defined in Appendix A.

# 4. An Estimation Procedure for the Covariance Structure

Besides inefficiency, a disadvantage of the estimation procedure is that no direct estimates are available for the covariance structure of the error components. From the residuals of the random effects model, the (marginal) variances of the individual specific and of the residual error component of the wage equation for men can be estimated by moments methods (Mátyás, 1992, p. 56). But, for women, residuals are only available for the subsample of working women. As a consequence, the variances of the error components of the wage equation for women and the covariances between the error components of the male and female wage equation must be estimated from this subsample.

For the purpose of microsimulation, the conditional error structure for couples is of most interest, since the correlation between the individual-specific error components should enter into the matching process independent of whether the individuals are working or not. Thus, the conditional covariance structure for couples must be estimated that is marginal to labor force participation. In principle these estimates can be obtained from the sample of all couples. However, estimation is complicated by the effects of selective labour force participation, since the conditional covariance structure in the observed subsample of double earners in general will differ from the marginal covariance structure for all couples.

Starting from the conditional covariance matrix of the residuals in the subsample of double earner couples, an estimator for the covariance structure can be obtained. If the residuals of the male and female wage equation for all couples are stacked into a vector $u' = (u^{m}{}', u^{f}) = (u_{11}^{m}, \ldots, u_{nT_n}^{m}, u_{11}^{f}, \ldots, u_{nT_n}^{f})$, the conditional covariance matrix can be derived using results by Johnson and Kotz (1972) and by McGill (1992) as:

$$\begin{aligned} V(u|d) &= V(u) - Cov(u,v)\Lambda Cov(v.u) \\ &= (\Sigma_{\mu\mu} \otimes diag(J_i) + \Sigma_{\varepsilon\varepsilon} \otimes diag(I_i)) \\ &\quad - (\Sigma_{\mu\gamma} \otimes diag(\tilde{J}_i) + \Sigma_{\varepsilon\eta} \otimes diag(\tilde{I}_i)) diag(\Lambda_i)(\Sigma_{\mu\gamma} \otimes diag(\tilde{J}_i) + \Sigma_{\varepsilon\eta} \otimes diag(\tilde{I}_i))' \end{aligned} \tag{7}$$

Here, $\Sigma_{\mu\mu}, \Sigma_{\varepsilon\varepsilon}, \Sigma_{\mu\gamma}, \Sigma_{\varepsilon\eta}$ are the variance-covariance matrices of the error components, $\Lambda = diag(\Lambda_i)$ is the block-diagonal matrix of the hazard gradient of the probit model for all observations, $J_i, \tilde{J}_i$ are matrices of appropriate dimension with all elements equal to one and $I_i, \tilde{I}_i$ are unit matrices.

The matrix $Q = I_2 \otimes diag(I_i - 1/T_i J_i) = I_2 \otimes diag(Q_i)$ can be used to perform the so-called within-transformation of the residuals, ie. to subtract the mean of the residuals over the observations for a couple from the residuals for the different periods. As $Q_iJ_i = 0$ and $Q_i\tilde{J}_i = 0$, the variance-covariance matrix of the transformed residuals $\tilde{u} = Qu$ simplifies to:

$$\begin{aligned} V(\tilde{u}|d) &= (\Sigma_{\varepsilon\varepsilon} \otimes diag(Q_i)) \\ &\quad - (\Sigma_{\varepsilon\eta} \otimes diag(Q_i\tilde{I}_i))diag(\Lambda_i)(\Sigma_{\varepsilon\eta} \otimes diag(Q_i\tilde{I}_i))' \\ &= (\Sigma_{\varepsilon\varepsilon} \otimes diag(Q_i)) - (\Sigma_{\varepsilon\eta}\Sigma'_{\varepsilon\eta}) \otimes diag(Q_i\tilde{I}\Lambda_{ii}\tilde{I}'Q_i) \end{aligned} \tag{8}$$

Now consider the sum of cross-products of the transformed residuals, $\tilde{u}^{g\prime}\tilde{u}^k = \sum \tilde{u}_{it}^g\tilde{u}_{it}^k$. Its expectation is given by:

$$\begin{aligned} E(\tilde{u}^{g\prime}\tilde{u}^k|d) &= trace\left\{E(\tilde{u}^k\tilde{u}^{g\prime}|d)\right\} \\ &= \sigma_{\varepsilon\varepsilon}^{gk} trace\{diag(Q_i)\} - \sigma_{\varepsilon\eta}^{g}\sigma_{\varepsilon\eta}^{k} trace\left\{diag(Q_i\tilde{I}_i\Lambda_i\tilde{I}_iQ_i)\right\} \end{aligned} \tag{9}$$

In this expression the expected value on the left side can be estimated consistently from the regression residuals by the corresponding sum of cross-products. An estimate for the variance and the covariance of the residual error component in the female wage equation with the residual error in the probit model is available from the parameter estimates. The corresponding covariance for the male wage equation can be obtained from a auxiliary regression, $u_{it}^m = \sigma_{\mu\gamma}^m\bar{\lambda}_i + \sigma_{\varepsilon\eta}^m\lambda_{it} + \zeta_{it}$, of the residual of the wage equation on the hazard rate and the sum of the hazards, that is obtained from the probit model. Given these estimates, the value of $\sigma_{\varepsilon\varepsilon}^{gk}$ can be estimated from (9) as:

$$\hat{\sigma}_{\varepsilon\varepsilon}^{gk} = \frac{1}{\Sigma(T_i - 1)}\left[\tilde{u}^{g\prime}\tilde{u}^k + \hat{\sigma}_{\varepsilon\eta}^g\hat{\sigma}_{\varepsilon\eta}^k \sum_i trace\{Q_i\tilde{I}_i\Lambda_i\tilde{I}_i'\}\right] \tag{10}$$

In a similar way, estimates for the variance-covariance structure $\Sigma_{\mu\mu}$ of the individual-specific error components can be obtained from expression (7). The expectation of the sum of cross-products of the combined regression residuals, $u^{g\prime}u^k = \sum u_{it}^g u_{it}^k$, is given by:

$$\begin{aligned} E(u^{g\prime}u^k|d) &= (\sigma_{\mu\mu}^{gk} + \sigma_{\varepsilon\varepsilon}^{gk})\sum_i T_i \\ &\quad - trace\left\{(\sigma_{\mu\gamma}^g diag(\tilde{J}_i) + \sigma_{\varepsilon\eta}^g diag(\tilde{I}_i))diag(\Lambda_i)(\sigma_{\mu\gamma}^g diag(\tilde{J}_i) + \sigma_{\varepsilon\eta}^g diag(\tilde{I}_i))'\right\} \end{aligned} \tag{11}$$

Again, the term on the left hand can be estimated by the cross-products of the residuals. Solving for $\sigma_{\mu\mu}^{gk}$ the covariances can be estimated by:

$$\hat{\sigma}_{\mu\mu}^{gk} = \frac{1}{\Sigma T_i}\left[u^{g\prime}u^k - \sum_i\left\{\hat{\sigma}_{\varepsilon\varepsilon}^{gk}T_i + trace\left((\hat{\sigma}_{\mu\gamma}^g\tilde{J}_i + \hat{\sigma}_{\varepsilon\eta}^g\tilde{I}_i)\Lambda_i(\hat{\sigma}_{\mu\gamma}^k\tilde{J}_i' + \hat{\sigma}_{\varepsilon\eta}^k\tilde{I}_i)\right)\right\}\right] \tag{12}$$

In Table 3, estimates for the covariances and correlations of the error components of the wage equations for couples are presented. A positive correlation is estimated for both the individual specific and the residual error components of the wage equation.

Table 3: Estimated Covariances and Correlation Coefficients of the Error Components Across Individuals Forming a Couple*

| | Man | | Woman | |
|---|---|---|---|---|
| | $\mu^m$ | $\varepsilon^m$ | $\mu^f$ | $\varepsilon^f$ |
| $\mu^m$ | 0.0761 | - | 0.0263 | - |
| $\varepsilon^m$ | - | 0.0526 | - | 0.0090 |
| $\mu^f$ | 0.2735 | - | 0.1218 | - |
| $\varepsilon^f$ | - | 0.1533 | - | 0.0657 |

*Variances/Covariances in Upper Triangular Part, Correlations in Lower Triangular Part

The estimates appear to be reasonable. A moderate positive correlation results for the individual specific components and the residual error components across the individuals. However, since no standard errors have been computed for these estimates, it is difficult to assess the significance of these results.

From a substantive point of view, these results imply that, on average, the marriage markets results in positive assortive matching, in the sense that men with high earnings capacity tend to marry women who also have a high earnings capacity and vice versa. Apparently, the hypothesis of specialisation of spouses, that can be derived from simple neoclassical arguments and that would imply a negative correlation of earnings capacities, is not supported by the data. Probably preferences for partners with a similar education and social factors, like the social groups where people look for partners, are more important than such simple efficiency arguments.

## 5. Conclusions

A positive correlation of the error components across individuals forming a couple implies that a microsimulation approach based on the assumption of conditional independence of the wage rates of partners in a couple tends to underestimate the dispersion of family income. Since earnings depend also on the number of hours worked, this effect may be compensated to some extent if there is a negative correlation between wages and labour supply - but probably the effects will not cancel. Since such a correlation has important consequences especially for analyses of the personal income distribution, one should try to include such stochastic dependencies into microsimulation models.

However, experiences with the estimates that have been presented here show that some efforts should be made to improve the estimation procedures. In principle, models of this type could be estimated by maximum likelihood methods even if more than two wage equations are considered. But this requires improvements in the numerical integration techniques used in the probit model. Estimators based on a multi-step approach of the Heckman type are simpler to compute. However the estimation of standard errors of the estimates poses severe problems. Nonetheless, some progress in this direction should allow substantial improvements in simulation models.

# Appendix A

### Definition of the Variables in the Regression

| | | | |
|---|---|---|---|
| F_...,M_... | Variables for women and men respectively | F_un_last | Years of unemployment in last spell |
| F_age | Age (years) | F_un_before | Years of unemployment before last speil |
| F_educ/M_educ | Years of schooling (institutional) | M_un_tot | Total number of years of unemployment |
| M_exp_last | Years of work in last spell | F_no | Years not working in last spell |
| M_exo_last^2 | M_exp_last squared | F_no_before | Years not working during interruptions before last spell |
| M_exp_prior | Years of work before last spell | F_training/M_training | Years of training after end of education |
| M_exp_prior^2 | M_exp_prior squared | F_public/M_public | Dummy: Other public employee |
| F_full_perm | Years of full-time work in last spell, no interruptions in work biography | F_civil/M_civil | Dummy: Civil servant |
| F_full_perm^2 | F_full_perm squared | F_self/M_self | Dummy: Self employed |
| F_full_inter | Years of full-time work in last spell, interruptions in work biography | Marriage | Dummy: married with spouse present |
| F_full_inter^2 | F_full_inter squared | partner | Dummy: Consensual union with partner present |
| F_full_prior | Years of full-time work before last spell | Children | Number of children under 16 years |
| F_full_prior^2 | F_full_prior squared | Child_03 | Dummy: youngest child 0-3 years |
| F_full*age | Interaction: (Age-F_full_inter)*F_full_inter | Child_06 | Dummy: youngest child 3-6 years |
| F_full_5 | Rebound-effect: 10*min(5,F_full_inter)-min(5,F_full_inter)^2 | M_unemp | Dummy: currently unemployed |
| F_part_last | Years of part-time work in last spell | M_working | Dummy: currently in labour force |
| F_part_prior | Years of part-time work before last spell | Year_yy | Dummy: Year equal to yy |

## References


Butler, J.S. and R. Moffitt (1982), 'A computationally efficient quadrature procedure for the one factor multinomial probit model', Econometrica, Vol. 50, No. 3.

Chamberlain, G. (1982), 'Multivariate regression models for panel data', *Journal of Econometrics*, Vol. 18, No. 1.

Greene, W. (1993), *Econometric Analysis*, 2nd edition, Macmillan Publishing Company, New York.

Johnson, N.L. and S. Kotz (1972), *Distributions in Statistics: Continuous Multivariate Distributions*, John Wiley and Sons, New York.

Judge, G.G., C. Hill, W. Griffith, and T. Lee, (1985), *The Theory and Practice of Econometrics*, 2nd edition, John Wiley and Sons, New York.

Killingsworth, M. and J. Heckman (1986), 'Female labour supply: A survey', in: O. Ashenfelter and R. Layard (eds.), *Handbook of Labour Economics, Vol. I*, Elsevier Science Publishers, Amsterdam.

Krupp, H.J., H.P. Galler, H. Grohmann, R. Hauser, and G. Wagner, (1981), *Alternativen der Rentenreform '84'*, Campus, Frankfurt.

Mátyás, L. (1992), 'Error components models', in: L. Mátyás and P. Sevestre (eds.), *The Econometrics of Panel Data, Handbook and Applications*, Kluwer Academic Publishers, Dordrecht.

McGill, J.I. (1992), 'The multivariate hazard gradient and moments of the truncated normal distribution', *Communications in Statistics - Theory and Methods*, Vol. 21, No. 11.

Mincer, J. and S. Polachek (1978), 'Women's earnings re-examined', *Journal of Human Resources*, Vol. 13, No. 2.

Nakamura, M., A. Nakamura and D. Cullen (1979), 'Job opportunities, the offered wage, and the labour supply of married women', *American Economic Review*, Vol. 69, No. 5.

Rendtel, U. and F. Buechel (1994), Tests for Non-Ignorable Panel Attrition and their Application on Wage Estimation from the German Socioeconomic Panel, Discussion Paper No. 89, Deutsches Institut fuer Wirtschaftsforschung, Berlin

Verbeek, M. and Th. Nijman (1992a), 'Incomplete panels and selection bias', in: L. Mátyás and P. Sevestre (eds.), *The Econometrics of Panel Data, Handbook and Applications*, Kluwer Academic Publishers, Dordrecht.

Verbeek, M. and Th. Nijman (1992b), 'Testing for selectivity bias in panel data models', *International Economic Review.*, Vol. 33, No. 3.

Weiss, Y. (1986), 'The determinants of life cycle earnings: A survey', in: O. Ashenfelter and R. Layard, (eds.), Handbook of Labour Economics, Vol. I, Elsevier Science Publishers, Amsterdam, 1986.

Willis, R.J. (1986), 'Wage determinants: A survey and reinterpretation of human capital earnings functions', in: O. Ashenfelter and R. Layard (eds.), *Handbook of Labour Economics, Vol. I*, Elsevier Science Publishers, Amsterdam.

Bengt Eklind
Ingemar Eriksson
Joakim Hussénius
Mikael Müller

Ministry of Finance
Sweden

# Chapter 14: Pension Analysis in a Static Model With Lifetime Income Distribution: Initial Results


### Abstract

*Initial results from a static model alternative to dynamic modelling of future old age pensions are presented in this chapter. A cross-section tax-and-benefit model population is expanded by adding a maximum 30 years of lifetime income, using exact matching from a public pension data file. The historical individual income profiles for the working population are projected to pension age, using an internal statistical matching procedure. These lifetime income profiles, with actual and simulated income, are then donated to the pensioners in the cross section model population. Future demographic structure and macroeconomic changes are introduced through conventional reweighting and multipliers. This new hypothetical cross-section income distribution for any future year allows full scale simultaneous simulation of taxes, benefits and public pension program rules producing net cost estimates, distributional effects and comparisons between the working age population and old age pensioners.*


## 1. Introduction

Expenditures upon public pensions have been a subject of intense discussion in most western countries in recent years. The problems are common: demographic pressure, program maturation and budget and financing constraints because of slow economic growth and high unemployment. Many governments and other institutions have invested substantial efforts in searching for reliable projections of future pension costs and financial burdens (Hancock et al., 1992).

Since the late 1970s, Sweden has had a dynamic pension projection model. This 'ATP' projection model is developed and run by the National Social Insurance Board (NSIB). The model projects future gross expenditure, with different assumptions about economic growth, death risks etc. It has recently been re-designed to allow microsimulation of policy changes. Connected, but not integrated, are other models that calculate funding and trends for other benefits.

The model projections have shown that program maturation and demographic change will increase the need for payroll taxes to the public supplementary pension program from 14.2 per cent to 20.8 per cent in the next 20 years, using historically reasonable macroeconomic assumptions (1.5 per cent growth).

However, the expansion of microsimulation tax-and-benefit models within government, to estimate program costs and the effects of program changes, has promoted a shift towards measuring *government net spending,* rather than gross expenditures. Consequently, pension expenditure and the distribution between the elderly and the working population should be measured after-tax and with regards to the interaction with income tested programs.

The closer a change in public supplementary pensions, the more wide-ranging are the questions asked in the policy-making process. The policy experiments must allow the testing of future design of minimum pensions and other programs for income guarantees. The experiments should also cover combinations of changes in pensions and taxes. The distributional effects of proposed new pension systems should be studied, both within pensioners and among the total population. The incentive effects of alternative methods of financing and delivering benefits must be measured (eg. lifetime marginal effects on earnings and savings).

The costs and time spent by many other governments to improve pension modelling is impressive but the urgent need for pension reform in Sweden does not allow time for such projects. Therefore, in order to improve distribution, incentive and net expenditure analysis, we are exploring and testing methods in our tax-and-benefit model to supplement the ATP-projection model. This chapter presents the first results of this project.

## 2. Choices in Long-term Pension Modelling

The literature on pension modelling specifies a number of decisions that must be taken before designing a model. Since the policy issues cover both expenditure, lifetime income distribution, and incentive effects, one basic conclusion is that any comprehensive model must generate complete individual life histories. The present ATP-projection model does not include rules for taxes, housing allowances etc.

There are no households in the model and at the moment it cannot calculate net pensions and disposable income.

As noted by many others (Eriksen, 1973; Ståhlberg, 1990; Merz, 1991; Harding, 1993; Pudney, 1993) long-term pension modelling can be based either on actual lifetime earning profiles or on synthetic estimated profiles. Most modellers must rely on completely synthetic earning profiles, due to the lack of longitudinal income history registration.

As described by Merz (1991), Harding (1993) and Pudney (1993), most existing micro-level dynamic simulation models concentrate heavily on simulating the development of the sample individuals, relegating the weighting system to a minor role. Monte Carlo methods seem to dominate in the generation of life events and earning profiles. The advantages and problems of probabilistic methods are well documented, especially when it comes to validation and simulation stability.

The policy issues in Sweden, as well as cost and time constraints, suggest that alternative model strategies should be examined. Policy analyses of public pensions seldom just include one part of the system. In Sweden, a program change usually includes many elements, such as supplementary pensions, minimum guarantees, tax allowances and means-tested benefits. Consequently, the 'Burden of Pensions' on the working population is by definition a result of the combined impact of many factors.

The following tentative conclusions emerge from both the methodological discussions of microsimulation and the public pension debate:

- any policy analysis on pensions would focus not only on long term future outcomes but also on transitional problems. Thus, the model must include acquired pension rights, their transformation to new systems, as well as simulated future earnings and pension eligibility;
- in order to allow analyses of the 'Burden of Pensions' and 'Net Financing Requirements in Pay-As-You-Go pensions, the model should include successive complete income distributions (including non-pensioners) and the full economic structure, including income and wealth;
- stochastic simulation of a complete income distribution necessarily must be preceded by many theoretical simplifications and technical complications -

and large investments of time and resources. A very simple method should be explored, based on the assumption that the future income distribution will resemble the present distribution more than it will be different, and that any trends and projected changes can be introduced as controlled marginal changes; and

- pension models that in the same simulation include lifetime earnings, a series of stochastic elements, tax and benefit regulations and re-weighting etc. appear to be difficult to validate and control. One way of avoiding these problems is to construct a model that allows all steps to be simulated separately.

The pressure to produce pension analyses has forced a choice of the simplest method possible. At present the method is:

- through exact matching from the Pension Administration Database, the complete yearly record of earned income from 1960 to 1991 is added to each adult individual in our cross section model database of 1991. By using personal number we can find almost every person in the Pension Database;
- a complete lifetime earning profile is generated for each person aged 20-64 by statistical matching;
- pensioners in the model population are donated a new, modern lifetime earning profile from people aged 31-64, this also through statistical matching;
- demographic re-weighting is used to establish a future population that corresponds with Census projections;
- conventional and straightforward adjustments are externally applied to reflect different assumptions of growth, interest rates etc.; and
- pensions, taxes and other transfers are calculated with different pension systems.

Thus, the aim of the modelling strategy is to retain the infinite and basically unknown number of complex interrelations between population characteristics and to introduce microdata changes externally in a controlled manner. Conceptually, this method illustrates a future in which income structure, employment etc. are basically quite similar to their present structure, with the exception of clearly identified marginal changes (eg. in the sex and age composition of the elderly, the acquired pension rights from the public pension system etc.). This *'frozen micro structure'* does not internally expand present trends into the future, and does not try to anticipate plausible but unseen new trends.

The model is an embryo, not a complete version. Furthermore, the model does not include any behavioral responses or any equilibrium assumptions.

## 3. The Model Database

The model prototype uses a composite database constructed from two sources: the Swedish Income Distribution Survey from 1991 (IDS91) and the Pension Administration Database. Added to this are aggregated pension profiles from the ATP-projection model.

The IDS is carried out by Statistics Sweden every year and consists of about 10,000 households. The main purpose of the study is to analyse the distribution of disposable income among different types of households and throw more light on income structure and redistribution.

The data have been collected directly from the households through a telephone inquiry and by using different registers from the authorities. The questions to the households are about the composition of the household, work and occupation for the members of the household, child-care costs and housing costs. Almost all economic data are collected from the registers of different authorities (eg. income tax returns, social insurance registers, benefit registers etc.).

The IDS91 has been supplemented with data from the National Social Insurance Board (NSIB). These data consist of the individual income from work for every year from 1960-1991. Data are added by exact matching. For the purpose of analysis, income is transformed to 'pension-points', which form the basis for calculating the pensions. For validation, future pension profiles are added to the IDS with statistical matching. These profiles are generated in the ATP-projection model.

## 4. Supplementing Earning Profiles

The public pension is divided into two components - a basic pension and an earnings-related supplementary pension.

The basic pension is 96 per cent of the basic amount (for singles). The yearly basic amount in 1991 was Swedish Crowns (SEK) 32,200, indexed to inflation. To get a full public supplementary pension one has to work 30 years. The pension is based on the mean of the best 15 years. If one has low or no supplementary pension at all, one receives a special additional basic pension. The public pension system also includes housing allowances.

Thus, public supplementary pensions depend on lifetime earnings and also on the level of earnings every year. Therefore, the NSIB yearly register all individual earnings. This started in 1960 when the public supplementary pension was introduced.

The first step of the model is to include lifetime earnings for the working age population. To be able to simulate future pensions and the public net costs of present or alternative supplementary pensions, it is necessary *both* to add acquired pension points *and* to project future earnings. By exact matching, the basis of future pensions for economically active people is known in various degrees. For already retired individuals, the basis is completely known.

The IDS database is from 1991, and the Pension Administration Database provides the income history from 1960 to 1991 for each person. The aim is to create the incomes for future years. The simulation of future earning profiles uses a simple cross-section statistical matching procedure, by sex and age, which is illustrated in the following example:

> *Receiver is a male 40 years of age in 1991. Donator will be a male of 41 in 1991, who has income at age 40 in 1990 as close as possible to the Receiver's income at age 40. The Donation is the income at age 41 from the Donator. The Receiver now has got his income at age 41. Using the same method, the Receiver will be donated his income at age 42 from a male 42 years of age in 1991, and whose income at age 41 is as close as possible to the Receiver's income at age 41, and so on.*

The donation is repeated individually for males and females in every age class. The completion of earning profiles is based on the income trend for 1990-1991. If a person in the statistical matching is donated a disability pension, his or her status and earnings are not changed further. Hence, the earning profiles consist of historic (actual) data for 1960-1991 and simulated earnings after 1991.

To achieve profiles free from trends in the Swedish economy, various corrections are made to the 1991 income data. The baseline is generated with the assumptions of zero per cent growth and correction for other structural trends. The historic growth has been calculated from the National Accounts, as the wages paid to employees divided by the number of employees. The rising real wages reflect changes for full-time employees and changes in the distribution of full-time and part-time work. The consumer price index has been used as a deflator. When simulating future growth scenarios, real wages are increased by, for example, 1.3 per cent yearly, and lifetime earnings are changed accordingly.

Unemployment is the most important trend correction. The unemployment rate increased from 1.7 per cent to 2.9 per cent between 1990 and 1991. Individual earnings in 1990 and 1991 are corrected so that the average is the same for both years. If the donation example is examined, it is essential that the effect of growth in the economy is eliminated in the matching procedure. If the general growth was 2 per cent, and was not taken away, the chosen donator would not be a person in the same income situation as the receiver in real terms.

The results of these statistical matching experiments suggest that this combination of actual and generated data produces reasonable lifetime earning profiles. As Figure 1 shows, the lifetime earning profile for men at age 35 at 1.3 per cent growth is similar to the historic earning profile for men at age 64. The level is higher for the 35-years cohort, which reflects their higher real wages. The income scale is in thousands of Swedish Crowns (SEK), and the income measure includes income from work and work-related transfers like sick pay.

Figures 1 and 2 give some other examples of actual and generated lifetime earning profiles for males and females at ages 35, 45 and 64 years in 1991. The generated profiles are estimated with zero per cent growth in real wages and with 1.3 per cent growth, which corresponds to the historic growth from 1963 to 1991.

The number of sample individuals in each cohort is relatively small (about 150), which explains the random variation in both actual and generated earnings. However, the variation is larger in the generated profiles than in the historic ones. The main reason is probably that the matching method only uses a one year memory. Another explanation is that - depending on the pension rules - all observations in the historic series have registered income only between 1 and 7.5 basic amounts, while the generated profile partly includes income over and below that level.

Figure 1: Actual and Generated Lifetime Earnings Profiles for Men Aged 35, 45 and 64 Years in 1991, Assuming Zero and 1.3 Per Cent Real Wage Growth

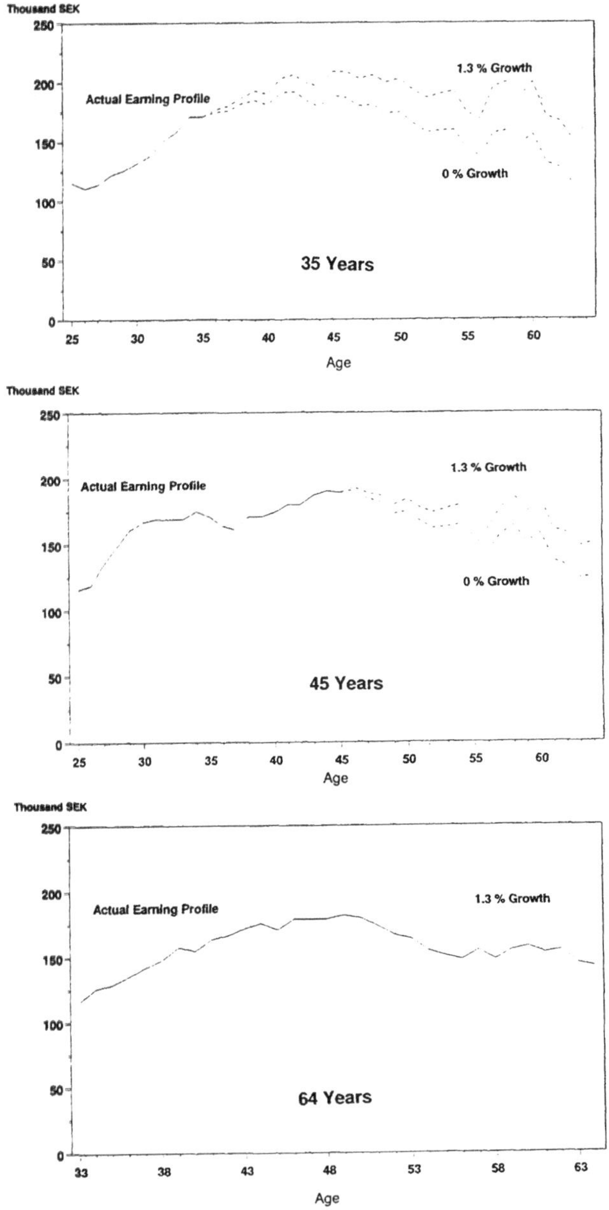

Figure 2: Actual and Generated Lifetime Earnings Profiles for Women Aged 35, 45 and 64 Years in 1991, Assuming Zero and 1.3 Per Cent Real Wage Growth

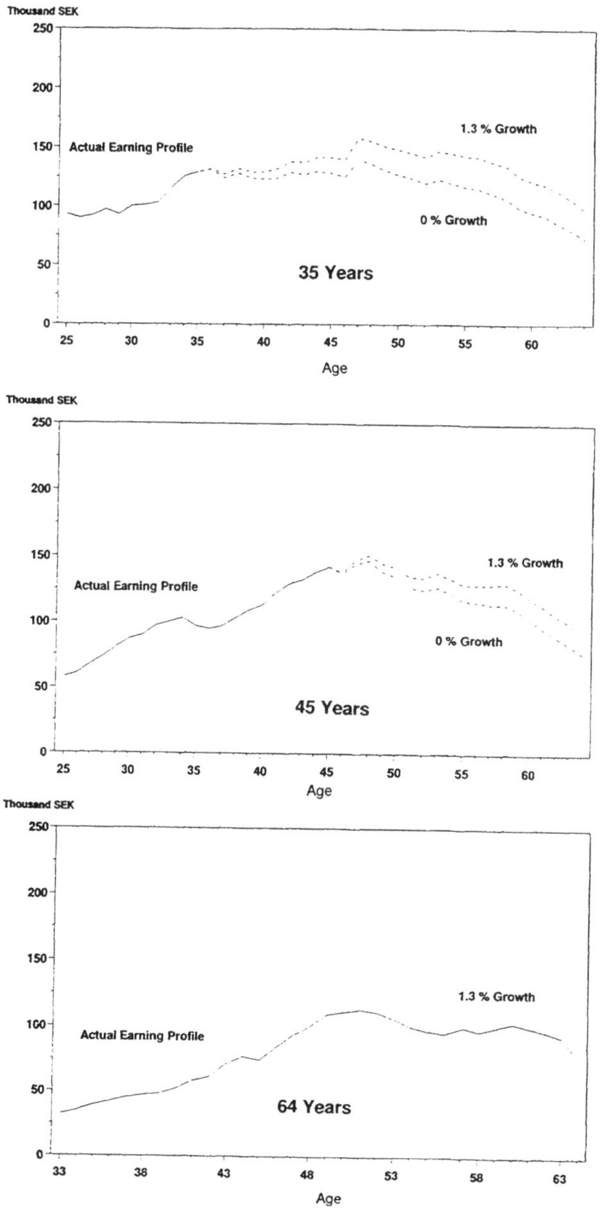

# 5. Donation of Earning Profiles - Macroeconomic Adjustment

After including and completing lifetime earnings for the working age population, these modern profiles are donated to the pensioners in the cross-section database, replacing the old profiles. This donation also uses statistical matching.

A pensioner at age 70 years in 1991 in the IDS91, is assumed, for instance, to still be a pensioner aged 70 years in twenty years time (in 2011). The pensioner is still living in the same type of dwelling and has the same composition of other income and wealth. The minimum change strategy is to replace the lifetime earning profile for the 70 year old pensioner with a new profile from a person aged 50 years.

As mentioned earlier, the assumption is that the future world - the composition of households, socioeconomic groups, dwellings etc. - will be essentially the same, except for a higher real income level from economic growth. The assumption of a 'constant world' is used in the prototype, because it is very difficult to foresee whether present trends will expand or diminish, or if new trends will appear. Furthermore, it is understandable and explainable to use the world of today as a baseline.

The *donation from the working population to pensioners* is based on a straightforward ranking method. Lifetime earning profiles from men at age 50 years are in this example donated to men aged 70 years, so that the working age person with the highest lifetime earnings donates his profile to the pensioner with the highest pension in 1991. The pensioner with the highest pension in 1991 is also supposed to have the highest pension in 2011. There are more individuals of age 50 years than age 70 years, so rather than donating from one person to another, the donation is made from percentile to percentile. For people aged 90 years in 2011, estimated lifetime earning profiles are not needed: instead actual profiles from pensioners aged 70 years in 1991 are used.

The *macro-economic adjustment* of the income distribution is implemented as a marginal change to the 1991 income distribution. Structural composition can be assumed constant or changed. If a constant change is assumed, all income sources are increased by conventional and equal economic multipliers. Wages, entrepreneurial income, capital income and wealth are increased at the same percentage as for the economy as a whole. Unemployment is assumed to be at the same level as in 1991. The percentage of pensioners who are entitled to supplementary pension will increase only as a result of changes in expected life earning profiles.

In the next steps, sensitivity analysis is used to measure the effects of different assumptions about gender, wage levels, socioeconomic group or growth levels of work and capital income.

With different growth assumptions, *tax and benefit regulations* also have to be adjusted. The static tax-and-benefit model includes almost all direct taxes, pensions, social insurances, housing and family benefits. When calculating taxes and benefits for a future income distribution, both tax allowances and tax brackets are adjusted to appropriate levels.

To illustrate the outcome, the following graphs and tables use the result from a validation matching done from the 2025 supplementary pension distribution in the NSIB's model for supplementary pensions. In this matching, future pensions have been donated from an NSIB pensioner to an IDS91 pensioner instead. The NSIB pension distribution is simulated with 1.5 per cent growth in wages per year, and also with an assumption that the supplementary pension can be based on wages greater than 7.5 basic amounts, which is not possible today.

The future income distribution will change substantially as a result of this experiment with donated increased pensions, additional growth assumptions, and adjusted taxes (Figures 3 and 4). Because of the combination of increased supplementary pensions and the marginal impact upon these pensions from minimum guarantees, taxes and means-tested benefits, the distribution among pensioners seems to become more equal. The large number of pensioners in 1994 with a yearly pension of around SEK 50,000 is much reduced in 2025. Basic pensions will for most pensioners be replaced with a supplementary pension (Figure 3). A larger proportion of the pensioners will have a substantial supplementary pension.

The distribution of disposable income for all households shows clearly the move upwards produced by all these effects (Figure 4). This graph includes the effects of direct taxes, housing allowances and other transfers to the households. It also includes the effect of increased real wages for the working population. The distribution of disposable income seems to become more equal as a result of these

Figure 3: Projected Pension Distributions in 1994 and 2025*

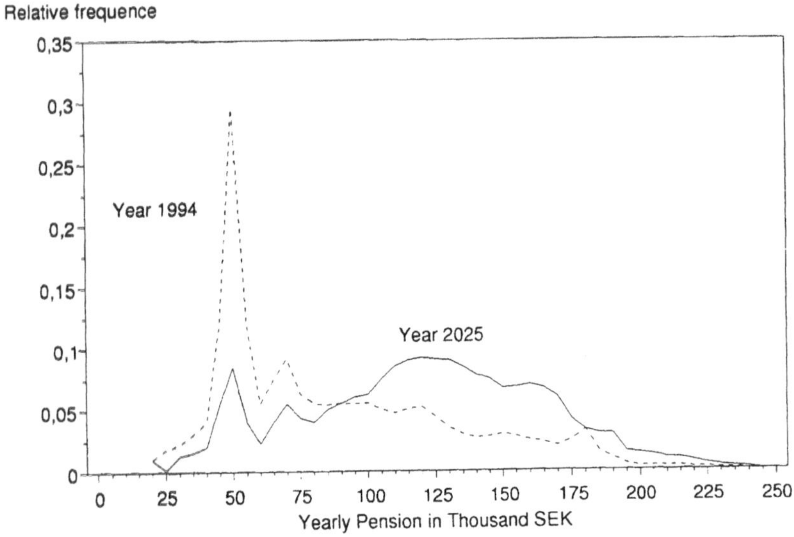


Figure 4: Projected Disposable Incomes of All Households in 1994 and 2025*

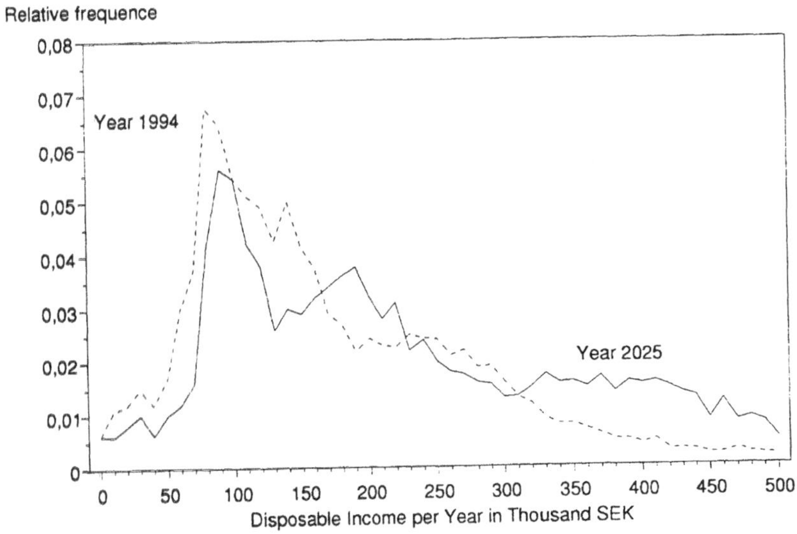


*1.5 per cent wages growth assumed in both cases.

experiments. All other things being equal, the increased supplementary pensions, although much reduced by the impact of taxation and other programs, will increase income at the lower end of the distribution.

## 6. Demographic Re-weighting

The 1991 IDS database, of course, reflects the demographic structure of 1991. Every individual in the database has a weight that can be viewed as the number of people the individual represents. The future distribution of age is projected by Statistics Sweden. This projection is used to reweight the IDS sample into a future year. The weights are proportionally changed for each sex and each five year age class, so that the sample reflects the projected distribution of age. In all other aspects, the sample is unchanged: a child is a child and a pensioner is still a pensioner.

The change in the age and sex distribution between 1994 and 2025 is rather small (Table 1). When demographic reweighting is simulated on the donated and adjusted 2025 pension distribution and income distribution, the changes are very small.

Table 1: Age Distribution by Sex in 1994 and in 2025

| | Percent in 1994 | | | Percent in 2025 | | |
|---|---|---|---|---|---|---|
| Age | Male | Female | All | Male | Female | All |
| 0-19 | 25.6 | 23.8 | 24.7 | 24.8 | 23.2 | 24.0 |
| 20-24 | 35.3 | 32.9 | 34.1 | 32.7 | 30.5 | 31.6 |
| 45-64 | 24.1 | 23.4 | 23.7 | 24.4 | 23.7 | 24.1 |
| 65-69 | 4.4 | 4.8 | 4.6 | 5.1 | 5.4 | 5.3 |
| 70-74 | 4.3 | 5.1 | 4.7 | 4.6 | 5.1 | 4.8 |
| 75-79 | 3.1 | 4.1 | 3.6 | 4.1 | 5.1 | 4.6 |
| 80-84 | 2.0 | 3.3 | 2.7 | 2.7 | 3.9 | 3.3 |
| 85-89 | 0.9 | 1.9 | 1.4 | 1.1 | 2.0 | 1.6 |
| 90-94 | 0.3 | 0.7 | 0.5 | 0.3 | 0.9 | 0.6 |
| 95- | 0.0 | 0.1 | 0.1 | 0.1 | 0.3 | 0.2 |
| ALL | 100.0 | 100.0 | 100.0 | 100.0 | 100.0 | 100.0 |

# 7. Gross and Net Pension Expenditures

If all these simulations are combined into one experiment, it is possible to describe the total effects of supplementary pensions, the macroeconomic assumptions, the adjusting of tax regulations, etc. and demographic reweighting.

The calculations are done with today's prices, but with an annual growth of 1.5 per cent. According to the rules for income tax, that would mean the tax parameters (except for a minor part) would be unchanged. The assumption here is that the tax schedules are unchanged.

The initial results from the experiments of macro estimates from the 1994 and 2025 distributions show some surprising results (Table 2). The demographic reweighting increases the share of pensioners in the population from 21.2 per cent to 24.5 per cent, which is projected by Statistics Sweden. The assumption of 1.5 per cent yearly growth and demographic development increases work income by 69 per cent. The pensioners' share of total wealth increases from 53 per cent to 56 per cent in 2025.

Gross supplementary pensions as a percentage of work income increase from 12.1 per cent to 15.9 per cent, which is one major reason for the present restructuring of Sweden's pension system. However, the net pension share decreases as a result of the effects of progressive income taxes, means tests etc. If these first model results can be believed this suggests that the *net pension financing requirement is decreasing, especially if it is considered that the share of the elderly is increasing.*

This result is quite possible, because the pension system now is at a stage where older pensioners with low or no supplementary pensions at all are replaced by younger pensioners with high supplementary pensions. Because pensioners with no or low supplementary pension receive a special basic pension, they cost the taxpayer more. In contrast, those with higher supplementary pensions pay more income tax and receive less basic pension because of means-testing.

# 8. Conclusions

Most of the pension long-term simulation problems are still unsolved. All experiments so far suggest that the level of unemployment and invalidity pensions are more important to all interaction and net financing effects than demographic

Table 2: Preliminary Estimates of Public Pension Gross and Net Expenditures by Components in 1994 and in 2025 (Billion SEK)

| | 1994 | 2025 | |
|---|---|---|---|
| | | Including donation + economic adjustment | Also including demographic reweighting |
| *Basic Macro-estimates* | | | |
| - '000: | | | |
| Total Population | 8 617 | 8 617 | 9 328 |
| Old Age Pensioners | 1 437 | 1 437 | 1 817 |
| Disability Pensioners | 386 | 386 | 455 |
| *- Billion SEK* | | | |
| Work Income, | 755 | 1 230 | 1 273 |
| Taxable Wealth | 600 | 951 | 1 142 |
| - among Pensioners | 317 | 473 | 634 |
| Capital Income | 36 | 52 | 59 |
| - among Pensioners | 14 | 19 | 25 |
| *Public Pension Expenditures* | | | |
| Supplementary Pensions | 94 | 161 | 200 |
| Basic Pensions | 65 | 58 | 72 |
| Total Gross Expenditures | 159 | 219 | 272 |
| Net Expenditures (after tax) | 121 | 153 | 191 |
| Gross Public Supplementary Pensions as % of Work Income | 12.1 | 13.1 | 15.9 |
| Net Public Pensions as % of Work Income | 15.6 | 12.4 | 15.2 |

* Assumes 1.5 per cent growth per year from 1995 to 2025.

development, pension indexation and formulas. However, simulations have not yet been designed that allow these trends to be controlled and varied.

Furthermore, the future actual total effects will depend largely on the coverage and level of non-public pensions. From other sources, it is known that these pensions are increasing even faster than public supplementary pensions. Due to lack of microdata - and even of reliable macrodata - it has not yet been possible to model the effects of non-public pensions. This means that future net costs and pension payments are overestimated.

If anything, the most important conclusion of this alternative model strategy is that it may at least be faster to initiate and more easy to understand and control than stochastic alternatives. Statistical matching is easy to implement and it produces individual event series with both variation and individual consistency. Generated earnings data on an individual level show a mix of stability and variation that looks very close to the observed actual pensions.

## References

Eriksen, T.E. (1973), 'En prognosmodell för den allmänna tilläggspensionering', National Social Insurance Board, Stockholm.

Hancock, R., J. Mallender and S. Pudney (1992) 'Constructing a computer model for simulating the future distribution of pensioners' incomes for Great Britain', in: R. Hancock and H. Sutherland (eds.), *Microsimulation Models for Public Policy Analysis: New Frontiers*, STICERD, London School of Economics, London.

Harding, A. (1993), *Lifetime Income Distribution and Redistribution*, North Holland, Amsterdam.

Merz, J. (1991), 'Microsimulation - A survey of principles, developments and applications', *International Journal of Forecasting*, Vol. 7, No. 1.

Pudney, S. (1993), 'Dynamic simulation of pensioner's incomes: Methodological issues and design for a computer model for Great Britain', mimeo, Department of Applied Economics, University of Cambridge, England.

Ståhlberg, A-C. (1990), *Generating Equality in the Welfare State - Lifecycle Income Redistribution of the Public Sector: Inter- and Intragenerational Effects*, English Series; 1990:42, Swedish Institute for Social Research, Stockholm.

Swedish National Social Insurance Board (NSIB) (1991), *The ATP Projection Model*, Department of Research and Analysis, Stockholm.

Leif Andreassen
Dennis Fredriksen
Olav Ljones

Statistics Norway
Norway

# Chapter 15: The Future Burden of Public Pension Benefits: A Microsimulation Study

## Abstract

*In this paper, the microsimulation model MOSART is used to analyse future long-run developments in disability and old-age pensions covered by the Norwegian National Insurance System. The number of pensioners will increase relative to the number of workers, leading to a growing tax burden on future generations. It is found that among those born before the year 2000, early generations possess a larger pension wealth (the discounted value of payments and received benefits) as a percentage of life time labour income than later generations. The chapter discusses changes in the rules for determining pension benefits and the consequences of shifting from a Pay-As-You-Go System to a funded system.*

## 1. Introduction

In many countries there is an ongoing debate about the modern welfare state. In a welfare state, the government takes explicit responsibility for creating social justice, reducing poverty and creating a stable economic environment. This usually entails active government intervention in the economy, a large government sector providing goods and services, and comprehensive redistributive programs.

To some extent, the welfare states developed in the Scandinavian countries have achieved these aims. However, the recent past has also revealed some of the negative effects of a comprehensive welfare state. It necessitates relatively high taxes (and/or large public sector deficits) and leads to negative labour supply effects. It has also been argued that the total redistributive effect of the welfare state is in reality fairly small.

* We are grateful to Marie W Arneberg for valuable comments and to the National Insurance Administration for letting us use their registers on pension entitlements. Financial support from Norges Rad for Anvendt Forskning (NORAS) is acknowledged.

One effect of the welfare state is to divide the population into receivers and contributors of government funds. If the number of receivers (or the value of the transfers to the receivers) grows rapidly, the tax burden on the contributors can become politically unacceptable. In particular, social security systems in countries facing low fertility rates and an ageing population represent a growing burden on government budgets.

This paper analyses the future burden of public pension benefits in Norway. It focuses on the public pensions received by the disabled and the retired compared to the contributions to the social security system from those in the labour force. The pension system is formally a Pay-As-You-Go system, where pensions are determined by legal rules based on each individual's income history. The pension system also has some built-in indexation.

The analysis is based on the dynamic microsimulation model MOSART.[1] One of the advantages of using a microsimulation model in the present context is that it produces individual life histories, which are important when calculating pension benefits and entitlements. The paper discusses how different immigration rates, labour force participation rates and disability propensities in the long run influence the labour force and the social security system. The paper also considers how the social security system redistributes income across generations (cohorts).

## 2. Model Description

The MOSART model is a dynamic cross-sectional stochastic microsimulation model which projects population size and composition, labour force status, educational level and future pension benefits. It includes the events of immigration and emigration, birth, death, marriage, divorce, education, disability, retirement, labour force participation, and labour market earnings. An overview of the MOSART model is given in Figure 1, while Table 1 lists the events that are simulated, together with the variables that each event depends on.

The present version of the model simulates life histories for a one per cent sample of the Norwegian population aged 16 years and older. This gives a model population of 30,000-35,000 individuals in each simulation year. It is possible to simulate a

---

[1] MOSART is a Norwegian acronym for model for Microsimulation of Schooling, Labour Supply and Pensions.

Figure 1: Structure of MOSART Model

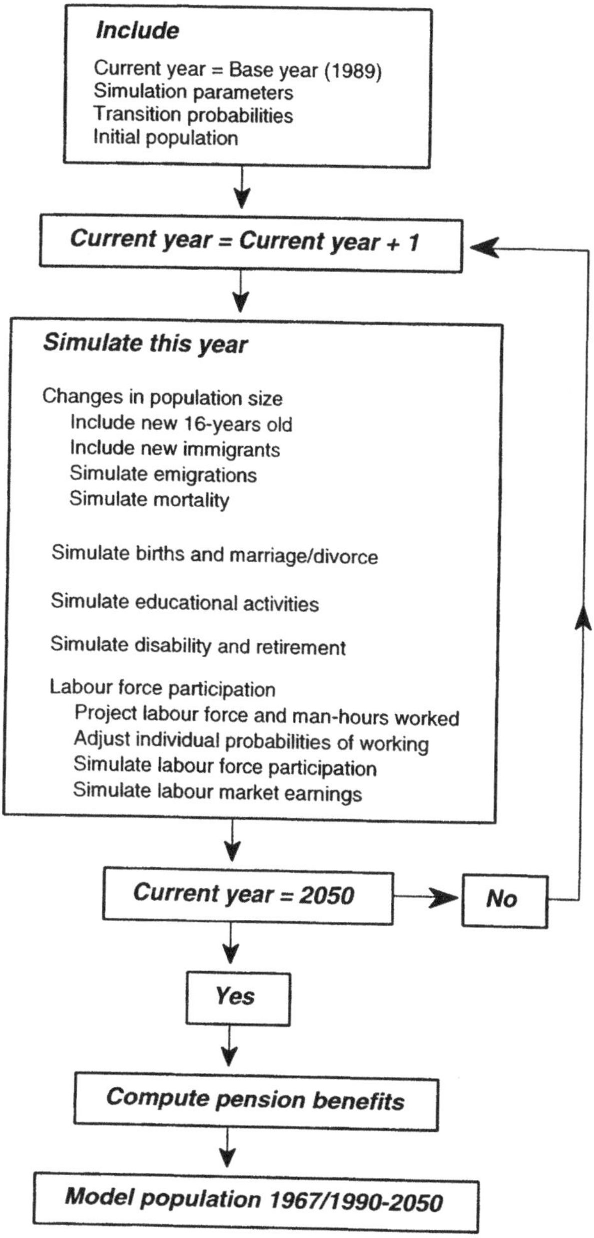

Table 1: Processes Represented in the MOSART Model

| Process or Event | Conditional Upon |
|---|---|
| Death | Age, sex, disability |
| Fertility (women) | Age, parity, age of youngest child |
| Marriage (female dominant) | Women's age and marital status (unmarried, divorced or widowed) and whether the woman has children |
| Matching with a husband | Husband's age depends upon woman's age |
| Divorce (female dominant) | Wife's age |
| Education | Age, sex, educational attainment, current educational activity |
| Entry into disability | Age, sex, educational attainment, marital status, labour force participation in the preceding year |
| Retirement | Happens at age 67 for all individuals (approximately) |
| Labour force participation and labour market earnings | Age, sex, educational attainment, whether an individual is a student, disability, labour force status in the preceding year, labour force attachment (a measure of the stability of labour force participation in the preceding years), marital status<br>In addition for women: the number of children and age of youngest child |

larger sample, but this level was chosen to achieve acceptable computing times (a total of about 1.5 hours of CPU-time for a simulation of 50 years). This version of the model is an expanded version of the model presented in Andreassen (1993) and Andreassen et al. (1993).

MOSART is a discrete time model, with one year intervals and a recursive structure. The simulation proceeds one year at a time, simulating all events that occur for all individuals during that year. This procedure enables the link between spouses to be tracked and constraints on some aggregate totals to be set for each year. As can be seen from Figure 1 and Table 1, the simulation is sequential, with later events to a large degree depending on earlier events. Transition probabilities are estimated as conditional probabilities, dependent on the events simulated earlier. This enables competing risk, to a certain degree, to be taken into account.

For each year, the model first supplements the model population with new 16 year olds and immigrants, and simulates emigration and mortality. Then births are simulated, followed by marriage and divorce. After these demographic events, the model then simulates educational activity, disability and retirement. Finally, labour force participation and labour incomes are simulated, depending on most of the events simulated earlier that year.

The simulation in the present version of the model goes from 1990 to 2050, with historical data going back to 1967. This results in a very large output, which is sorted and organised into individual life histories after the simulation is finished. These post-simulation routines also include calculation of pension benefits, and production of tables with aggregate figures.

MOSART is mainly a demographic model, where educational transition probabilities, the transition into disability, labour force participation rates, and labour market earnings depend only on demographic variables. An important feature of the model is the large initial population with no synthetically generated (simulated) variables. A model having many similarities with MOSART is the Dutch dynamic cross-sectional model, NEDYMAS, described in Nelissen and Vossen (1989) and Nelissen (1991).

## 2.1 The initial population

The initial population is based on a 10 per cent sample of the Norwegian population 16 years and older in the base year 1989. The sample is drawn stratified by age, sex and marital status, and married couples are linked together. This sample is used when estimating many of the transition probabilities in the model and as a starting point for simulations. To avoid estimation bias, the sample includes 10 per cent of all individuals who have died or emigrated since 1967. Each simulation uses only one-tenth of the sample (amounting to one per cent of the total population or about 33,000 people) while simulation of the entire original sample is carried out by combining ten simulations, each using a different part of the original sample.

The initial population and the larger 10 per cent sample are the result of merging different data registers by a unique personal identification number. The Central Population Register supplies female birth histories and information on each individual's sex, age, marital status and, for immigrants, year of entry into Norway. The Educational Register in Statistics Norway supplies educational attainment and activities since 1985. Data from registers in the National Insurance Administration show whether a person is a recipient of an old age, disability or widow pension for each year since 1985, and also existing pension entitlements based on annual labour market earnings since 1967 (first year of the National Insurance System).

## 2.2 Demographic transitions

The number of new 16 year olds added to the model population each simulation year can be determined in two ways. One can use the number of births simulated in the model, and thus make an endogenous population projection. Alternatively, one can use an exogenous number of new 16 year olds (for example, from the official population projection from Statistics Norway). New 16 year olds are assigned the starting-values of unmarried, childless, non-disabled, non-working and just left primary school.

The total number of net immigrants is specified exogenously in the model and divided into groups by age and sex. Women's birth histories and marriage among immigrants are also simulated, using probabilities that characterise the Norwegian population as a whole. Immigrants in the model are assumed to face the same probabilities as non-immigrants. This is done for the sake of simplicity, since data on immigrants at present is limited. The educational attainment for immigrants is set to 'unknown', as in the educational statistics. If net emigration occurs in some age groups, the model draws individuals from the sample who are then assumed to emigrate.

Mortality varies by age and sex. If a married person dies, the surviving spouse is made a widow or widower. Disability pensioners have substantially higher mortality rates than others, and the model provides an option of increasing their mortality by a fixed percentage. The mortality of the non-disabled is in such a case reduced endogenously so as to leave aggregate mortality unchanged. In the present version, mortality is independent of marital status and educational attainment - assumptions that are open to future improvement.

After mortality and migration, the model simulates births and changes in marital status. Giving birth is simulated using probabilities that depend on the woman's age, the number of children, and the age of the youngest child. These fertility probabilities are independent of marital status, although unwed mothers are exposed to increased probabilities of marrying. This assumption is to some extent a reflection of the large proportion of couples cohabitating without being married.

Simulation of marriage and divorce is female-dominated, based on marriage and divorce probabilities for females. Marriage is simulated with probabilities that depend on the woman's age, her marital status (unmarried, divorced or widowed), and whether she recently had a child. If a woman marries, the model draws the age of her spouse (with probabilities depending on the woman's own age), and then randomly draws a spouse among all unmarried males of this age. This implies that the marriage probability of a man will be the same whether he is unmarried, divorced or a widower. If a woman is drawn to divorce, her husband will also be divorced. It is assumed that it is only possible to change marital status once a year.

## 2.3 Education

The MOSART model simulates, for each individual, year by year educational activities and the resulting influence on educational attainment. The transition probabilities through the educational system depend only on age, sex, educational attainment and educational activity in the previous year. Education is described by level and subject. Educational level is divided into one-year steps, and educational subjects are gathered in relatively large groups, where a separation is made between health subjects, technical subjects and administrative subjects at all levels. The transition probabilities have been estimated on all educational transitions in Norway between the school years 1986-87 and 1987-88.

The number of pupils and students in Norway has increased dramatically since 1988, perhaps related to the rise in unemployment. To take this into account, the estimated probabilities have been adjusted so that the model simulates the observed number of pupils and students in 1991.

About 90 per cent of those leaving primary school continue directly into secondary school, with 40 per cent choosing further general education that prepares them for university studies. The flow through the educational system seems slow, and is characterised by movements across subjects and fairly low progression from one level of education to a higher level. This is especially the case for higher education and vocational training. Low progression in vocational training is caused by a combination of good capacity in basic courses and low capacity in final courses. On the basis of estimated transition probabilities for education from 1987, about 40 per cent of all individuals experiencing these probabilities will attain higher education (post-secondary) during their lifetime, and about 40 per cent will end up with three-year vocational training. The increase in the number of pupils and students since 1987 will probably result in a substantially higher proportion of the population attaining higher education in the future.

## 2.4 Old age and disability pensions

Public old age pensions, disability pensions and pensions for widows and widowers are in Norway organised through the *National Insurance System*. This version of MOSART simulates the number of recipients under these three programs and also calculates the corresponding pension benefits. The intention of these programs is to secure a minimum level of income and to ensure pensions are related to previous labour market earnings.

The key to understanding the National Insurance System is the so-called *basic pension unit* (BPU), an amount of nominal Norwegian Kroner (Nkr). The BPU is increased every year to compensate for inflation and, to some degree, real income growth. The BPU is also used to calculate the upper limit for earning pension entitlements (see Andreassen et al, 1994, for further information).

The main features of the pension system are that all pensioners are guaranteed a *minimum pension*, which is about one-third of average full-time labour market earnings. Most pensioners also receive an *additional pension*, depending on average labour market earnings during their working years. These pension entitlements are roughly calculated by dividing labour income each year by the BPU that year. The additional pension is calculated by multiplying a predetermined fraction of the current pension entitlements by the current BPU.

The after-tax pension/wage ratio decreases with rising income and is, with stable labour market earnings, around 80 per cent for an income of 100,000 Nkr and 50 per cent for an income of 300,000 Nkr. The National Insurance System was established in 1967, and only earnings obtained after that year qualify for the additional pension. This implies that average entitlements will keep on growing well into the next century.

People aged 70 years and older receive old-age pensions independent of other sources of income, while people aged from 67 to 69 have their pensions means-tested against labour market earnings. The fact that 90 per cent retire at the age of 67 simplifies the simulation of retirement[2]. Widow pensions cover mainly widows with

---

[2] Certain occupational groups have a lower retirement age, but not through the National Insurance System. During recent years a scheme for general retirement at age 64 for members of the main labour market organisations has been established.

low labour market earnings[3]. The simulation of such pensions is therefore closely connected with the simulation of marital status and labour force participation.

Disability pensions are granted to all people with a permanently reduced working ability of at least 50 per cent caused by disease or accident. Once a person is granted a disability pension, it is common to remain a disability pensioner until she or he reaches the age of 67 (and is transferred to an old-age pension) or dies.

Working ability is measured relative to a person's opportunities in the labour market. At the end of the 1980s, medical claims had become rather open and included more or less any kind of long-term problems in the labour market. This included problems such as alcoholism, stress syndromes and long-term unemployment. In many cases, disability pension has been (mis)used to enable elderly workers to retire earlier than the official retirement age of 67. In 1990, about 8.5 per cent of the population in the age group of 16 to 66 years received a disability pension and, by the age of 65 to 66 years, about 45 per cent received a disability pension. During the last few years, the criteria for being disabled have been tightened and the number of new disability pensioners has been reduced. It is too early to see the long-term effects of this, since there has been an increase in the number of persons in situations that commonly lead to a disability pension (for example, work rehabilitation and long-term unemployment).

The transitions into disability in MOSART depend on sex, age, educational attainment, marital status and labour force participation. The probabilities are represented by a logit function and are estimated on the same 10 per cent sample of the population as the initial population is based on. Data on transitions into the state of being a disability pensioner are available for the years 1986 to 1989.

## 2.5 Labour force participation

Simulation of yearly labour force participation is intended to give projections of the labour force and of the pension entitlements earned in the future. Lack of good data covering both the labour force and labour market earnings has led to the use of two different types of simulation of labour force participation based on two different definitions. One definition is the one usually used by labour force surveys, namely the yearly average number of persons in the labour force. The other definition is the

---

[3] In principle, men can also receive this type of pension, but few men become widowers before the age of 67. In addition, the widow and widowers pension is means-tested, and most men earn more than the maximum limit.

total number of persons who have been in the labour force at any point of time during the year. This last definition corresponds to the data for labour income[4] held in the data registers of the National Insurance System.

These two definitions of labour force participation are related by the probability that a person with labour income during the year will also be a part of the labour force at any given time during that same year.[5] This probability can be interpreted as the proportion of the year a person works. The two definitions vary in the extent to which they capture pupils, students and pensioners participating only in seasonal work during holidays and vacations.

Simulation based on the first definition of labour force participation is calculated from data from the Norwegian Labour Force Surveys. The model computes the expected labour force participation rate and working hours of each individual. By adding across all individuals in the model population, the model produces a projection of the labour force and man-hours worked. The labour force participation rates and average working hours are estimated from the Labour Force Surveys in 1991, and depend on sex, age, educational activities and attainments, disability, marital status, number of children, and age of youngest child. Labour force participation rates are in the present version assumed constant throughout the simulation period. Thus the projection shows the effect on the labour force of changes in the size and composition of the population.

Simulation based on the second definition of labour force participation simulates exits from and entries into the state of having positive labour income during the year. These transition probabilities are dependent on age, sex, educational activities and attainments, disability, stability of labour force participation in preceding years, marital status, number of children, and age of youngest child. They are estimated using a logit function on the 10 per cent sample of the Norwegian population for the years 1985 to 1988.

Consistency between the two types of simulation is ensured by letting the first type determine the level of the total labour force and man-hours worked, and letting the second type determine the distribution of labour force participation across the

---

[4] This is the sum of wages, income of the self-employed, sick leave benefits, unemployment benefits and maternity leave benefits during each year.

[5] The possiblity of a person without labour income being a part of the labour force has been excluded (for example, young unemployed without previous work experience and therefore no unemployment benefits cannot be part of the labour force).

population. The transition probabilities used in the second simulation are adjusted to make the expected number of individuals with labour income consistent with the projected labour force of the first simulation. This is done by ensuring that the individual differences in the probability of having labour income during the year are maintained.

## 2.6 Distribution of labour income

Simulation of labour income is based on the distribution of income among individuals and is estimated by using an ordinary least squares model on the model's 10 per cent sample of the Norwegian population during the years 1985 to 1988. The estimates are based only on persons with positive labour income during the year (ie. more than 1,000 Nkr). The model can be written:

$$\log Y_{it} = X_{it}\beta + U_{it} \tag{1}$$

where $Y_{it}$ represents labour income during year t for individual i, and $U_{it}$ is a disturbance term capturing unexplained deviations from the mean across both individuals and time. The vector of explanatory variables, $X_{it}$, includes age, sex, educational activity and attainment, disability, whether an individual the same year has entered or left[6] the state of having labour income, stability of labour force participation in the preceding years, marital status, number of children, and the age of the youngest child.

Total labour income is determined outside the model while its distribution across individuals is determined by the estimates derived from equation (1) using both the explained variation across individuals (captured by the estimated valued of $\beta$) and the unexplained variation found in the empirical residuals $\varepsilon_i$. An adjustment term k is used to calibrate the income distribution so that it is consistent with the assumed level of total labour income. Simulated labour income for individual i at the time t can thereby be written:

$$Y_{it}^{s} = k \, . \, Exp(x_{it}\hat{\beta} + \varepsilon_i) \tag{2}$$

where $\varepsilon_i$ is a fixed residual specific to each individual and $\hat{\beta}$ is the OLS estimate from equation (1). The residual $\varepsilon_i$ does not vary over time. This specification implies that each individual, given their age, sex, education, etc. is a consistent amount above or below the mean. The chosen specification of the residual $\varepsilon_i$ gives

[6] It is assumed that an individual leaves the state of having labour income in the year before labour income becomes zero.

an income distribution similar to the observed income distribution in the model's 10 per cent sample of the Norwegian population over the years 1967 to 1989. The adjustment term k is set so that the following condition is satisfied:

$$\sum_i Y^s_{it} = W_t \cdot H_t \qquad (3)$$

where $W_t$ is an exogenous given wage level and $H_t$ is projected man-hours from the simulation. This ensures that the income distribution given by (2) will be consistent with the average full-time labour income set outside the model.

# 3. The National Insurance System in the Long Run

In this analysis, a base line projection is used together with several alternative projections. These alternative projections are simulations using different levels of the most important transition rates and other exogenous variables such as net immigration. This gives an indication of how sensitive the results are to changes in current behaviour.

## 3.1 The baseline projection

In Table 2 are summarised the main assumptions underlying the base line projection. The most important demographic assumptions concern immigration, death, and fertility. Net immigration is assumed to be 5,000 persons per year during the whole projection period. (The total population in Norway is just above four million people.) Since 1985, actual net immigration has varied from -1,500 (in 1989) to 14,000 (in 1987). Death rates are at the outset assumed to equal those observed in 1989, and to then decrease until 2010 at a rate equal to that observed in the period 1965 to 1988. After 2010, death rates are assumed to be constant. This will lead to an increase in expected lifetime at birth of 1-2 years. Fertility rates are assumed to equal the period fertility rates observed for 1989, implying a total fertility rate of 1.89. By comparison, the fertility rate for the period 1981 to 1985 was 1.68, while in 1991 it was 1.92. In 1967, when the National Insurance System was established, the total fertility rate was well above 2.5.

Transitions into disability in the base line alternative are from the year 1989, when they were at a historically high level. They were much lower in 1986, having increased by 60 per cent in the period to 1989. Labour force participation rates are set at the 1991 level, when 68.5 per cent of the population aged 16 to 74 years was in the labour force. By comparison, 71.3 per cent of the population aged 16 to 74 years was in the labour force in 1987 (highest observed level in post-war Norway).

Table 2: The Main Assumptions in the Baseline Projection

| Process or Event | Assumption |
|---|---|
| Net immigration | 5,000 per year |
| Death rates | 1988-1989 data, with death rates decreasing to 2010 at a rate equal to that observed in the period 1965 to 1988 |
| Fertility | 1989 data (period fertility) implying a total fertility rate of 1.89 |
| Marriage | 1984 data |
| Education | Data from the school years 1986-1987 and 1987-1988, with an ad hoc adjustment so that the model simulates the observed number of pupils and students in 1991. |
| Disability | 1989 level (a historically high level) |
| Labour force participation | 1991 level (a relatively low level compared with the preceding 5 years) |
| Labour market earnings | 0.75% growth in real wages |
| Social security benefits | The basic pension unit BPU (which determines individual pension benefits) increases at the same rate as wages |

In the base line alternative, the BPU, and thus additional pensions, increase in line with wages. Real wage growth is assumed to be 0.75 per cent a year, although this growth rate is by itself not very interesting in the model. It is the difference between the growth in benefits and in wages that is of interest, and that is the main focus of this chapter.

## 3.2 Population, labour force and the number of pensioners

Figure 2 shows five different projections that illustrate how sensitive the base line projection is to changes in the assumptions about immigration and fertility. As mentioned earlier, the base line alternative assumes a total fertility rate of 1.89 and a net immigration of 5,000 persons. The other four alternatives are based on changing either the net immigration assumption (to 0 and 10,000 persons a year) or the fertility assumption (to a total fertility level of 1.68 and 2.06). The figure shows that in the base line alternative, the population starts declining just before 2030, while with a net immigration of 10,000 or a total fertility rate of 2.06 (approximately the replacement level of fertility), the population levels off at about five million persons. A drop in net immigration to zero or a fall in the total fertility rate to 1.68 will lead to a significantly smaller population, approaching four million in 2050. This illustrates the fairly dramatic consequences arising from changes in immigration or fertility - but it also illustrates how it is possible to compensate for a fall in fertility by increasing immigration.

In Figure 3, developments in the labour force are similar to the population developments shown in Figure 2. High net immigration or high fertility levels lead to an increasing labour force during the whole period. Figure 3 also illustrates that changes in immigration and fertility have different effects on the age profile of the population. Increased immigration feeds directly into the labour force (because immigrants are mainly adults under the age of 40), while increased fertility only affects the labour force when the newborn reach maturity. This explains why a net immigration of 10,000 individuals a year in the short run increases the labour force by more than a total fertility rate of 2.06, but has about the same effect in the year 2050.

Figure 4 illustrates the increase in the number of pensioners implied by the different demographic assumptions. It includes pensioners on both public disability and old-age pensions. The base line alternative shows a peak in 2040. As expected, the effects from a shift in assumptions are not as dramatic for the number of pensioners as for the population as a whole. This is largely because fertility assumptions have little effect upon the number of pensioners before the year 2050. Changes in immigration have larger consequences because, as mentioned earlier, most immigrants arrive in the country as adults. In 2050, the difference between the alternative with high and low immigration is almost 200,000 persons.

One of the main determinants of the burden of pension benefits is the number of pensioners compared with the number of workers. Figure 5 gives the number of pensioners per person in the labour force under different assumptions, and shows little change in this dependency ratio until the year 2010. It then starts increasing and reaches a peak around 2040. In the case of a total fertility rate of 1.68, it peaks at a level 51 per cent higher than in 1990, in the base line alternative it is 39 per cent higher, and in the case with high participation rates and low disability rates, it is 23 per cent higher.

If the two most extreme cases are disregarded, this dependency ratio in 2040 lies between 43 per cent (if there is a net immigration of zero) and 31 per cent (with a

Figure 2: Population Projections Under Different Demographic Assumptions

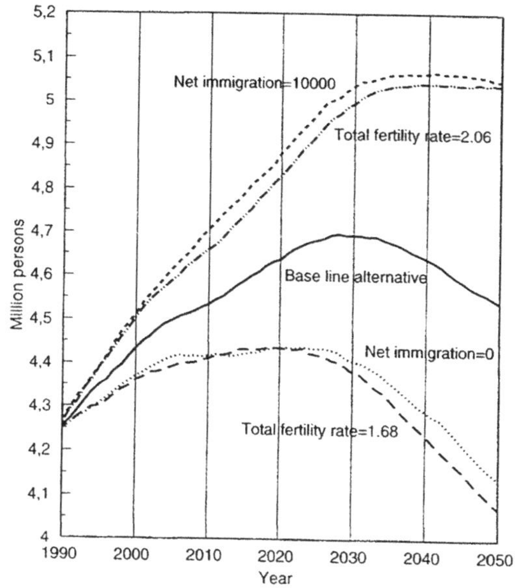


Figure 3: Labour Force Projections Under Different Demographic Assumptions

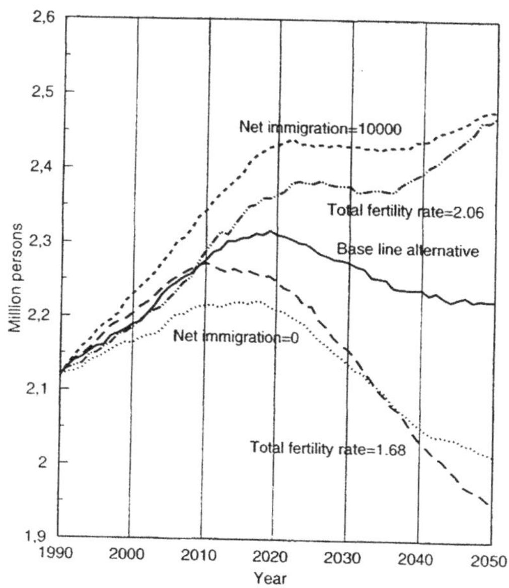

Figure 4: Estimated Number of Pensioners Under Different Demographic Assumptions

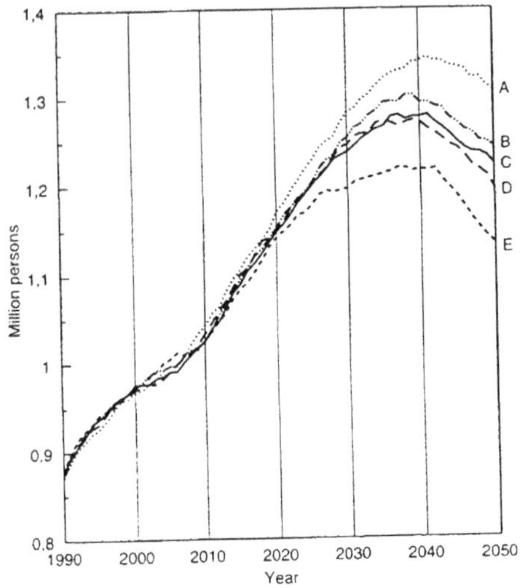


A. Net immigration = 10000
B. Total fertility rate = 2.06
C. Base line alternative
D. Total fertility rate = 1.68
E. Net immigration = 0

Figure 5: Estimated Number of Pensioners Per Worker Under Different Demographic Assumptions

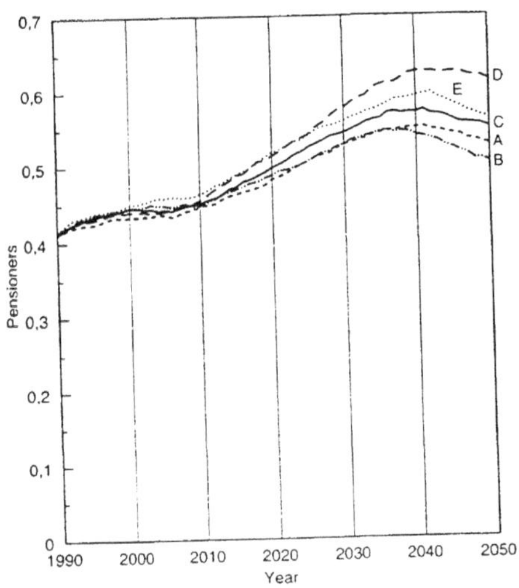


A. Net immigration = 10000
B. Total fertility rate = 2.06
C. Base line alternative
D. Total fertility rate = 1.68
E. Net immigration = 0

total fertility rate of 2.06) higher than in 1990. It is therefore a fairly robust conclusion that, in the first half of the next century, Norway will experience a very large increase in the number of pensioners relative to the number of workers.

Figure 5 also illustrates that changes in fertility have a greater effect on the dependency ratio than immigration does, because fertility mainly affects the labour force rather than the number of pensioners in the period until 2050. Toward the end of the period, increased immigration will have increased both the labour force and the number of pensioners, leading to a relatively small decrease in the number of pensioners per worker. During the later part of the century, increased fertility will also lead to an increasing number of pensioners.

## 3.3 Pension benefits

Figure 6 shows the growth in average old-age pensions when wages and the BPU are held constant. The Figure thus shows the growth in average pension benefits relative to the average wage level. This gives basically the same results as letting wages and the BPU grow at the same rate. The growth in average old-age pensions for men until the year 2020 is due to the increase in average entitlements as more and more people earn full pension rights. Average pension benefits for men reach a maximum just before 2020, when most male pensioners will have received pension entitlements from their entire working careers.

In 1992 the possibility of earning pension entitlements was reduced for those on high incomes, (but with no effect upon already earned pension entitlements[7]). As Figure 6 shows, this reform will reduce male old-age pension benefits after 2020. Stronger growth in the educational level for women than for men produced a reduction in male/female differences in labour income in the simulation. This also contributes to the reduction in male pensions relative to the average wage level.

Old-age pensions for women continue to grow for the next 60 years, and end up close to the pension level for men. The main reason for this long-lasting growth is a strong increase in female labour participation rates throughout the last 25 years in Norway. Only female cohorts born after 1970 will have these high labour

[7] Before 1992, each unit of income over 8 x BPU (289,600 Nkr in 1992) resulted in a smaller increase in pension entitlements than each unit of income under this level (only a third as large). After 1992, this threshold was reduced to 6 x BPU (217,200 Nkr in 1992). In both cases, any income over 12 x BPU (434,000 Nkr in 1992) led to no extra pension entitlements. In 1992, the additional pensions were also reduced by approximately 7 per cent for a given level of pension entitlements.

participation rates throughout their full working careers. Even though average labour income for women born after 1970 in the simulation is about 70 per cent of that for men in the same generations, the differences in male/female pensions are small. Two important reasons for this are that the pension/wage ratio decreases with income and that persons without labour income receive pension entitlements if they have extensive family care obligations (mothers with young children). Another important contribution comes from the fact that women on average live longer than men. As widows, they receive higher basic pensions than when they were married, and they also inherit pension entitlements from their late husbands. The magnitude of the latter is shown in Figure 6.

Figure 6: Estimated Average Old Age Pensions (Assuming Constant Wage Level)

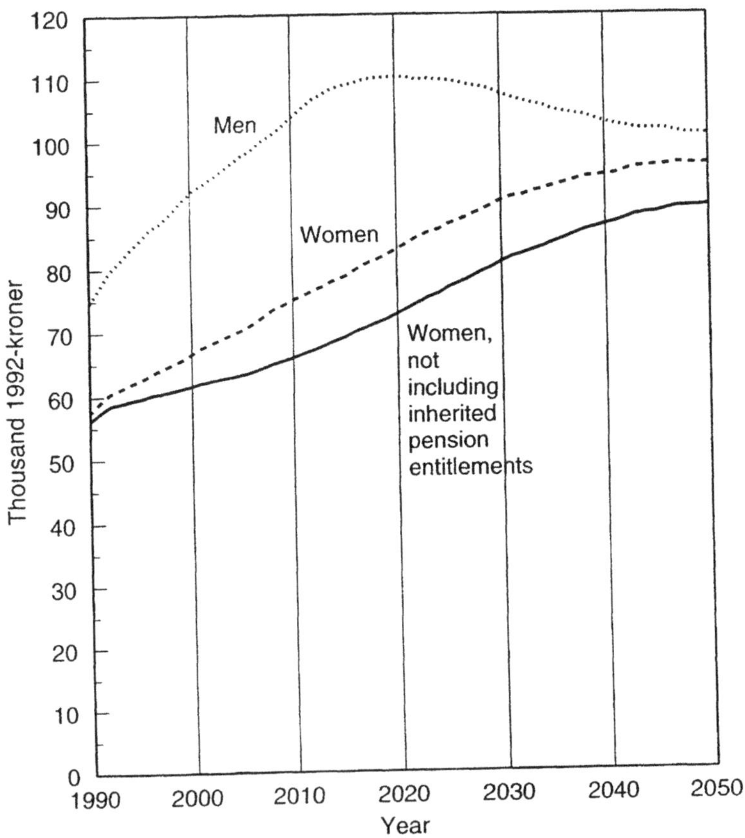

# 4. Contribution Rates

An indicator of the burden of public pension benefits is the ratio between total pensions and total income. This ratio can be thought of as a general tax rate or as the required contribution rate to the social security system. These discussions of contribution rates and income distribution incorporate economic variables that are not explicitly modelled in MOSART. One way of interpreting this is to think of MOSART as representing a closed economy producing a non-durable commodity. This means there is no saving or investment, and future pension payments cannot be based on foreign capital income. Labour is the only production factor and labour income is the only source of income. Total income in the model will then be equal to the sum of labour income, while consumption will be equal to yearly output. This leads to an interpretation of the MOSART model close to the overlapping generations model in Samuelson (1958).

The elderly are, in this type of model, completely dependent on sharing a part of the workers' income. Old-age pensions represent an implicit contract, where those in the labour force (the young) care for the elderly, in return for being cared for by the next generation when they themselves become old. Assuming that the government is only concerned with this type of redistribution, the government budget constraint will be:

$$\Sigma_i p_i = t\Sigma_i w_i h_i \tag{4}$$

where $p_i$ is the pension, $w_i$ is the wage, and $h_i$ is the number of hours worked by individual i. The variable t is the contribution rate (tax rate) necessary for the social security system to be able to honour its obligations. If pensions are also taxed, this relationship becomes:

$$\Sigma_i p_i = t\Sigma_i (w_i h_i + p_i) \tag{5}$$

This budget balance can be interpreted as an implicit social contract. The contribution rate t will vary over time, depending on variations in average income, the size of the labour force, and the number of old-age and disability pensioners. The future developments in the number of pensioners per worker shown earlier in Figure 5, and in average old-age pensions in Figure 6, are thereby important factors in determining the contribution rate.

The contribution rate t has been calculated for the two cases where only workers contribute, as in equation (4), and where pensioners contribute at the same rate as workers, as in equation (5). In 1992, Norwegian firms paid an average of 15.2 per cent of wages to the National Income System (covering both pensions and health insurance), and workers contributed 7.8 per cent of labour income towards old-age and disability pensions. In contrast, pensioners only contributed 3.0 per cent of pension income. It therefore seems that the Norwegian situation is closest to that described in equation (4), which is therefore designated the base line alternative. Figure 7 shows that the contribution rate in the base line alternative will grow dramatically for the next 50 years. From a level of 17-18 per cent today, the contribution level will stabilise at a level just below 30 per cent.

The growth in the contribution rate during the first 20 years is caused by the maturing of the National Insurance System and a major increase in the number of disability pensioners. In this period an increase in the labour force and a (slight) reduction in the number of people older than 67 years are both expected. These socio-demographic changes reduce the contribution rate, but increases in average entitlements are stronger, with a resulting increase in the contribution rate. Just after 2010, the first cohorts of the post-war baby boom reach retirement age, and this will lead to a strong growth in the base line contribution rate.

Figure 7 also shows the effect of equal taxation of labour income and pension benefits. This tax rate is estimated as the ratio between total pensions and the sum of total pensions and total wages each year. The reduction in relation to the base line contribution rate is large, especially after 2010 when pension benefit expenses increase rapidly.

The contribution rate is sensitive to changes in the underlying assumptions. Historically, the basic pension unit (BPU) has been increased at a lower rate than the increase in wages. The BPU, and thus the additional pensions, have mainly been increased with the general inflation rate, while the minimum pension has increased at a slightly higher rate than wage growth.

Figure 8 illustrates how changes in the growth of the politically determined BPU will affect the contribution rate. The base line alternative assumes that the BPU increases at the same rate as wages. In the M:0.75 alternative, the BPU is increased at a rate 0.75 per cent lower than the rate of wage growth (implying that BPU increases at the same rate as prices), and in the M:2.0 alternative, the BPU is increased at a rate 2 per cent lower than wages. Minimum pensions are held constant relative to the wage

Figure 7: Estimated Future Contribution Rates

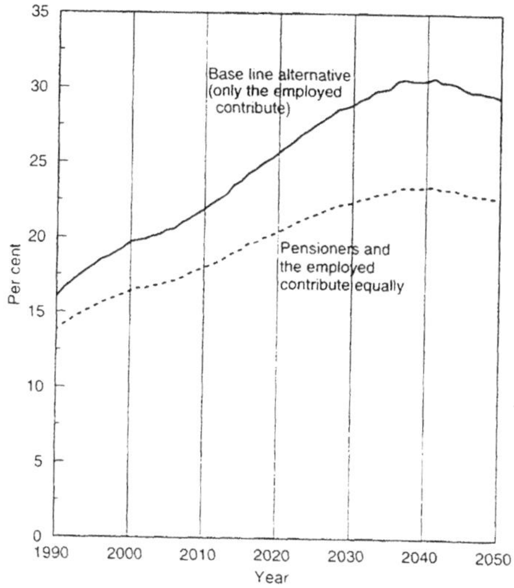


Figure 8: Estimated Future Contribution Rates Under Different Assumptions About Disability and Growth in the BPU

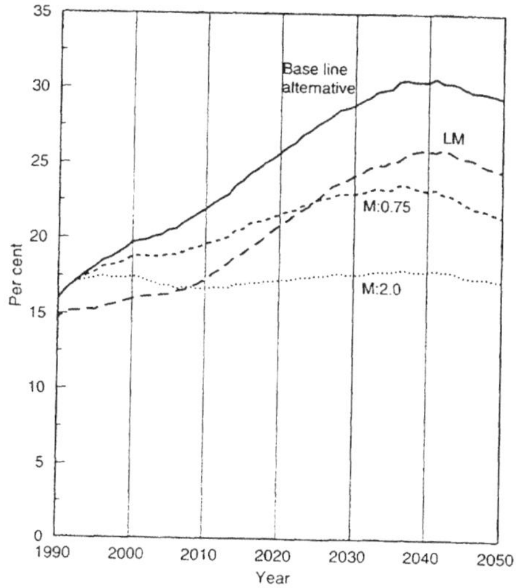


LM: Entry into disability pension reduced by one-third, 1987 labour force participation rates.

M:0.75,M:2.0: BPU increases at a rate which is 0.75 and 2.0 % lower than that of wages.

level in all the above cases. Compared to the base line alternative, where additional pensions are increased at the same rate as wages, the impact of lower growth in the BPU on the contribution rate is considerable.

Since the base line alternative is based on the relatively high disability propensities observed in 1989 and the low labour force participation rates observed in 1991, calculations have been done on where entry into disability is reduced by one third and the high labour force participation rates observed in 1987 are used (71.3 per cent of persons aged 16 to 74 years). The LM-alternative in Figure 8 shows the results of these calculations. Decreased entry into disability and higher labour force participation rates lead to a substantial fall in the contribution rate.

## 4.1 Income Transfers Between Generations

Social justice is one of the basic ideas behind the welfare state. Social justice concerns both income distribution and legal principles about fair and equal treatment by the government. The welfare state consists of a complex set of rules and laws that cover the concerns of a very heterogeneous population. Some of the effects of welfare state institutions are visible immediately, but others have to be judged in a lifecycle perspective. In a social insurance system, a lifecycle perspective must be used to compare the contributions paid with the transfers gained. Such comparisons can be made for different socio-demographic groups, by comparing high versus low income, male versus female, or different birth cohorts. Comparing the contributions and gains of different birth cohorts is to some extent also the aim of the generational accounting approach, such as taken in the study done for Norway by Auerbach et al. (1993).

The complexity of the welfare state and the heterogeneity of the population make dynamic microsimulation a very effective tool for analysing how different population groups are treated by the welfare state. The microsimulation approach makes it possible to take into account in great detail the socio-demographic structure of the population and to use a relatively precise description of the laws and rules that apply. The weak point in the present model is the modeling of economic behaviour.

In the following, an attempt is made to analyse how lifecycle income is influenced by a transfer system, such as the National Insurance System, based on a Pay-As-You-Go social contract. Whether the present system has any redistributive effects across generations is also examined.

In the closed economy interpretation of MOSART discussed earlier, it was established that net income $y_i$ for individual i in a given year will be

$$y_i = w_i h_i + p_i - t(w_i h_i + p_i) \quad (6)$$

where $w_i$ is the wage, $h_i$ is the number of hours worked, $p_i$ is the pension and *t* is the contribution rate. For most people either $p_i$ or $w_i h_i$ equals zero. The distribution of net income over each individual's life time depends on gross income, the tax structure and, of course, the rules that apply in the pension system.

In the simple economy with only a non-durable good, lifetime income can be seen as simply the sum of yearly incomes. This is a similar assumption to that used in calculating lifecycle earnings in Wolfson (1988) and Harding (1993). If one abandons this simple interpretation of MOSART and assumes that there are financial markets where it is possible to save at a given interest rate, it is natural to discount income. A certain income earned today is of greater value than income earned later. The interest rate is an exogenous variable in MOSART in the same manner as average wages. In the base line projection, these variables were determined in a separate macroeconomic model.

In the following, an attempt is made to analyse the effect the National Insurance System has on the distribution of income, both between generations and within each generation. The key concept is individual pension wealth, calculated as the difference between the discounted value of pension benefits and payments into the National Insurance System (pension premiums) over the life course:

$$\text{Pension wealth} = \sum t[(\text{pension benefit - pension premium})\,(1+r)^{-t}] \quad (7)$$

where

t = the number of years since this person was 16 years old, and

r = the discount rate

It is assumed that the discount rate is equal to the real interest rate. It is further assumed that that the pension premium is the estimated contribution rate discussed earlier, multiplied by each individual's labour income. This probably gives a reasonable representation of how the National Insurance System actually has been financed. Full lifetime income series are available only for persons born after 1950, making it difficult to compare the pension wealth of those born before 1950 with those born after that year. An extra complication is the fact that there were some public pension schemes prior to 1967.

The premiums paid into the National Insurance System may be considered as an alternative to saving (bank deposits, private pension funds). The size of these premiums depends upon the number of persons who contribute to the insurance system and the size of the total pensions being paid out. Figure 9 shows the development in pension wealth as a percent of lifetime labour income, using different interest rates (discount rates) and assuming that wages grow by 0.75 per cent a year from 1989 onwards.

All three cases illustrated in Figure 9 are based on the base line alternative (see Table 2 for the main assumptions). The only difference between them is that the resulting pension wealth has been calculated using three different discount rates. As before, the basic pension unit BPU (which determines individual pension benefits) increases at the same rate as wages. It is only the difference between the interest rate and the growth rate of wages which matters in this model. If these are equal, it can be seen from Figure 9 that individuals born after 1980 receive a return on their contributions about equal to what they could have received by saving privately. Those born before 1980 receive more than they otherwise could achieve, because there were relatively few pensioners when they paid their premiums and the pensions paid were fairly low. Because of the increase in additional pensions, average pension benefits will continue to increase faster than wages until the year 2020.

If the discount rate is higher than wage growth, future generations may lose a substantial part of their potential lifetime incomes (what they could achieve by saving privately instead of being a member of the National Insurance System). In this case, the National Insurance System gives a lower return than the market, as shown by the lower lines in Figure 9. The higher the discount rate compared with the growth in wages, the larger the loss for future generations. Another interpretation is that current pension premiums are too low to cover the pension entitlements earned today, and that the interest on the accumulated 'deficit' will inflict a loss on all future generations. This redistribution is not necessarily unjust, because future generations may still achieve higher incomes than current generations, as can be seen from Figure 10.

Other ways of financing the National Insurance System are shown in Figure 11 (which assumes a discount rate of four per cent and wages growth of 0.75 per cent

Figure 9: Estimated Pension Wealth as a Percentage of Lifetime Labour Income, Using Different Discount Rates

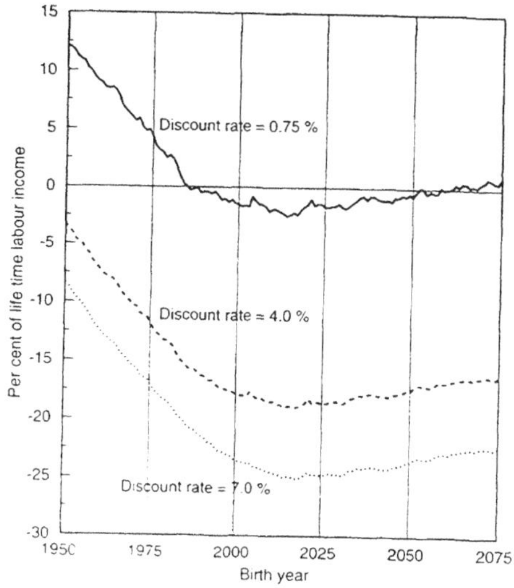


Smoothed time series

Figure 10: Estimated Lifetime Income for Different Birth Cohorts

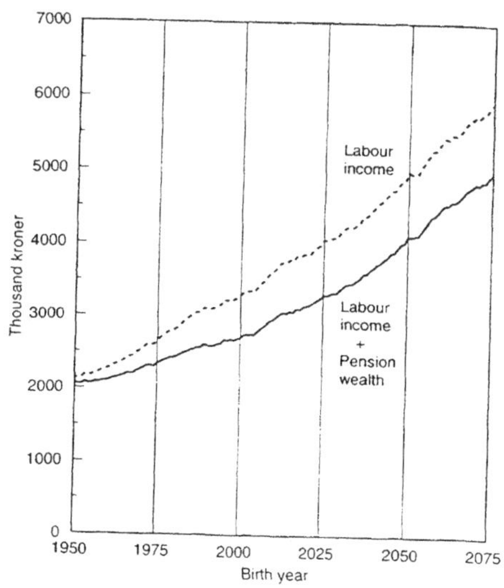


Smoothed time series

per year). The baseline alternative in Figure 11 is the same as the line in Figure 9 showing pension wealth as a percentage of lifetime income with a discount rate of 4 per cent. The M: 0.75 alternative shows the movement in relative pension wealth if BPU increases at a rate 0.75 per cent lower than that of wages (implying that the BPU increases at the same rate as inflation), all other assumptions being equal to the base line alternative. A lower increase in BPU leads to a less generous pension system, reducing the relative loss experienced by future generations. The F67 and F95 alternatives show the development in relative pension wealth if the old-age disability pension system were required to be self-financing from 1967 and 1995 onwards respectively.

Figure 11: Estimated Pension Wealth as a Percentage of Lifetime Labour Income, Under Various Funding Scenarios

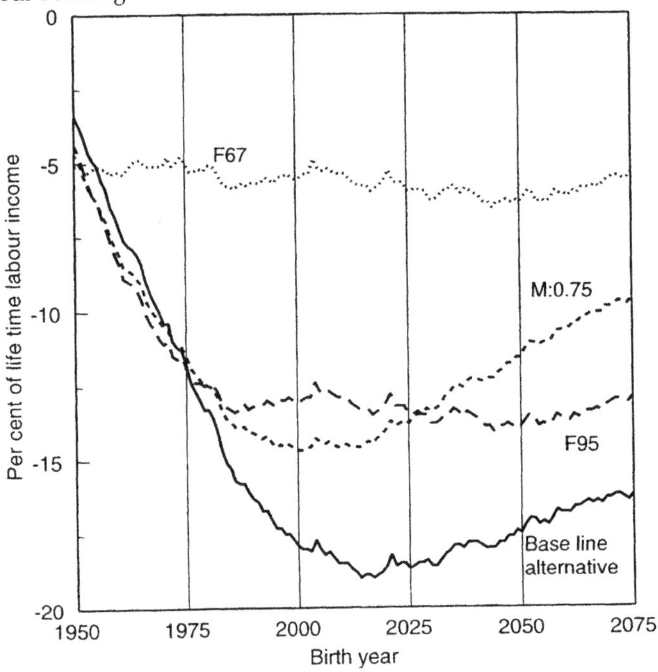


F67,F95: Funded, with a constant contribution rate from 1967 and 1995 respectively.

M:0.75: BPU increases at a rate which is 0.75 % lower than that of wages.

Smoothed time series.

The year 1967 was when the National Insurance System was started. In both these alternatives, the contribution rate is assumed to be constant in all future years, requiring the building of large funds in the early years to fund the later increase in the number of elderly.[8] If properly funded since the beginning in 1967, the contribution rate could have been 17.5 per cent, assuming a discount rate of 4 per cent and a wage growth of 0.75 per cent. Related to a long-run contribution rate of 29 per cent (see Figures 7 and 8), this would have reduced the loss of future generations substantially. The loss of about 5 per cent shown in Figure 11 is mainly a result of the need to pay the pensions of generations that in 1967 had few or no years left of their working careers.

If the National Insurance System were to be funded starting in 1995, the pension premium rate would have to be 25 per cent. Related to a current contribution rate of just below 20 per cent, this leads to only small reductions in future losses. Another alternative is to reduce future benefits by allowing the BPU, and thus the additional pensions, to increase at a lower rate than wages growth, without changing the minimum pension. Letting additional pensions grow by 0.75 per cent less than wages will slightly reduce the pension wealth of generations born before 1975, and increase that of all future generations.

## 4.2 Further distribution analysis

Income distribution within generations is a complicated matter, since the redistributive effects of pension wealth depend on differences in purely actuarial elements (how long one lives and the probability of being disabled) as well as income inequalities. It would be most appropriate to measure expected pension wealth, while the model generates actual pension wealth.

When the latter is examined with this reservation in mind, the redistributive effects are found to be large. Gini-coefficients for discounted lifetime income for a synthetic generation are reduced by 15 per cent in the base line alternative compared with a situation without a pension system. High discount rates reduce the redistributive effects, because pension benefits are received late in life. The present system, with additional pensions, has a greater redistributive effect than a system where an equal pension is paid to everyone (at today's low level of the minimum pension). The reason is that additional pensions increase the average pension, and this is more

8 This was the intention when the National Insurance System was established in 1967, but it was abandoned in the seventies. The size of these funds is today far too small to meet the increase in the contribution rate after 2010. But total public wealth, mainly future taxes on petroleum production, may change this conclusion.

important than the differences in pension benefits. In terms of pension wealth, the National Insurance System benefits women (at the cost of men). One important reason is that women receive higher pensions relative to previous income. Two other important aspects are that women become pensioners at a lower age (through disability pension) and they live longer.

## 5. Conclusion

This paper has described the construction of MOSART, a dynamic microsimulation model which projects population size and composition, labour force, educational level and future pension benefits. The model was used to analyse possible long-run developments in disability and old-age pensions covered by the Norwegian National Insurance System.

In the early part of the next century, Norway will see a large increase in the number of pensioners relative to the number of workers. Since the Norwegian pension system is based on a Pay-As-You-Go principle, this implies that workers in the future will have to contribute an increasing portion of their labour income to the pension system. This, in turn, means that early generations possess a larger pension wealth as a percentage of lifetime labour income than later generations. Even so, the assumption that real wages will grow by 0.75 per cent a year implies that future generations will be better off than earlier generations. Making the pension system less generous, by not letting the basic pension unit BPU increase at the same rate as wages, would lessen the differences in pension wealth as a percentage of lifetime labour income, but would increase differences in income levels between generations. Changing to a funded system in 1995 would also lessen the differences in relative pension wealth.

## References

Andreassen, L. (1993), *Demographic Forecasting with a Dynamic Stochastic Microsimulation Model*, Discussion Paper No. 85, Statistics Norway, Oslo.

Andreassen, T., G. Spurkland and Y. Vogt (1992), 'MOSART, a microsimulation model', Paper prepared for the 1992 Conference on Computing for the Social Sciences, University of Michigan, Ann Arbor, May.

Andreassen, L. T., G. Andreassen, G.S. Fredriksen, G. Spurkland and Y. Vogt (1993), 'Framskriving av arbeidsstyrke og utdanning'. Mikrosimuleringmodellen MOSART 1, *Rapporter*, Vol. 93, No. 6, Statistics Norway, Oslo.

Auerbach, A.J., J. Gokhale, L.J.Kotlikoff and E. Steigum Jr. (1993), 'Generational accounting in Norway: Is Norway overconsuming its petroleum wealth?', Unpublished paper presented to the Conference of Quantitative Aggregate Economics at Norwegian School of Economics and Business, Bergen, August.

Harding, A. (1993), *Lifetime Income Distribution and Redistribution - Applications of a Microsimulation Model*, North-Holland, Amsterdam.

Nelissen, J.H.M. (1991), 'Household and education projections by means of a microsimulation model', *Economic Modelling*, Vol. 9, No.4.

Nelissen, J.H.M. and A.P. Vossen (1989), 'Projecting household dynamics: A scenario-based microsimulation approach', *European Journal of Population*, Vol. 5, No. 3.

Samuelson, P.A. (1958), 'An exact consumption-loan model of interest with or without the social contrivance of money', *The Journal of Political Economy*, Vol. 66, No. 6.

Wolfson, M.C. (1988), 'Homemaker pensions and lifetime redistribution', *Review of Income and Wealth*, Vol. 34, No. 3.

PART FOUR

# Micro/Macro Links and Firm Behaviour

G. A. Meagher

Centre of Policy Studies, Monash University
Australia

# Chapter 16: Forecasting Changes in the Distribution of Income: An Applied General Equilibrium Approach

## Abstract

*This paper is concerned with the development of techniques for applying an existing set of detailed economic forecasts for the Australian economy to the question of income distribution. The centrepiece of the forecasting system is a large dynamic applied general equilibrium model, the MONASH model. Via the markets incorporated in the MONASH model, a range of otherwise intractable information is brought to bear on a forecast of the incomes of various groups of individuals over the period 1990-91 to 1996-97. The range includes detailed scenarios on macroeconomic developments in both the world and domestic economies, changes in the world prices of internationally traded commodities, changes in protection and primary factor saving technical change. To generate the income forecasts, projections of changes in factor prices and factor employment levels from the MONASH model are interfaced with micro data from the 1990 Australian income survey. The key to the interface is an allocation of the ownership of the relevant factors among the individuals identified in the survey. Results are reported for sixty income groups differentiated by region and size of income.*

## 1. Introduction

Changes in the incomes of individuals or groups of individuals over time depend on changes in their personal characteristics (such as age and marital status) and changes in the characteristics of the economy of which they form a part (such as wage rates and employment levels). The forecasting of changes in these two general categories has tended to be undertaken independently - the first by analysts trained in demography and the second by analysts trained in economics. However, in recent times, demographers have begun to link their dynamic microsimulation models with simple macroeconomic models, and economists have begun to employ microsimulation techniques to enhance the range of their forecasting models in the area of distribution. This paper describes a contribution to the development of such an integrated forecasting capacity, beginning with a well-established forecasting system for the Australian economy.[1]

A forecast of changes in the distribution of income can be regarded as consisting of the construction and linking together of four major components:

- a forecast of changes in factor incomes (ie. wages and profits) over time;
- a description of the distribution of factor incomes between individuals to give personal incomes at a particular point of time;
- a description of the redistribution of personal incomes between individuals at a particular point of time, particularly via the tax/benefit system; and
- a forecast of changes in the individuals' ownership of factors, tax liabilities and eligibility for benefits over time.

The first of these components is the province of economic modelling, and a wide variety of relevant models are already in existence. This study relies on the extensive set of economic forecasts for the period 1990-91 to 1996-97 recently released by the Centre of Policy Studies at Monash University, in association with the commercial forecasting agency Syntec Economic Services. The forecasts are derived, in part, by solving a large dynamic applied general equilibrium (AGE) model of the Australian economy, the MONASH model. As far as distributional analysis is concerned, an important feature of the AGE approach is that it explicitly models the determination of factor incomes on an industry basis. In the case of MONASH, the level of disaggregation is quite high, with 112 industries being identified in each of six regions. This industry dimension matters, because different individuals earn their factor incomes in different industries - so variations in factor incomes across industries translate into variations in personal incomes. Less obviously, the industry disaggregation allows *a priori* structural information to be built into the forecasts - information which by its nature has a differential impact on industries and hence on personal incomes. This kind of influence on distribution would go unaccounted for in an aggregate macroeconomic model.

---

[1] For a recent Australian example of an integrated distributional model beginning from the demographic side, see Antcliff (1993).

The second and third components are associated with static microsimulation models and are based on unit record data from a sample survey recording the amounts of income individuals obtained from a range of sources in a particular year. Such models have been used extensively to analyse the first-round impact or 'morning after' effect of changes in the tax/benefit system. By assumption, the changes do not affect factor incomes so there is no need to employ an economic model. So popular have the static microsimulation models become that, by sheer weight of analyses, distribution is threatening to turn into a branch of tax policy. The difficulty with analysing changes in other government policies, changes that would appear on the surface to be no less important or interesting, is that their effect on personal incomes is mediated by their effect on factor incomes. While suitable economic models are available for analysing the latter, methods for linking them with microsimulation models are not well developed.

The same considerations apply to forecasting as to policy analysis, and the primary concern of the present chapter is the establishment of appropriate linkages between the economic and microsimulation models. The key ingredient is not so much a good description of income sources of individuals as a good description of their ownership of the factors of production (or, more correctly, the employed factors). Information on ownership is not generally included in an income survey and hence it must be inferred from the survey data and other sources.

The fourth component is associated with dynamic microsimulation models. The demographic techniques characteristic of the genre are particularly well suited for handling changes in income that result from an individual's position in the life cycle. Examples are changes in eligibility for various government benefits, changes in the full-time wage rate as work experience is accumulated, and changes in labour force participation as children are born and raised. Given the focus of this paper, only the most rudimentary attention is paid to the demographic ageing of the population over the forecast period. However, it is envisaged that the methods developed here will be adapted to link the MONASH model with a properly specified dynamic microsimulation model in due course.

Two other limitations of the present study are worth noting. The first concerns the role of financial assets, which can be divided conceptually into bonds (which earn interest), equities (which earn dividends and capital gains) and cash (which earns nothing). A comprehensive treatment would require the incorporation of markets for financial assets in the MONASH model, and the specification of the savings behaviour and asset portfolios of all the individuals in the income survey. The treatment here is much simpler than this and ignores capital gains altogether (ie. the effect on distribution of the difference between corporate retained earnings and dividend payments in each year is ignored).[2] The other limitation concerns implicit incomes such as rent on owner-occupied housing, interest on life and superannuation funds, and payment in kind for household production. With the exception of the first, these income sources are also ignored. However, the MONASH modelling framework is quite flexible and more sophisticated treatments could be adopted at a later date.[3]

In the balance of the paper, the economic forecasts are discussed in somewhat more detail in Section 2, both in terms of methodology and results. The same kind of discussion is then repeated for the distributional forecasts in Section 3. Section 4 contains concluding remarks.

## 2. The Economic Forecasts

### 2.1 Methodology

The MONASH model is a development of the well known ORANI model (Dixon et al., 1982) which has been widely used for policy analysis in Australia by economists located in universities, the government and the private sector[4]. The original version of ORANI was only suitable for comparative-static analysis, but the more recent ORANI-F version (Horridge et al., 1993) includes enough accumulation relations to allow the average annual growth rates of macroeconomic and structural variables to be forecast over the medium run. In MONASH, the dynamic specification has been refined so that year-by-year time paths for the variables are now provided (Adams et al., 1994). The MONASH model is the centrepiece of a forecasting system that currently proceeds in four stages.

The fundamental data requirement for an applied general equilibrium model is an input-output table. In Australia, input-output tables are produced by the Australian

---

[2] Note that, in our stylized representation of the forecasting process, dividends and capital gains comprise that part of profits retained by corporate enterprises, and interest payments and receipts comprise a redistribution of personal incomes.

[3] For example, Lim and Meagher (1988) have added markets for financial assets to an antecedent of the MONASH model.

[4] A recent review of the application of the ORANI model to policy analysis in Australia can be found in Powell and Snape (1993).

Bureau of Statistics (ABS), with a lag of about five years. Hence the first stage in the forecasting process is to update the latest available input-output data to the base year for the forecast (from 1986-87 to 1990-91, in the present case) via an ORANI simulation. The method (Dixon and McDonald, 1993) results in updated data which are consistent with all relevant published data about the structure of the economy in 1990-91.

The second stage begins with the determination of the values of the exogenous variables of the MONASH model over the forecast period. Forecasts for most exogenous macroeconomic variables are provided by Syntec Economic Services (1993). Syntec believes there will be a slow recovery in the world economy, with increased activity in the United States eventually stimulating recovery in Japan and Europe. As demand picks up, the prices of primary commodities will gradually strengthen, leading to an improvement in Australia's terms of trade and an average GDP growth of about 2.5 per cent per annum. The share of international trade in economic activity is forecast to increase sharply, with aggregate imports and aggregate exports both growing at nearly three times the rate of GDP.

For the exogenous structural variables, the main sources of information are the Australian Bureau of Agricultural and Resource Economics, the Bureau of Tourism Research and the Industry Commission. According to forecasts by these institutions, industries in the transport and recreation sectors will benefit from continued strong growth in international tourism, while heavily protected industries (particularly automotive vehicles, textiles, clothing and footwear) will suffer from the Australian government's ongoing program of phased reductions in barriers against manufactured imports.

To complete the second stage, the implications of these exogenous forecasts for a range of structural variables are computed using the MONASH model. The endogenous variables include outputs by commodity and industry, employment and capital formation by industry, and all components of domestic usage (ie. usage by firms, households and the government) by commodity for imports and domestically produced goods.

At stage three, the ORANI regional equation system ORES is used to convert economy-wide forecasts into regional forecasts for the Australian States (for a description of ORES see Dixon et al., 1982, Chapter 6). In this calculation, a commodity is classified as national if it is traded extensively across State boundaries, or as local if each State satisfies most of its own demand for the commodity. For national commodities, ORES allocates the Australia-wide output growth rate to each State. For local commodities, ORES imposes regional-balance constraints which generate regional income multipliers. This treatment implies that, if a State is specialised in the production of national commodities with relatively rapid output growth rates, the producers of local commodities in that State will experience growth rates above the Australia-wide average.

The final stage of the forecasting process converts growth rates in employment by industry at the national and regional levels into growth rates in employment by occupation. The forecasts are produced at the major, minor and unit group levels of the Australian Standard Classification of Occupations, the last of which identifies 283 different occupations. The database for the occupational disaggregation facility is a cross-classification of employment by industry, occupation and State, compiled from the 1986 Census of Population and Housing.

## 2.2 Results

Table 1 shows results of the forecasting process for a selection of variables relevant to the calculation of changes in the distribution of income. Although the process generates forecasts separately for each year between 1990-91 and 1996-97, only average annual rates of growth are considered in this paper. The results have also been heavily aggregated across industries and occupations to bring the classifications into line with those employed in the Australian Bureau of Statistics (ABS) income survey. This involves a significant loss in distributional information, but it is one that cannot be avoided with the current mismatch between the levels of detail in the survey and in the other data sources used by the MONASH model. In the ORES regional classification, and hence in Table 1, the Australian Capital Territory is included with New South Wales and the Northern Territory is included with South Australia.

Consider first the economy-wide forecasts of output by industry. According to the table, the output of the *Mining* industry in Australia is forecast to grow at an average rate of 3.55 per cent per annum over the forecast period. This is the largest growth rate of all the industries shown and is 1.09 percentage points more than the growth in GDP. The *Mining* industry is distinguished by the large share of its output exported to foreign countries and its good growth prospects derive from the forecast increase in international trade as a share of GDP. Two other industries , namely *Transport, storage and communication* and *Recreation, personal and other services*, have growth rates of more than 3 per cent per annum. The first benefits from substantial linkages with international trade (both imports and exports in this case), while the second owes its favoured position to the strong growth forecast for international tourism. The prognosis for *Construction*, which has the slowest growth, is crucially affected by the excess stocks of commercial buildings that accumulated in Australia's major cities during the 1980's. It is worth noting here that *Residential construction* is forecast to grow at the quite healthy rate of 3.27 per cent per annum but, as far as the distributional analysis is concerned, this potentially important distinction between the two types of construction activity is lost in the industry aggregation. The *Manufacturing* industry does relatively poorly because it suffers most from the planned reductions in protection.

Now consider the relative output growth rates across the regions. Queensland and Western Australia do well because they are relatively specialised in national industries which grow rapidly, especially *Mining*. Victoria and South Australia, on the other hand, are relatively specialised in the slow growing *Manufacturing* industry and have correspondingly poor prospects. Recall that local industries are stimulated in regions with rapidly growing national industries. As a result of this multiplier effect, all industries in Queensland (at least at the level of aggregation shown in Table 1) have growth rates greater than or equal to the national average. Similarly most industries in Victoria and South Australia grow more slowly than the national average.

The second panel in Table 1 shows forecasts of employment growth by industry. The output and employment forecasts are related by production functions which determine the increase in output associated with given increases in inputs (capital or labour) and a given rate of technical progress. In the present forecasts, factor-saving technical change of one per cent per annum is assumed. Hence, when output growth rates are converted into employment growth rates, the first one per cent is absorbed by technical change and the latter rates tend to be less than the former.

The change in capital inputs depends critically on whether an industry was under- or over-capitalised in the base period, ie. on whether the rate of return in the industry was above or below the average across industries. An industry with a relatively high rate of return attracts investment and enjoys a relatively high rate of capital growth. For a given rate of output growth and technical change, this implies a relatively low rate of employment growth. Conversely, an industry which is relatively over-capitalised in the base period will tend to have a relatively high rate of growth in employment. It is these differences in the degree of capitalisation that are primarily responsible for the differences in the industry rankings for output and employment growth.

Table 1: MONASH Forecasts (Average Annual Growth Rates, 1990-91 to 1996-97 Per Cent)

| Variable | New South Wales | Victoria | Queensland | South Australia | Western Australia | Tasmania | Australia |
|---|---|---|---|---|---|---|---|
| **Output by industry:** | | | | | | | |
| Agriculture, forestry, fishing and hunting | 1.59 | 1.85 | 2.01 | 1.72 | 1.62 | 1.40 | 1.75 |
| Mining | 3.84 | 3.65 | 3.78 | 3.68 | 3.13 | 3.57 | 3.55 |
| Manufacturing | 1.29 | 0.92 | 1.73 | 1.13 | 1.78 | 1.35 | 1.25 |
| Public utilities | 2.43 | 2.33 | 2.51 | 2.36 | 2.47 | 2.42 | 2.42 |
| Construction | 0.42 | 0.53 | 1.16 | -0.48 | 1.33 | 0.04 | 0.58 |
| Wholesale and retail trade | 2.89 | 2.71 | 3.06 | 2.77 | 3.04 | 2.77 | 2.87 |
| Transport, storage and communication | 3.31 | 3.14 | 3.39 | 3.10 | 3.43 | 3.36 | 3.27 |
| Finance, property and business services | 2.60 | 2.29 | 2.83 | 2.40 | 2.78 | 2.56 | 2.56 |
| Ownership of dwellings | 2.37 | 2.10 | 2.60 | 2.28 | 2.34 | 2.25 | 2.32 |
| Public administration and defence | 2.80 | 2.80 | 2.80 | 2.80 | 2.78 | 2.76 | 2.80 |
| Community services | 3.13 | 2.06 | 3.18 | 2.08 | 3.12 | 3.13 | 2.75 |
| Recreation, personal and other services | 3.25 | 3.00 | 3.38 | 3.05 | 3.22 | 3.19 | 3.18 |
| All industries | 2.45 | 2.22 | 2.72 | 2.16 | 2.63 | 2.39 | 2.46 |
| | | | | | | | |
| **Employment by industry:** | | | | | | | |
| Agriculture, forestry, fishing and hunting | 0.44 | 1.02 | 1.23 | 0.76 | 0.46 | 0.41 | 0.78 |
| Mining | 4.28 | 2.02 | 3.54 | 1.27 | 0.67 | 0.47 | 2.44 |
| Manufacturing | 0.06 | -0.50 | 0.53 | -0.01 | 0.74 | -0.22 | -0.04 |
| Public utilities | 1.37 | 1.24 | 1.43 | 1.27 | 1.39 | 1.36 | 1.34 |
| Construction | -0.53 | -0.40 | 0.26 | -1.49 | 0.47 | -0.92 | -0.35 |
| Wholesale and retail trade | 1.48 | 1.31 | 1.65 | 1.37 | 1.62 | 1.37 | 1.47 |
| Transport, storage and communication | 2.15 | 2.00 | 2.25 | 1.98 | 2.31 | 2.37 | 2.14 |
| Finance, property and business services | -0.01 | -0.31 | 0.23 | -0.19 | 0.22 | -0.02 | -0.04 |
| Ownership of dwellings | 0.00 | 0.00 | 0.00 | 0.00 | 0.00 | 0.00 | 0.00 |
| Public administration and defence | 1.89 | 1.89 | 1.89 | 1.89 | 1.89 | 1.88 | 1.89 |
| Community services | 2.23 | 1.14 | 2.28 | 1.17 | 2.22 | 2.23 | 1.84 |
| Recreation, personal and other services | 2.71 | 2.46 | 2.82 | 2.50 | 2.65 | 2.63 | 2.63 |
| All industries | 1.16 | 0.73 | 1.48 | 0.87 | 1.37 | 1.16 | 1.09 |

| Variable | New South Wales | Victoria | Queensland | South Australia | Western Australia | Tasmania | Australia |
|---|---|---|---|---|---|---|---|
| **Gross operating surplus by industry:** | | | | | | | |
| Agriculture, forestry, fishing and hunting | 3.94 | 4.47 | 4.81 | 4.40 | 4.27 | 3.89 | 4.36 |
| Mining | 6.33 | 6.11 | 7.44 | 6.06 | 5.40 | 5.54 | 6.28 |
| Manufacturing | 3.24 | 4.85 | 5.71 | 5.05 | 5.05 | 4.54 | 4.43 |
| Public utilities | 5.82 | 6.46 | 6.74 | 6.47 | 6.38 | 6.40 | 6.31 |
| Construction | 2.35 | 3.08 | 3.73 | 2.24 | 3.58 | 2.55 | 2.93 |
| Wholesale and retail trade | 2.98 | 4.88 | 5.42 | 4.87 | 4.54 | 4.50 | 4.22 |
| Transport, storage and communication | 5.25 | 6.12 | 6.50 | 6.08 | 6.07 | 6.12 | 5.84 |
| Finance, property and business services | 3.38 | 4.16 | 4.76 | 4.23 | 4.31 | 4.12 | 3.88 |
| Ownership of dwellings | 4.68 | 4.41 | 4.91 | 4.59 | 4.64 | 4.56 | 4.62 |
| Public administration and defence | 7.44 | 7.44 | 7.44 | 7.44 | 7.44 | 7.44 | 7.44 |
| Community services | 4.00 | 4.80 | 5.79 | 4.79 | 5.10 | 5.18 | 4.65 |
| Recreation, personal and other services | 4.11 | 5.95 | 6.51 | 5.94 | 5.56 | 5.65 | 5.30 |
| All industries | 3.99 | 4.87 | 5.59 | 4.80 | 4.87 | 4.63 | 4.66 |
| **Employment by occupation:** | | | | | | | |
| Managers and administrators | 1.00 | 0.94 | 1.34 | 0.84 | 1.08 | 0.88 | 1.03 |
| Professionals | 1.55 | 0.80 | 1.77 | 1.01 | 1.66 | 1.60 | 1.32 |
| Para-professionals | 1.72 | 1.08 | 1.87 | 1.11 | 1.69 | 1.92 | 1.53 |
| Tradespersons | 0.70 | 0.37 | 1.16 | 0.60 | 1.17 | 0.65 | 0.73 |
| Clerks | 0.98 | 0.68 | 1.39 | 0.79 | 1.16 | 1.15 | 0.98 |
| Salespersons and personal service workers | 1.55 | 1.02 | 1.67 | 1.27 | 1.60 | 1.54 | 1.41 |
| Plant and machine operators, and drivers | 1.14 | 0.53 | 1.42 | 0.81 | 1.40 | 0.86 | 1.02 |
| Labourers and related workers | 1.05 | 0.59 | 1.53 | 0.74 | 1.42 | 0.95 | 1.03 |
| All occupations | 1.16 | 0.73 | 1.48 | 0.87 | 1.37 | 1.16 | 1.09 |
| **Other variables:** | | | | | | | |
| Nominal wage rate | | | | | | | 3.66 |
| Consumer price index | | | | | | | 2.96 |

The third panel shows the forecast changes in gross operating surplus (GOS), an industry's GOS being the product of its capital input and its gross rate of profit.[5] These changes are generally larger than the output growth rates because they are expressed in nominal, rather than real, terms. The rate of inflation incorporated in the GOS forecasts is approximately equal to the forecast for the consumer price index given at the bottom of the table.

The forecasts of employment by occupation shown in the fourth panel are obtained by first converting the MONASH employment forecasts for 112 industries into forecasts for 283 occupations region-by-region using Census data, and then aggregating across occupations. Averaging occurs at both stages of the process with, to take the case of Victoria as an example, the range of the forecasts being reduced first from (-7.91 : +7.74) to (-6.69 : +3.44) and then to (+0.37 : +1.08). The narrowness of the final range again illustrates the limitations imposed on distributional analysis by the classifications of industries and occupations adopted in the ABS income survey.

The final panel of Table 1 shows forecasts for the nominal wage rate and the consumer price index. Given the highly centralised system of wage fixation that has traditionally existed in Australia, the MONASH model currently assumes any variations over time in relative industry-specific and/or region-specific average wage rates may be safely ignored in comparison with variations in the economy-wide average wage rate. Recently there has been some pressure for a shift to enterprise bargaining which, if it comes to fruition, may require a change in this modelling strategy. For the time being, however, all nominal wage rates are taken to move together and hence only a single forecast is required.

# 3. The Distributional Effects

## 3.1 Methodology

The distributional projections rely primarily on data from the ABS 1990 Survey of Income and Housing Costs and Amenities (or IHS data, for short), which shows the amounts of income received from various sources in 1989-90 by about thirty

---

[5] In practice, part of GOS represents the wages of owner-operators. In the simulations reported here, wages are imputed to owner-operators in rural, construction, retail trade, transport, insurance and health industries and modelled as a return to labour rather than as a return to capital.

thousand persons. The income sources fall into five broad categories: wages and salaries, income from unincorporated enterprises, property income, government benefits and other income. Using population weights produced as part of the survey, the individual income profiles can be combined to form profiles for various broad income groups of interest. This paper considers sixty such groups, differentiated by size of income and region of residence. In defining the groups, income recipients are first classified into deciles and then, for each decile in turn, into six State regional groups. Income deciles are defined using equivalent net (ie post-tax) incomes and Kakwani's (1986) equivalence scale, which gives a single adult a value of 1, second and subsequent adults a value of 0.7 and children a value of 0.4.

The distributional model draws on the MONASH/ORES results in Table 1 to compute the implied growth rates in the incomes of the selected groups. In the remainder of this subsection, the methods used to determine the income changes are discussed for each of the five broad categories of income in turn.

(a) Wages, salaries and supplements

For labour incomes, the computation involves the following steps.

(i) Among the 30,444 respondents to the income survey, 10,849 did not work at all during 1989-90. All persons belonging to the latter group, including 461 who collected some unemployment benefits during the year, are assumed to remain outside the employed labour force during the forecast period.

(ii) The respondents who did work at some time during 1989-90 are divided first into seven groups according to their region of residence. The last of these groups, representing residents of the Northern Territory and the Australian Capital Territory, is excluded from the distributional calculation as the regions concerned are not separately identified in the ORES model.[6] Each of the remaining regional groups is then divided into 90 subgroups differentiated by size of income and occupation, the latter choice reflecting the view that occupation is more

---

[6] In Table 1, the Northern Territory is included with South Australia and the Australian Capital Territory with New South Wales, introducing a small discrepancy between the IHS and ORES regional classifications. This discrepancy is ignored.

important than industry in determining a person's employment opportunities.

(iii) For the included respondents, the income survey records the number of weeks during 1989-90 that each was employed, unemployed or not in the labour force. Hence the average labour force status over the year, measured in person-years, can be calculated separately for the 540 groups defined at step (ii).

(iv) The forecast average annual rates of growth of employment by occupation and region are available from Table 1. According to Syntec, the working age population will grow at an average annual rate of 1.41 per cent during the forecast period. Hence, given the assumption of a uniformly expanding workforce, 1.41 percentage points of the forecast growth of 1.72 per cent per annum for *Para-professionals* in New South Wales, for example, is absorbed by population growth. In other words, employment of the pre-existing population of *Para-professionals* in New South Wales is forecast to increase by only 0.31 per cent per annum. Changes in employment are calculated in this way for each of the 54 groups identified by occupation and region (the OR groups). Note that employment actually contracts for many of these groups.

(v) An increase in the number of employed persons belonging to an OR group must be matched by a corresponding decrease in the number of persons without a job. Such changes are distributed between unemployed persons and persons not in the labour force on the assumption that the ratio of the number of unemployed to the number without a job remains constant for each group. This assumption amounts to the specification of implicit labour supply functions for the individuals involved. It turns out that the changes in employment are not sufficient to exhaust the stocks of potential workers in any of the OR groups.

(vi) The changes in labour force status by occupation and region must now be allocated across income groups. If employment for an OR group increases, it is assumed that the number of persons without a job decreases in such a way that each income group maintains its base period share of unemployed persons and of persons not in the labour force. The distribution of employment between income groups is then determined as a residual. Thus, if no *Para-professionals* belonging to a particular income decile were without jobs in New South Wales in 1989-90, that income group does not participate in the 0.31 per cent average annual growth rate of employment for the OR group. If employment for an OR group decreases, it is assumed that each income group maintains its base period share of employed persons and that unemployed persons account for a constant share of persons without a job for each income group. The method is designed to ensure that, for each income group, the stock of persons in any labour force category is never exhausted.

(vii) At the conclusion of these machinations, the changes in the numbers of employed and unemployed persons by occupation, region and income decile are determined. The changes in wages, salaries and supplements are then obtained by adding into the calculation the increase in the nominal wage rate of 3.66 per cent per annum.

(b) Income from unincorporated enterprises

The computation for this category involves four steps.

(i) The income or net profit $Y_n$ derived by the nth person in the income survey from the operation of an unincorporated enterprise can be written:

$$Y_n = G_n - D_n - I_n$$

where $G_n$ represents gross operating surplus or gross profit, $D_n$ represents consumption of fixed capital or depreciation, and $I_n$ represents interest payments. The equation can also be written as:

$$y_n K_n = g_n K_n - d_n K_n - i_n r_n K_n$$

where $K_n$ is the amount of capital owned by the nth person, $y_n$ is the net rate of profit, $g_n$ is the gross rate of profit, $d_n$ is the rate of depreciation, $i_n$ is the rate of interest and $r_n$ is the debt-to-equity ratio.

To prepare a database suitable for forecasting, it is assumed that the rate of depreciation, the rate of interest and the debt to equity ratio are the same for all unincorporated enterprises. Then:

$$Y_n = y_n K_n = (g_n - \bar{d} - \bar{i}\,\bar{r}) K_n,$$

ie. the variations in $Y_n$ between individuals recorded in the income survey arise only from variations in the amount of capital $K_n$ they own and in the gross rate of profit $g_n$ they earn on that capital.

(ii) Next, an estimate $\bar{y}$ of the average net rate of profit in 1989-90 is deduced from a range of rates published by the Australian Bureau of Statistics. Given $\bar{y}$ and the total values of net profits, gross profits, depreciation and interest payments obtained from the national accounts, the average amount of capital $\bar{K}$ employed in unincorporated enterprises can be determined, along with $\bar{d}$, $\bar{i}$ and $\bar{r}$.

(iii) The net profit rates $y_n$ implicit in the income survey are then assumed to be drawn from a normal distribution with mean $\bar{y}$ and standard deviation $\bar{y}$ (ie. with about two-thirds of the observations in the range zero to $2\bar{y}$), and with no $y_n$ greater than 2.5 times $\bar{y}$. Apart from the mean, these parameters are chosen as a matter of economic judgement. The capital stocks $K_n$ are assumed to be drawn from a normal distribution with mean $\bar{K}$ and standard deviation $3\bar{K}$, with no $K_n$ less than zero. This time the standard deviation is chosen to cover the range of the $Y_n$'s reported in the income survey. Twenty thousand values of the product $y_n K_n$ are then selected at random and the nth individual is assigned the capital stock corresponding to the product that most closely approximates the value of $Y_n$ reported by that individual.

(iv) The forecast average annual rate of growth of the gross operating surplus earned by all enterprises operating in a particular industry and region is available from Table 1. Incorporated and unincorporated enterprises are not distinguished in the MONASH model, so the same growth rate is assumed to apply to both types. As for employment, 1.41 percentage points of the growth in gross operating surplus earned by unincorporated enterprises are accounted for by new operators, and the growth rate for pre-existing operators is correspondingly reduced.

After making this adjustment, Table 1 yields the average annual growth rate (in per cent):

$$100\Delta G_{js} \; / \; G_{js} \, ,$$

where:

$$G_{js} = g_{js} K_{js}$$

is the total gross profit earned by all pre-existing unincorporated enterprises in industry j in region s (or in industry js for short). Now the average annual change in total net profit is given by:

$$\Delta Y_{js} = \Delta G_{js} - (\bar{d} + \bar{i}\,\bar{r})\,\Delta K_{js} \, ,$$

where $\Delta K_{js}$ is the average annual change in the amount of capital owned by the pre-existing enterprises. Thus we have enough information to determine the average annual rate of growth $\hat{Y}_{js}$ in the total income earned by all pre-existing unincorporated enterprises in industry js. However, the $\hat{Y}_{js}$ cannot be simply assigned to all the individual enterprises operating in that industry, because typically some made profits and some made losses in the base period (ie. in 1989-90). If $\hat{Y}_{js}$ is positive, the incomes of the corresponding individual enterprises ought to increase. But the application of a positive rate of growth to the income of an enterprise originally making losses would increase its losses and reduce its income. Consequently we define the average annual change:

$$\Delta y_{js} = \Delta Y_{js} \; / \; K_{js}$$

in net profit expressed as a share of base period capital in industry js. The average annual rate of growth (in per cent) in the income of an individual enterprise n operating in that industry is then given by:

$$\hat{Y}_n = 100\Delta y_{js} K_n \; / \; Y_n \, .$$

### (c) Property income

Property income consists of interest, dividends and rent, with the last category being further subdivided into landlord rent and imputed rent on owner-occupied dwellings. The method for imputing rent contains the following three steps.

(i) The IHS data records the value $K_n$ of an owner-occupied dwelling and the amount $M_n$ of any outstanding debt incurred in its purchase for each income unit n in the survey. If income unit n is not an owner-occupier, both variables have zero values. Hence the total value of the owner-occupied stock of dwellings is given by:

$$K = \Sigma_n K_n$$

and the total value of associated debt by:

$$M = \Sigma_n M_n.$$

(ii) Values for total gross operating surplus G, total depreciation D, total interest payments I and total net income Y associated with owner-occupied dwellings are available from the national accounts. Hence we can obtain the corresponding average rates $\bar{g}$, $\bar{d}$, $\bar{i}$ and $\bar{y}$ from:

$$\bar{g} = G \; / \; K, \text{ etc.}$$

(iii) The average values $\bar{g}$, $\bar{d}$, and $\bar{i}$ are then assumed to apply to all owner-occupiers, so that the rent imputed to the nth income unit is given by:

$$Y_n = (\bar{g} - \bar{d}) K_n - \bar{i} M_n .$$

If the income unit contains two parents, the rent is allocated equally between them.

For landlords, rental income is given directly in the income survey and the task is to impute ownership of the housing stock. It is assumed that the debt-to-equity ratios $r_n$ of the landlords are drawn from a normal distribution with the same mean and standard deviation as the distribution for owner-occupiers. Values of $r_n$ equal in number to the number of landlords are chosen at random from this distribution and ranked in descending order. The landlord with the largest loss in the base period is then assigned the first value, ie. the largest debt ratio. The same values of g, d and i are assumed to apply to landlords and owner-occupiers, yielding the value:

$$K_n = Y_n \; / \; (\bar{g} - \bar{d} - \bar{i} \, r_n)$$

for the housing stock rented out by the nth income unit.

At this stage, the changes in the rental incomes of both landlords and owner-occupiers can be calculated in the same manner as the rates for incomes from unincorporated enterprises. In the present case, the calculations are based exclusively on the change in the gross operating surplus of the *Ownership of dwellings* industry in Table 1.

The two remaining categories of property income, namely, interest and dividends, are handled much more simply. Interest income is assumed to grow at the same rate as the yield on 5-year Treasury bonds, as forecast by Syntec. The growth rate for dividend income is governed by the economy-wide growth in gross operating surplus available from Table 1.

(d) Government benefits

Government benefits are divided into four types: unemployment benefits, other taxable benefits, non-taxable benefits and benefits from overseas. Income from unemployment benefits for a selected group depends on the number of benefit recipients and the size of the benefit per recipient. The latter is assumed to be indexed to the national consumer price index while the change in the former is obtained as a by-product of the calculation for wages and salaries already described. For other taxable benefits and non-taxable benefits, the number of recipients is assumed to remain constant and the benefit per recipient to vary with the national consumer price index. Benefits from overseas are assumed to be unaffected by developments in the domestic economy. Clearly, the treatment of this part of the distributional calculation would be greatly enhanced if it were informed by a dynamic microsimulation model.

(e) Other income

The final category, ie. other income, is divided into taxable and non-taxable components, and includes superannuation, alimony and income provided by relatives. Both components are assumed to remain constant.

Summing over the five broad categories gives gross income. Taxable income is obtained by subtracting from gross income the imputed rent of owner-occupiers, non-taxable government benefits and non-taxable other income. The total amount of income tax paid is taken from the national accounts, while its distribution is determined by IHS data. In calculating the growth rates in net incomes, the average rate of income tax for each group is assumed to remain constant.

## 3.2 Results

Table 2 presents the results for the sixty income groups. The most striking feature of the table is the strong growth in income for the lowest decile in every region. The relevant characteristic of that decile is that it contains many income recipients who reported large losses, either from landlord rent or from unincorporated enterprises, in 1989-90. The method used for imputing ownership allocates large amounts of capital to persons who made large losses or large profits in the base period (ie. large amounts of capital are allocated to persons at either end of the income scale). More specifically, the tenth income decile ends up with 25 per cent of the capital stock in unincorporated enterprises, but the first decile (with more than 20 per cent) gets significantly more than any of the other eight. The ownership of the housing stock follows the income scale much more closely, but the first decile is still allocated more than the second and third. The general thrust of the economic forecasts is for a mild recovery, and persons with large holdings of capital are in a position to benefit most in absolute terms. This can be seen clearly in Table 3, which sets out the contributions to net income growth made by each of the five broad income categories. Of the 8.88 per cent per annum growth recorded nationally for the first decile in Table 2, 5.02 percentage points are accounted for by income from unincorporated enterprises.

At first glance, the results in Table 3 do not appear to reflect the large holdings of unincorporated capital by persons in the top decile, income from this source accounting for only 1.06 percentage points of net income growth. However, from the average income figures in Table 2, this represents $432 per income recipient on average, while the 5.02 percentage points for the first decile represents only $283 per recipient. Thus income from unincorporated enterprises increases by comparable absolute amounts for the two income classes, the large difference in the contributions to net income growth arising from the large difference in the two income bases.

If the first decile is excluded, the forecast changes in income across deciles are mildly progressive on balance. From Table 3, this general result can be partly attributed to the influence of government benefits, but the other income sources also contribute in a non-systematic way. Note that, compared with the seventh, eighth and ninth deciles, the tenth decile relies much less heavily on wages and salaries (and much more heavily on property income) as an income source. Hence the contribution of wages and salaries to net income growth is relatively low for the tenth income decile, a deficiency that is not made up for by the contribution of property income. Closer inspection of the database reveals that interest receipts account for the relative importance of property income for the tenth decile. Hence the poor forecast for that

Table 2: Nominal Net Income by Decile Class and Region

| Variable | New South Wales | Victoria | Queensland | South Australia | Western Australia | Tasmania | Australia |
|---|---|---|---|---|---|---|---|
| **Average income, 1989-90, $** | | | | | | | |
| 1. First income decile | 5725 | 5061 | 5529 | 6332 | 6268 | 6021 | 5632 |
| 2. Second income decile | 12015 | 12112 | 12194 | 11981 | 12252 | 12074 | 12092 |
| 3. Third income decile | 13398 | 13231 | 13109 | 12672 | 13032 | 12527 | 13164 |
| 4. Fourth income decile | 14410 | 14306 | 14178 | 14191 | 14371 | 13697 | 14292 |
| 5. Fifth income decile | 15903 | 16040 | 15225 | 15465 | 16145 | 15940 | 15808 |
| 6. Sixth income decile | 18007 | 18190 | 17837 | 18036 | 17642 | 17823 | 17986 |
| 7. Seventh income decile | 20704 | 20727 | 20210 | 20160 | 20407 | 20534 | 20550 |
| 8. Eighth income decile | 23250 | 22962 | 22368 | 22576 | 22281 | 21717 | 22839 |
| 9. Ninth income decile | 27141 | 26627 | 25458 | 25372 | 26412 | 25619 | 26494 |
| 10. Tenth income decile | 40688 | 40498 | 41362 | 40043 | 41818 | 36914 | 40726 |
| 11. All income deciles | 19712 | 19265 | 17971 | 18067 | 19173 | 17340 | 19029 |
| **Average annual income growth 1990-91 to 1996-97, per cent** | | | | | | | |
| 1. First income decile | 8.29 | 8.87 | 10.37 | 8.33 | 9.11 | 7.49 | 8.88 |
| 2. Second income decile | 3.73 | 3.67 | 4.34 | 4.12 | 3.99 | 3.76 | 3.90 |
| 3. Third income decile | 3.75 | 3.65 | 4.05 | 3.49 | 4.01 | 3.46 | 3.78 |
| 4. Fourth income decile | 3.62 | 3.64 | 3.99 | 3.48 | 3.83 | 3.57 | 3.70 |
| 5. Fifth income decile | 3.72 | 3.32 | 3.72 | 3.33 | 4.03 | 3.84 | 3.62 |
| 6. Sixth income decile | 3.64 | 3.37 | 4.05 | 3.32 | 3.74 | 3.45 | 3.62 |
| 7. Seventh income decile | 3.61 | 3.32 | 4.06 | 3.22 | 3.80 | 3.32 | 3.58 |
| 8. Eighth income decile | 3.45 | 3.07 | 3.89 | 3.41 | 3.87 | 3.65 | 3.46 |
| 9. Ninth income decile | 3.52 | 3.22 | 3.79 | 3.36 | 3.93 | 3.60 | 3.50 |
| 10. Tenth income decile | 3.10 | 2.70 | 3.55 | 2.83 | 3.31 | 3.34 | 3.07 |
| 11. All income deciles | 3.60 | 3.33 | 4.08 | 3.47 | 3.91 | 3.66 | 3.63 |

Table 3: Income Components Expressed as Shares of 1989-90 Net Income, Australia, Per Cent

| Variable | Income Decile | | | | | | | | | | All |
|---|---|---|---|---|---|---|---|---|---|---|---|
| | 1st | 2nd | 3rd | 4th | 5th | 6th | 7th | 8th | 9th | 10th | |
| **Components of net income, 1989-90** | | | | | | | | | | | |
| 1. Wages, salaries and supplements | 69.4 | 61.5 | 58.1 | 61.7 | 68.9 | 78.8 | 93.9 | 100.0 | 100.1 | 78.6 | 81.1 |
| 2. Income from unincorporated enterprises | -9.0 | 7.0 | 6.4 | 5.0 | 5.9 | 9.6 | 7.3 | 7.4 | 10.9 | 18.9 | 9.7 |
| 3. Property income | 4.3 | 4.0 | 4.6 | 8.8 | 11.1 | 12.0 | 10.7 | 10.7 | 12.9 | 32.9 | 14.8 |
| 4. Government benefits | 48.7 | 37.0 | 40.4 | 34.9 | 25.8 | 15.1 | 8.6 | 3.9 | 2.1 | 1.1 | 14.6 |
| 5. Other income | 2.1 | 1.3 | 1.1 | 2.2 | 2.6 | 2.4 | 1.4 | 1.5 | 1.5 | 2.0 | 1.8 |
| 6. Gross income | 115.4 | 110.7 | 110.6 | 112.6 | 114.3 | 117.8 | 121.9 | 123.4 | 127.6 | 133.5 | 122.1 |
| 7. Income taxes | 15.4 | 10.7 | 10.6 | 12.6 | 14.3 | 17.8 | 21.9 | 23.4 | 27.6 | 33.5 | 22.1 |
| 8. Net income | 100.0 | 100.0 | 100.0 | 100.0 | 100.0 | 100.0 | 100.0 | 100.0 | 100.0 | 100.0 | 100.0 |
| **Contributions of components to income growth, 1990-91 to 1996-97** | | | | | | | | | | | |
| 1. Wages, salaries and supplements | 2.31 | 2.02 | 1.91 | 2.01 | 2.25 | 2.53 | 3.02 | 3.21 | 3.23 | 2.56 | 2.63 |
| 2. Income from unincorporated enterprises | 5.02 | 0.66 | 0.57 | 0.49 | 0.52 | 0.73 | 0.57 | 0.48 | 0.69 | 1.06 | 0.83 |
| 3. Property income | 1.29 | 0.46 | 0.43 | 0.54 | 0.54 | 0.47 | 0.41 | 0.38 | 0.38 | 0.31 | 0.44 |
| 4. Government benefits | 1.48 | 1.13 | 1.22 | 1.07 | 0.77 | 0.47 | 0.28 | 0.12 | 0.07 | 0.03 | 0.45 |
| 5. Other income | 0.00 | 0.00 | 0.00 | 0.00 | 0.00 | 0.00 | 0.00 | 0.00 | 0.00 | 0.00 | 0.00 |
| 6. Gross income | 10.10 | 4.27 | 4.14 | 4.11 | 4.07 | 4.20 | 4.28 | 4.19 | 4.37 | 3.98 | 4.34 |
| 7. Income taxes | 1.21 | 0.37 | 0.37 | 0.42 | 0.46 | 0.57 | 0.70 | 0.73 | 0.87 | 0.91 | 0.71 |
| 8. Net income | 8.88 | 3.90 | 3.78 | 3.70 | 3.62 | 3.62 | 3.58 | 3.46 | 3.50 | 3.07 | 3.63 |

Table 4: Income Components Expressed as Shares of 1989-90 Net Income, All Income Deciles, Per Cent

| Variable | New South Wales | Victoria | Queensland | South Australia | Western Australia | Tasmania | Australia |
|---|---|---|---|---|---|---|---|
| **Components of net income, 1989-90** | | | | | | | |
| Wages, salaries and supplements | 81.7 | 85.6 | 76.7 | 74.9 | 80.5 | 76.3 | 81.1 |
| Income from unincorporated enterprises | 9.3 | 8.1 | 10.8 | 11.0 | 12.7 | 9.9 | 9.7 |
| Property income | 16.0 | 14.7 | 14.9 | 13.7 | 13.0 | 11.2 | 14.8 |
| Government benefits | 14.2 | 12.6 | 16.6 | 17.6 | 14.2 | 20.2 | 14.6 |
| Other income | 1.7 | 2.0 | 1.5 | 2.4 | 1.8 | 1.7 | 1.8 |
| Gross income | 122.8 | 123.1 | 120.4 | 119.5 | 122.2 | 119.4 | 122.1 |
| Income taxes | 22.8 | 23.1 | 20.4 | 19.5 | 22.2 | 19.4 | 22.1 |
| Net income | 100.0 | 100.0 | 100.0 | 100.0 | 100.0 | 100.0 | 100.0 |
| **Contributions of components to income growth, 1990-91 to 1996-97** | | | | | | | |
| Wages, salaries and supplements | 2.72 | 2.45 | 2.78 | 2.28 | 2.86 | 2.53 | 2.63 |
| Income from unincorporated enterprises | 0.63 | 0.75 | 1.14 | 0.99 | 1.08 | 0.88 | 0.83 |
| Property income | 0.54 | 0.41 | 0.44 | 0.26 | 0.35 | 0.27 | 0.44 |
| Government benefits | 0.43 | 0.41 | 0.48 | 0.54 | 0.41 | 0.63 | 0.45 |
| Other income | 0.00 | 0.00 | 0.00 | 0.00 | 0.00 | 0.00 | 0.00 |
| Gross income | 4.32 | 4.01 | 4.85 | 4.08 | 4.71 | 4.32 | 4.34 |
| Income taxes | 0.71 | 0.68 | 0.76 | 0.61 | 0.80 | 0.66 | 0.71 |
| Net income | 3.60 | 3.33 | 4.08 | 3.47 | 3.91 | 3.66 | 3.63 |

decile (compared with the seventh, eighth and ninth) can be traced to the low interest rate environment predicted for the forecast period.

The earlier discussion of the relative prospects of the regions carries over to the distributional results of Table 2, with Queensland and Western Australia doing generally better than the national average and Victoria and South Australia doing generally worse. In the more disaggregated results of Table 4, the role of the forecasts for gross operating surplus from Table 1 is evident in the regional pattern of the contributions to net income growth of income from unincorporated enterprises.

## 4. Conclusion

Most individuals receive most of their income in the form of factor payments (ie. wages and profits). A forecast of changes in the distribution of income, therefore, will be improved if it is informed by a well constructed forecast of changes in factor markets and their associated commodity markets. Forecasts that rely entirely on modelling changes in the personal characteristics of individuals, as is typically the case in microsimulation studies, have a limited capacity to take into account developments that affect individuals only indirectly through the operation of markets. Furthermore, for purposes of distribution, the level of disaggregation at which the markets are modelled is important. Particular individuals derive their factor incomes from particular industries and occupations, and the prospects for different industries and occupations may deviate quite widely from the economy's general macroeconomic prospects.

This paper has been primarily concerned with the development of techniques for applying an existing set of detailed economic forecasts for the Australian economy to the question of income distribution. The centrepiece of the forecasting system is a large dynamic applied general equilibrium model, the MONASH model. Via the markets incorporated in the MONASH model, a range of otherwise intractable information is brought to bear on a forecast of the incomes of various groups of individuals over the period 1990-91 to 1996-97. The range includes detailed scenarios on macroeconomic developments in both the world and domestic economies, changes in the foreign demand for Australia's exports, changes in the world prices of all internationally traded goods, changes in protection, changes in indirect taxes and primary factor saving technical change. To generate the income forecasts, projections of changes in factor prices and factor employment levels from the MONASH model are interfaced with micro data from the 1990 Australian income survey. The key to the interface is an allocation of the ownership of the relevant factors among the individuals identified in the survey.

Results are reported for sixty income groups differentiated by region and size of income. They suggest that the change in distribution over the forecast period is likely to be mildly progressive. They also highlight the role played by the business cycle in determining the membership of the various income deciles over time. During recessions, the lowest income decile contains a significant number of individuals who own large amounts of capital and suffer large losses. When the economy picks up, and the return to capital increases, these individuals benefit more than those who own little or no capital, and hence they move rapidly up through the income deciles. Clearly, this circumstance should be borne in mind when determining social policies on the basis of income data from surveys carried out during recession years.

## References

Adams, P.D., P.B. Dixon, D. McDonald, G.A. Meagher and B.R. Parmenter (1994), 'Forecasts for the Australian economy using the MONASH Model', *International Journal of Forecasting*, Vol.10, No. 4.

Antcliff, S. (1993), 'An Introduction to DYNAMOD: A Dynamic Microsimulation Model', DYNAMOD Technical Paper No.1, National Centre for Social and Economic Modelling, University of Canberra.

Dixon, P.B., and D. McDonald (1993), 'Creating 1990-91 input-output tables for Australia by ORANI simulation', Paper presented to the Economic Society of Australia, 1993 Conference of Economists, Murdoch University, Perth.

Dixon, P.B., B.R. Parmenter, J.M. Sutton and D.P. Vincent (1982), *ORANI: A Multisectoral Model of the Australian Economy*, North-Holland, Amsterdam.

Horridge, J.M., B.R. Parmenter and K.R. Pearson (1993), 'ORANI-F: A General Equilibrium Model of the Australian Economy', *Economic and Financial Computing*, Vol.3, No. 2.

Kakwani, N. (1986), *Analysing Redistribution Policies: A Study Using Australian Data*, Cambridge University Press, Melbourne.

Lim, G.C. and G.A. Meagher (1988), 'Modelling taxation and financial flows in a general equilibrium framework', Paper presented to the Second International Conference on Economic Modelling, London.

Powell, A.A. and R.H. Snape (1993), 'The contribution of applied general equilibrium analysis to policy reform in Australia', *Journal of Policy Modelling*, Vol.15, No. 4.

Syntec Economic Services (1993), *Five Year Forecasts*, Melbourne.

Michael Wolfson

Statistics Canada and Canadian Institute for Advanced Research
Canada

# Chapter 17: XEcon: An Experimental/Evolutionary Model of Economic Growth


## Abstract

*The role of technical innovation in economic growth is currently a matter both of keen public policy interest and of active exploration in economic theory. However, formal economic theorising is often constrained by considerations of mathematical tractability. Evolutionary economic theories which are realised as computerised microsimulation models offer significant promise for transcending mathematical constraints and for addressing fundamental questions in a more realistic and flexible manner. This paper sketches XEcon, a microsimulation model of economic growth in the evolutionary tradition.*


## 1. Introduction

This paper provides a brief sketch of a new microsimulation model, XEcon, which is under development at Statistics Canada. As XEcon is a work in progress, there are no results to report. Moreover, in contrast to most other microsimulation work in the social sciences, XEcon is intended for theoretical exploration rather than practical empirical application. As XEcon is relatively unusual, the sketch of its broad architecture provided here may be of interest in its own right.

The objectives of XEcon are both substantive and methodological. The substantive objectives at the most general level are the same as those of much theoretical modelling in economics - to seek understanding of basic processes and phenomena in real economies, particularly long run economic growth, technical change, individual income inequality, and their inter-relationships. The use of abstract and

* I am indebted to my colleagues in the CIAR Economic Growth Program for helpful discussions; to discussants at the MERIT Conference on Convergence and Divergence (Maastricht, December 1992), especially Giovanni Dosi, Richard Nelson and Sidney Winter, to Francesca Chiaromonte, and to my colleagues at Statistics Canada, especially Steve Gribble and John Baldwin.

formal models in this endeavour helps to sharpen intuition and to construct logically coherent arguments. The typical style of argument is to posit formally a network of plausible causal pathways, then to derive further implications by logical methods (in this case computer simulation), and to assess the extent to which these derived implications hold in the real world.[1] Another substantive objective of the theorising is to inform and guide the development of new measurement systems, such as those of national statistical agencies.

As theoretical modelling in economics is perhaps overly valued (see, for example, Leijonhufved, 1973), XEcon has been conceived with the following criteria in mind. Subject to the inherent complexity of the phenomena under study, the model should be parsimonious and as transparent as possible. The model should rely on at least potentially measurable constructs. One model is 'better' than another, *ceteris paribus*, if it can account for a wider diversity of 'facts' or empirical patterns. Desirable features of theoretical models are that they address important public policy or social questions; the model structure provides a foundation whereby the model can be extended in the direction of greater realism (eg. via scenarios, not unconditional forecasts); and the model points to potentially efficacious 'policy levers'. Further details of the substantive objectives, beyond these broader characteristics desired of XEcon, are outlined later via a plan of simulation experiments.

The methodological objectives respond to concerns that the formalism of much contemporary economic theory, and the associated constraints of mathematical tractability, have excessively limited realism and have engendered an abstract research program increasingly divorced from empirical grounding (eg. Lipsey, 1993). XEcon shows (proof 'by construction') that many of these concerns can be readily and usefully transcended using modern computer microsimulation methods and that 'better' theoretical models (in the sense just defined) are feasible. In other words, there are important benefits from technical change, not only in real economies but also in the art of constructing theoretical economic models. Moreover, these benefits are not just from the application of new mathematical tools.

---

[1] This phrasing is deliberate in order to eschew the more commonly heard descriptions of this process as 'writing down a theoretical model in order to <u>explain</u> some phenomenon' and '<u>testing</u> the implications of the model'. 'Explain' is inappropriate insofar as it has a connotation of uniqueness when, in fact, many quite different theoretical models can usually be written down which accord at least moderately well with the same 'stylised facts'. 'Testing' also implies excessive precision for the empirical grounding of the theorising, certainly in economics.

## 2. Background

Neo-classical models dominate the literature on economic growth and form an essential backdrop to the development of XEcon. Recent work (eg. Romer, 1990) has emphasised not only the centrality of technical progress but also the idea that technical progress should be endogenous - an explicit part of the formal theorising. The advance of technology should not be seen as some magical process of creative lightning bolts of ideas beyond the scope of economic forces, nor as a 'given' that is appropriately described by the smooth exponential growth in some technical efficiency parameter. In fact, technical progress is uneven; it is the result of purposive search activity, and the scale and character of this activity is responsive to economic incentives. This view is not without controversy. For example, Mankiw et al. (1992) argue that growth in physical, and especially human, capital can account for observed national patterns of growth in the post WW II period and that technical change is not necessarily a major explanatory phenomenon. Lucas (1993) emphasises learning-by-doing, coupled with the continuous (and exogenous) introduction of new techniques of production, as yet another key explanatory factor.

Common to many of these recent theoretical growth models are the following components:

- a formal mathematical representation of the technology of production, typically a 'smooth', perfectly known, aggregate production function, including an abstract but explicit characterisation of technical progress;
- a set of agents - usually one or two kinds of (generally) homogeneous (hence simply aggregated) firms and workers - who jointly produce output according to the technology;
- a formally specified set of behaviours for the agents - typically omniscient inter-temporal profit or utility maximisation;
- a prose story about the institutional arrangements - such as perfect or monopolistic competition, and infinitely lived patents; and
- an assumption of steady-state growth equilibrium.

XEcon is similar insofar as it contains many of the same components - for example, a set of agents, formally specified behaviours and descriptions of the technology of production. However, XEcon is in the evolutionary tradition, so it is not 'solved' using formal mathematical deduction. Instead, the model's dynamic trajectory is 'revealed' using explicit computer simulation of the behaviours of the myriad micro-entities posited in the theoretical economy. Computer simulation approaches have the advantage that mathematical tractability need not be a constraint in formalising ideas. Instead, the only constraint is that the formal descriptions of agents, technologies, behaviours etc. must be well-defined procedures or algorithms.

Thus, XEcon need not and does not make any *a priori* assumption of equilibrium. Instead, the basic scenario is that of an economy operating in real calendar time, starting at some time $t_1$ (eg. 1900) and evolving either continuously or in discrete steps until time $t_2$ (eg. 2050). It may turn out that the path followed by the economy is a steady-state growth path, but it is much more likely (and realistic, of course) that the path will involve fluctuations with no clear equilibrium or asymptotic end-point. More interestingly, the economy may start out with a relatively simple structure and generally become more complex over time, or new kinds of arrangements or patterns might emerge. This is the essence of evolutionary models, and evolutionary theory.

Another aspect of the computer simulation approach of evolutionary models like XEcon, and of the consequential relaxation of the constraints of mathematical tractability, is that many assumptions in neoclassical models become optional. For example, production technology can, but need not, be represented by smooth aggregate production functions. Instead, individual firms can be explicitly represented by heterogeneous techniques of production that evolve over time in a groping, 'higgledy-piggledy' manner (again, arguably more realistically). Other neoclassical assumptions can become impossible in an evolutionary model. The main example is omniscient maximisation. The reason is that the optimisation problems faced by agents in a formal evolutionary model like XEcon are insoluble. Thus, agents' behaviour not only can, but must, be described by heuristics or rules of thumb. (These heuristics may adapt or evolve over time.)

Formal evolutionary models in economics are, of course, not new, but they are much less common than neoclassical models. The seminal model is that of Nelson and Winter (1982). Their model was designed to give an equally plausible but alternative and decidedly unorthodox account of 20th century technical progress in the US, compared with the account given by the Solow (1957) model, which is the seminal model in the neoclassical growth literature. A brief sketch of the Nelson and Winter model is useful because it provides an initial *point de depart* for XEcon.

The Nelson and Winter model is highly abstract and simplified. It starts with a finite population of firms, all of which produce a homogeneous output Q (or GNP). The i-th firm uses two inputs, capital ($K_i$) and labour ($L_i$), with fixed coefficient production functions - a labour productivity coefficient ($Q_i/L_i$) and a capital intensity coefficient ($K_i/L_i$). Each firm's output is a 'backwards L' function of labour input, with capital on-hand providing the upper constraint on output. The firms are heterogeneous because they each can have different fixed coefficients. Each firm's capital is subject to 'noisy' exponential depreciation. More importantly, there is an explicit stochastic process of technical change via R&D. Firms search for 'better' techniques of production, either incrementally in a local 'neighbourhood' of their existing technology, or imitatively by trying to copy the technique of a more successful firm.

A simulation of the Nelson and Winter model starts with a handful of firms endowed with technologies randomly distributed around the same 1909 average used in the Solow model. The population of firms then evolves by a random process of innovation according to the stochastic search process just mentioned and to the investment of any gross profits in new capital. A principal conclusion of this unorthodox evolutionary model is that it is equally capable of reproducing the macroeconomic 'stylised facts' as the aggregate profit-maximising parable in the 1957 Solow model. This is not necessarily a very strong test of the relative merits of the two models nor of the two methodological approaches more generally. But it clearly shows some measure of superiority of the evolutionary approach, as the Nelson and Winter model also accounts for the evolving size distribution of firms at the micro level (a feature which is entirely absent in aggregate growth models).

Lane (1993) describes a model under development with Dosi et al. which is more general than that of Nelson and Winter. The Dosi-Lane model, which contributed significantly to the initial conceptualisation of XEcon, has two sectors - one for producing heterogeneous machines (m), the other for producing a homogeneous consumer good (g). In the first sector, 'mfirms' make m's by employing workers; and search for new technologies (new m's) by employing scientists. In the other, consumer goods, sector, 'gfirms' use m's and workers to produce g.

The locus of technical change is in the first sector. Each machine is represented formally by a two element vector ($w_m$, $w_g$), where $w_m$ is the embodied number of FTE (full-time equivalent workers) needed to build one m, and $w_g$ is the number of FTEs required to make one physical unit of g using one m. Technical progress then involves reducing either or both $w_m$ and $w_g$.

Figure 1 gives a graphical representation of this technology. The vertical axis shows the labour required to produce the machine itself ($w_m$), and the horizontal axis is the labour required to produce one unit of the (homogeneous) consumer good using the machine ($w_g$). Each point in the graph therefore represents one kind of machine and, in turn, the one firm which knows how to manufacture that machine. The multiple dots in the figure reflect the coexistence of various kinds of machines, and hence the realistic coexistence in the economy of a set of heterogeneous firms using quite different techniques to produce the same commodity. Unambiguously 'better' machines lie to the south-west (SW) in this figure. Moving NW or SE produces an ambiguous result, as it depends on the relative prices of labour and of the consumer good. Somewhat artificially, Dosi-Lane identify westward movements in this space (ie. holding the labour content of the machine fixed) as technical progress along a 'technological trajectory'.

Using one of three strategies each firm invests in R&D to search for new technologies in this abstract space. The three strategies are (1) incremental (westward movements) or (2) imitative (copying 'nearby' firms) as in Nelson and Winter, or (3) radical (ie. 2-d) search, possibly moving south as well. As the model is still under development, no conclusions have been published.

Finally, two macroeconomic models with explicit microsimulation foundations should be noted. One is Eliasson's (eg. 1986, 1991) micro-to-macro MOSES model. This is much more than a pure theoretical evolutionary model. It is a large empirical model that has been under development since the mid 1970s. It builds on the actual historical behaviour of about 150 named large Swedish firms, and has explicitly reconciled their financial attributes with the Swedish national accounts and input/output data. The evolving behaviour of these firms is represented by simple but heterogeneous production possibilities and satisficing rules of thumb. Technological progress is exogenous, based on a simple exponential growth trend, though Ballot and Taymaz (1996) have recently extended MOSES to include an explicit learning process to make technical change substantially endogenous. The other macroeconomic model with explicit microsimulation foundations is the Bennett and Bergman 'microsimulated transactions model' of the US economy (1986). It is avowedly in the microsimulation tradition, while at the same time explicitly designed to fit 1970s US macrodata. It is relatively richly articulated with respect to the myriad kinds of transactions in an economy and its financial sector.

However, technical progress, as in Eliasson's MOSES model, is based on an exogenous exponential growth parameter, albeit with explicit vintage capital.

Figure 1: Technology and R&D Search in the Dosi-Lane Model

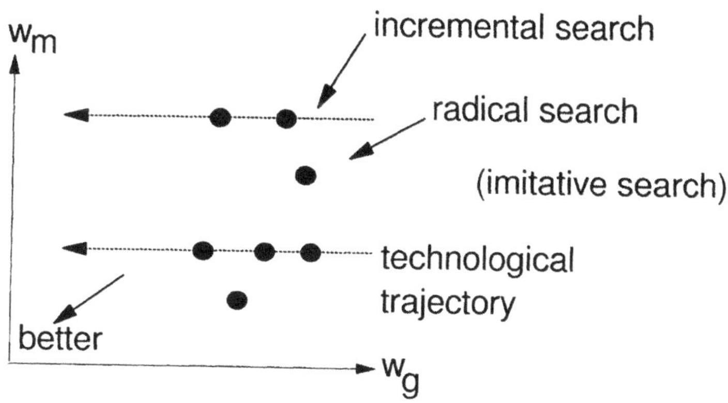


Given the diversity of approaches, both within and between the neoclassical and the evolutionary efforts, a central issue in the design of XEcon has been the formal representation of technology. The idea that a technology can be formally represented by a mapping from a set of inputs to a set of outputs seems elemental. At the same time, it would be useful if these 'atoms' of technology could be combined and recombined into 'molecules' that capture some of the historical evolution of firms - for example, as in Best's (1990) description of historical changes from early 19th century craft shops through the developments of interchangeable parts and standardisation, the specialisation and 'vertical disintegration' that characterised the emergence of the machine tool industry to the mass production 'flow processes' of Ford's production line.

## 3. Technology in XEcon

The 'atom' of production technology assumed in XEcon is a technique for producing one commodity output from a mix of inputs. We assume linear constant returns to scale, as in a column vector of a use matrix of an input/output table. (Non-linearity could easily be implemented but is avoided in the first instance to keep the initial version of XEcon simpler and easier to understand.)

In order to provide a sense of the possibilities opened up by a microsimulation approach rather than a purely mathematical approach, let us describe briefly some further detail of the posited production technology. Each commodity is uniquely identified by an index, say i or j, and it can be produced by one or more techniques. In turn, each technique is represented by a pair of column vectors for the j-th output:

g(i,j) = the amount of commodity i required 'to be present' to produce one unit of commodity j

d(i,j) = the amount of i 'used up' per unit of input requirement in producing j.

While this two-part linear representation is straightforward, it allows a richer range of production relationships than does a single input/output style column vector for each technique, and is much richer than usual neoclassical production functions. For example, if d(i,j) = 1, then the input (eg. electricity, petrochemical feed stocks) is used up one-for-one in production. If d(i,j) = 0, then the input is an infinitely durable capital good (eg. a catalyst, or the bit string for a new piece of software) for which there must be at least g(i,j) units present for each unit of output. If $0 < d(i,j) < 1$, then there must be at least g(i,j) units present for each unit of output and d(i,j) x g(i,j) units will be 'used up' in the production of one unit of the j-th output.

This d(i,j) formulation seems more realistic than the usual assumptions (the conventional accounting assumption of straight line depreciation, or the standard theoretical assumption of exponential decay) regarding durable goods. Both conventional assumptions imply that a fixed amount of the capital good must be around to 'support' any given level of output flow. But, unlike the d(i,j) formulation, the rate at which capital is physically used up is a function of time, rather than of the intensity with which the asset is actually used. (Depreciation in the sense of technological obsolescence is modelled separately and explicitly as the profitability of various techniques is assessed by firms' management or through market selection - see below.)

As we are using microsimulation methods, it is straightforward to populate XEcon's artificial or 'toy' economy with myriad heterogeneous firms. They will produce different commodities. Those producing the same commodity will generally use different techniques, ie. firms in the same industry will typically have different g(i,j) and d(i,j) coefficients. As a result, even simple versions of this abstract economy can exhibit 'economic growth' as the intensity of production shifts among the various available techniques. "Patterns of differential survival and growth in a population of firms can produce change in economic aggregates characterising that population even if the corresponding characteristics of individual firms are constant" (Nelson and Winter, 1982). Technical progress (ie. the discovery of new column vectors) is also possible in XEcon and is discussed later.

## 3.1 Agents

XEcon consists of two main types of agents - firms and worker/consumers. In addition, there is a rudimentary finance sector, a government tax/transfer branch, and an information sector. These agents are described briefly below.

*Firms*: Firms are central in XEcon and have considerable internal structure. They are organised collectivities of individuals engaged in the production of commodities using other commodities as inputs. The basic 'atom' of physical production (the transformation of inputs into outputs) is a production line or line of business or 'firm-line' for short. Firm-lines need not be identified with explicit industries or sectors; they are identified only by the commodities they produce. In addition, firms have a 'mind and management', denoted 'firm-hq' for headquarters, which makes strategic decisions with regard to levels and composition of production, human resource policy, financing, investment, pricing, marketing, R&D strategy, etc. This means that firms are best thought of as 'molecules' consisting of one firm-hq and one or more firm-lines.

Firms also tend to search continuously for new, more efficient methods of production. This activity is concentrated in one organisational sub-unit in each firm, the research and development department, or 'firm-rad' for short. (The R&D search process is typically described by relatively simple rules of thumb.) The internal structure of firms is illustrated in Figure 2.

Individual firms can not only grow and decline over time but also can be born as new entrants, and die or exit via bankruptcy. Growth (and decline) can take several forms. The most straightforward is changes in levels of activity. With explicit R&D, other forms of growth would be the discovery of more efficient techniques for producing commodities the firm is already producing (incremental or imitative R&D, or continuous improvement), and the introduction of techniques for producing entirely new commodities (radical R&D). This latter type of change allows firms to evolve from uni-cell to multi-cell (ie. multi-line). A further possibility is the acquisition (takeover) of the firm-lines belonging to other firms.

Figure 2: Firm Structure in XEcon

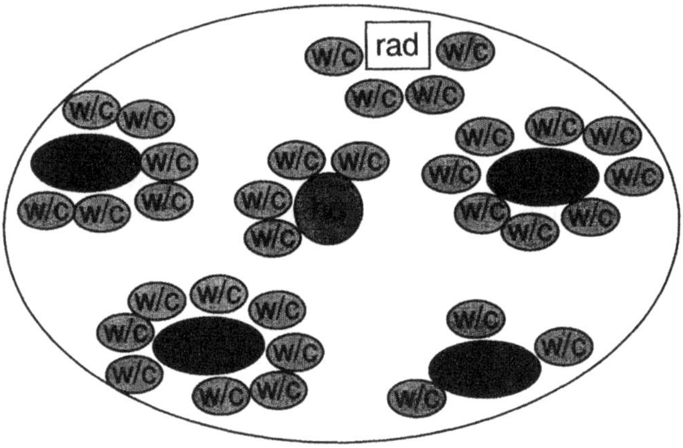


Firms are assumed to follow similar patterns of behaviour - such as formation of expectations, pricing, and production decisions. Even if a number of firms have identical technologies and formalised behavioural patterns, they will generally be heterogeneous in relation to, for example, numbers of employees, rate of output price adjustment based on (heterogeneous) past experience, and profitability.

To start, XEcon will assume limited heterogeneity in production techniques and no technical progress, even though the study of technical change is a primary objective of XEcon The reason is that we must first be able to generate plausible patterns of micro- and macro-level behaviour in the simpler case of static technology. Our intuition is that this will *not* be easy, because we shall be departing so radically from the usual neoclassical omniscient maximising assumptions.

*Worker/consumers*: Individuals in XEcon are considered in two main roles; workers who supply labour to firms, and consumers of final outputs. To begin, we assume homogeneous labour with no learning or skill heterogeneity, though extensions of these assumptions are anticipated. Such extensions would allow a more natural representation of learning-by-doing at the level of individual workers than is the case if it is disembodied at the firm or industry level. (Furthermore, with skills embodied in workers, firm bankruptcy need not imply complete loss of such specialised human capital, insofar as laid off workers are re-employed where they can still use their skills.) Initially, homogeneous behavioural patterns and preferences over consumption goods are also assumed. Again, such assumptions are easily relaxed in a microsimulation approach. However, they are useful not only for debugging the model but also in coming to understand its properties.

Worker/consumers offer their labour to firms based on a reservation wage, which, in turn, is a function of their previous employment experience. Workers who are already employed will only search for new work if their wage drops by more than a given percentage. Unemployed workers lower their reservation wages by another amount at some annual rate and search for work.

Simple consumer choice could be based on lexicographic preferences (eg. as in Pasinetti, 1982) over final outputs. For example, worker/consumers allocate all their income to necessities like food until they have met their basic requirements, and then consume as much 'luxury' goods (eg. TV) as they can afford. Such a hierarchical structure of preferences is no less realistic than those implied by conventional 'smooth' utility functions; however, it is virtually impossible to analyse in models using purely mathematical techniques.

Even though worker/consumers have homogeneous labour, preferences and behavioural rules of thumb, they will still be heterogeneous in employment, wages and consumption patterns because they will work for different firms which are likely to be heterogeneous in their wage rates, and will have the possibility of being unemployed. Also, a more complete description of the XEcon structure would make it clear that the optimisation problem for worker/consumers is intractable, so rules of thumb and gradual adaptation to evolving targets will characterise behaviour. (Optimisation is intractable for firms as well.)

*Government tax/transfer*: Simple income, payroll, profits, value added and retail sales taxes, demogrant, negative income tax, and unemployment insurance style transfers to worker/consumers are readily included, given the microanalytic foundations for XEcon.

*Finance*: The financial sector is highly simplified and represented by a single monolithic 'finance' agent whose deposit-taking and lending (investing, venture capital) policies are parameters of the XEcon model.

Finance has two main groups of actions. The first is the informational action of setting credit limits and interest rates for firms based on firms' financial performance (or business plans in the case of new entrants). The second is a set of cash flow actions related to lending money to firms, and taking deposits from firms and worker/consumers.

*Information*: The information sector can be thought of as an amalgam of a national statistical agency and market research firms. For now, it simply produces a set of statistics, like price and average wage indices, that are instantly and costlessly available to all the agents in XEcon (see Antonov and Trofimov, 1993 for an interesting analogous analysis using Eliasson's MOSES model.)

## 3.2 Visual metaphors for technology and technical change

Modelling processes of technical change, including evolving skills amongst workers, is a fundamental objective of XEcon. A central feature of technical change is increasing complexity of the economy, reflected in the growing variety and diversity of commodities and technologies. Thinking in terms of an input/output table, there are growing numbers of distinct inputs and outputs and, thus (at least conceptually), an increasing number of rows and columns over time, and more non-zero cells in the matrix. Labour is also becoming more heterogeneous with respect to skill, occupation, education, and tacit firm-specific knowledge.

Figure 3 draws on the concept of an input/output table use matrix as a visual metaphor for representing the technologies formalised in XEcon. Each row indicates a distinct input and each column represents the output of one firm-line. An empty cell corresponds to zero input requirements; a black cell indicates that the row commodity is the output of the given column vector. The intensity of shading roughly indicates the physical volume of the given commodity input required per physical unit of output. The diagram shows that there can be many heterogeneous commodities and that more than one technique for producing each commodity is possible. Recall that each column in the diagram actually has two underlying vectors, {g(i,j), d(i,j)}, to describe the technology fully.

In Figure 3 basic labour requirements are shown as homogeneous (but not of equal intensity) by virtue of there being only one row. These labour inputs could be re-interpreted as efficiency unit input requirements per physical unit of output. In turn, each technique of production or firm-line (ie. column in the matrix) could have a unique capacity for transforming labour of varying skill level(s) into efficiency units (according to, say, a simple linear firm-specific function E(s,f), where W(s,f) is the distribution of workers in firm f by skill level(s)). That is, firm-lines could vary

Figure 3: A 'Use' Matrix Visual Metaphor for Commodities and Technology in XEcon

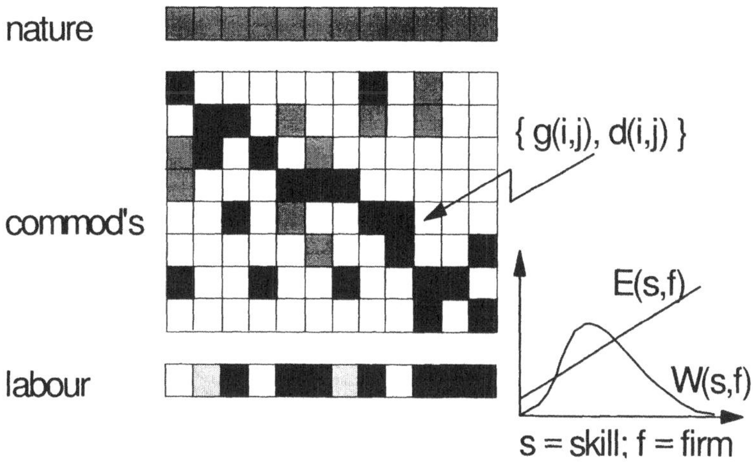


in their abilities to utilise skilled labour. Such variation is essential to another objective of XEcon, the exploration of the relationship between technical change and changing patterns of wage inequality from a microanalytic perspective. (The 'nature' row is a reminder that, at some point, use of natural resources should be explicitly considered.)

It may be noted that non-rival goods (Romer, 1990) are easily represented in this visual metaphor, and in the description of technology given above. Non-rival goods are those which have very low variable inputs; the major cost is in developing the 'recipe' (ie. R&D) or in making the first copy or a key ingredient. An example is the original bit string for the DOS 6.0 operating system, which was very costly, versus the shrink-wrapped package on sale at local software retailers. These latter packages are virtually costless in comparison to the original development costs of the software. The production of a commodity like this could be represented in the technology formalism of XEcon by two firm-lines, or two (pairs of) column vectors belonging to the same firm: one for producing the original 'design' which would have very large inputs per (physical) unit of output (an operating system), and one for producing the shrink-wrapped packages with very small inputs per unit of output.

Non-rival, though excludable via copyright and legal vigilance and hence private, commodities such as this are widespread, and can be seen increasingly to be at the heart of much that is considered technical progress. At the same time they exemplify indivisibilities and hence non-convexities in production possibilities. These in turn cause serious problems for neoclassical theory. The point for now is that such non-convexities and consequential increasing returns to scale are not at all problematic for an evolutionary microsimulation model like XEcon.

## 3.3 Running an XEcon simulation

XEcon, given a population of firms with these kinds of technologies as well as a population of workers, has the potential to become quite complex. Thus, it is important that other researchers, not just the authors, should find the simulation model easy to run and to use for experiments. One requirement is that the software should be modular; a second is that each module or component should itself be relatively simple and transparent; and a third is that the model should be open - a 'glass box' rather than a black box. This combination of characteristics, where overall complexity results mainly from the interactions amongst a large number of modules, each of which is relatively simple and easily understood, seems to have led to the widespread acceptance of simulation models in other fields such as cosmology.

The real power and interest of XEcon comes in the ability to view simulation results - in Lane's (1993) sense, the properties of the artificial world - as they unfold. It is therefore very important to have a variety of easy-to-use intuitive graphical displays. The image is of a researcher sitting at a PC with a Windows or equivalent operating system. Then, while XEcon is executing, the user can 'click' between a variety of views of the dynamic economy as it is evolving, as in SimCity and SimLife. Alternatively, the user could explore one or more parallel artificial worlds as an economic historian, albeit one who could assure that any data desired could be found in the archives. The views of the economy could include the examples of graphs described below.

*Labour input by sector* - numbers of employed workers by production sector, and unemployed over time.

*Factor incomes* - profits of the bottom 50 per cent of firms and the top 50 per cent of firms, wages of the bottom 50 per cent of worker/consumers and wages of the top 50 per cent (totals 100 per cent of factor income), all over time. (Note that the difference between the bottom and top 50 per cent is a quick indicator of skewness and hence of inequality of the underlying distributions.)

*Firm size distribution* - a 2-d scatter plot of firms' total sales against number of employees at a given point in time, and a histogram plot of value added per worker by firms sorted by this variable, with the width of each bar proportional to the firm's total employment.

*Firm micro-dynamics and volatility* - 3-d scatter plot of each firm's (1) average market share (ie. sales as a percentage of total sales), (2) trend in market share (represented by a slope coefficient from a simple linear regression), and (3) coefficient of variation of market share around the trend; all defined over an arbitrary time interval $t_1$ to $t_2$.

*Alternative views of productivity* - ingredients in typical aggregate total factor productivity calculations: wage and profit shares, index of labour inputs, index of capital inputs, and the 'Solow' residual, all against time; and the difference between actual GDP and 'microanalytic' potential GDP based on a linear programming style convex hull of the population of firms' actual techniques, capital, labour and inventories on hand, all over time.

*Financial sector* - total loans out, total deposits, interest payments and receipts, interest rates, debt-to-equity ratios by size group of firm, all over time.

*Basic SNA statistics* - components of the basic national income identity $Y = C + I + G$, as well as inventory versus capital investment, gross and net, all over time.

*Aggregate welfare measures* - GDP per capita, average wages, average disposable income, average 'utility' (given explicit underlying utility functions posited for worker/consumers), all over time.

*Distributional patterns* - quintile shares, polarisation shares for wages, incomes, and utility, all time, as well as Gini, and other summary inequality and polarisation measures (Wolfson, 1994) over time.

## 3.4 Program of experiments

Given a working version of XEcon, including the kinds of graphical 'views' just described, the research program with a theoretical microsimulation model consists of a series of simulation experiments. These experiments should start very simply (to assure that the model is functioning as intended) and are likely to be drawn from the following list.

*Reproducing equilibrium* - as a first set of 'debugging' experiments. Note that XEcon's input/output technology can be specified such that there is an 'equilibrium' steady-state path for the economy with a zero growth rate. One test of the model would be to give it a set of initial conditions at this equilibrium, and check that it endures.

*Disequilibrium dynamics* - once an equilibrium path is demonstrated, it would then be interesting to explore what happens to the economy when it is started at various points away from equilibrium. We conjecture that, in the absence of any intervention, the economy could spend long periods far away from equilibrium. In other words, there is no reason to believe that XEcon (or a real economy like Russia) has any tendencies to approach an equilibrium quickly or asymptotically when it is started with a highly distorted price vector .

*Conventional statistics versus 'unobservables'* - conventional statistics like GDP and disposable income per capita are readily computed for this artificial economy. So is an input/output table, and several measures of total factor productivity. At the same time, 'unobservable' items (like potential output and aggregate utility) can also be computed, as indicated in the proposed graphical outputs above. An obvious conjecture is that observable measured growth need not correspond well with growth in otherwise unobservable indicators of social welfare.

*Robustness and path dependence* - one question about an evolutionary path for an economy is its robustness. This can be assessed directly with XEcon by running several simulations with exactly identical inputs, with the exception of the starting point (seed) for the (pseudo) random number generator. A robust economy would then be one with small variance over such replicates. Non-robustness, or a high degree of path dependence, would correspond to high variance over replicates.

*Role of adaptive expectations* - the stability properties of XEcon probably depend critically on the rules of thumb used by firms and worker/consumers to adjust to faulty expectations, whether pessimistic or optimistic. This can be explored by trying various adjustment rules - for example, the rates at which wages, prices, and production are adjusted to evolving circumstances. One conjecture is that slower adjustments will give more short term volatility, but greater long-term stability.

*Role of finance* - the finance sector sets interest rates and credit limits. One obvious question is the role of interest rates in economic growth. Recent macro policy in Canada suggests that high interest rates have had serious depressing effects, while Romer's (1990) model suggests the opposite.

*Alternative market forms* - the microanalytic foundations of XEcon offer the possibility of exploring decidedly non-neoclassical but more realistic kinds of market structures - for example, abandoning the pervasive 'law of one price' (Smith and Williams, 1992) and assuming that firms follow 'posted offer pricing' with transactions continually occurring at a variety of posted prices . One question is how much difference these markets make, in comparison with, say, the conventional textbook Walrasian markets.

*Keynesian automatic stabilisers* - one conjecture is that XEcon, without redistributive government taxes and transfers, is more likely to enter a vicious spiral of rising unemployment and falling GDP per capita - a conjecture that can easily be explored with a series of simulations.

*Patterns of income inequality* - income inequality in XEcon derives in part from heterogeneity in the wages paid by firms. One question centres on the feedback from macro level policies, like demand stabilisation, to income inequality. Another question is whether or not tough credit limits which drive weak firms out quickly will have a beneficial impact on inequality.

*Firm dynamics* - Dosi et al. (1993) suggest, and Canadian longitudinal microdata support the view, that firm dynamics are characterised by a number of 'stylised facts'. These are, first, that the overall size distribution of firms is positively skewed and has a relatively stable shape (eg. Pareto or lognormal). Second, within the population of firms, there is a high and persistent degree of turbulence. Third, part of this turbulence is high rates of new entry and death. Fourth, the turbulence, however, is not entirely random as in Brownian motion; many firms have persistently above- or below-average performance (ie. their annual growth rates are autocorrelated). Finally, even though the overall size distribution has a stable and characteristic shape, this shape is not reproduced within sectoral sub-populations (eg. energy versus retailing).

Formal mathematical models have been constructed which can generate one or other of these properties, but not all together. Dosi et al.'s (1993) simulation model does. XEcon should similarly be able to account simultaneously for these patterns.

*Role of new firm entrants* - Dosi et al. (1993) also explore the importance of entry to economic growth. In particular, they experiment with the extent to which new technology is introduced only by new entrants rather than incumbents. Similar experiments would be possible with XEcon. Another set of experiments could relate the financing of new entrants to their birth and success rates. Jointly, these two sets of experiments might give some insight into the relative importance of venture financing to the overall pace of technical progress.

*Tax incidence in an evolutionary world* - the incidence of the corporate profits tax is a continuing subject of debate depending, for example, on alternative assumptions regarding capital mobility. However, this debate is still largely framed within a static neoclassical framework. It is likely that in XEcon imposition of such a tax will have widespread, complex and subtle effects that will vary with initial conditions and behavioural assumptions. Experiments to explore these impacts should have a salutary impact on the arguably simplistic analyses that dominate the literature.

## 4. Conclusion

As stated at the outset, XEcon is a work in progress. This paper has provided a sketch of the structure of the model, and illustrative simulation experiments. Work is currently focussed on developing an appropriate software environment along the lines sketched in Gribble and Wolfson (1992). The actual writing of code to implement XEcon is awaiting this new software environment.

*Postscript (November 1995):* The required software environment for writing XEcon and other microsimulation models, ModGen, has been successfully developed by Statistics Canada. An initial version of XEcon has also been completed. It is much simpler than the version sketched here, and is focussed more on consumer choice and shopping in a world where new commodities can come into being. In other words, developments have focussed on several essential prerequisites to modelling technical change in the way discussed above.

## References

Antonov, M. and G. Trofimov (1993), 'Learning through short-run macroeconomic forecasts in a micro-to-macro model', *Journal of Economic Behaviour and Organisation*, Vol. 21, No. 2.

Ballot, G. and E. Taymaz (1995), 'Firm-sponsored training, technical progress and aggregate performance in a micro-macro model', in: A. Harding (ed.), *Microsimulation and Public Policy*, North Holland, Amsterdam.

Bennett, R.L. and B.R. Bergmann (1986), *A Microsimulated Transactions Model of the United States Economy*, Johns Hopkins University Press, Baltimore and London.

Best, M. (1990), *The New Competition: Institutions of Industrial Restructuring*, Policy Press, Cambridge, England.

Dosi, G., O. Marsili, L. Orsenigo and R. Salvatore (1993), 'Learning, market selection and the evolution of industrial structures', mimeo, Department of Economics, University of Rome, 'La Sapienza', March.

Eliasson, G. (1986), 'The Swedish micro-to-macro model: Idea, design and application', in: G.H. Orcutt, J. Merz, and H. Quinke (eds.), *Microanalytic Simulation Models to Support Social and Financial Policy*, North Holland, Amsterdam.

Eliasson, G. (1991), 'Modelling the experimentally organised economy', *Journal of Economic Behaviour and Organisation*, Vol. 16, Nos. 1-1, July.

Gribble, S.F. and M.C. Wolfson (1992), 'Design issues for longitudinal microsimulation software', mimeo, Social and Economic Studies, Statistics Canada, Ottawa.

Lane, D. (1993), 'Artificial worlds in economics, Parts 1 and 2', *Journal of Evolutionary Economics*, Vol. 3, Nos. 2 and 3.

Leijonhufved, A. (1973), 'Life among the econ', *Western Economic Journal*, Vol. 11, No. 2.

Lipsey, R.G. (1993), 'Review of books by Blaug and Hutchison', *Canadian Journal of Economics*, Vol. 26, No. 3.

Lucas, R.E. (1993), 'Making a miracle', *Econometrica*, Vol. 61, No. 2.

Mankiw, N.G., D. Romer and N. Weil (1992), 'A contribution to the empirics of economic growth', *Quarterly Journal of Economics*, Vol. 107, No. 2.

Nelson, R. and S. Winter (1982), *An Evolutionary Theory of Economic Growth*, Harvard University Press, Harvard, Mass.

Pasinetti, L. (1982), *Structural Change and Economic Growth*, Cambridge University Press, Cambridge, England.

Romer, P. (1990), 'Endogenous technological change', *Journal of Political Economy*, Vol. 98, No. 5.

Smith, V.L. and A.W. Williams (1992), 'Experimental market economics', *Scientific American*, December.

Solow, R.M. (1957), 'Technical change and the aggregate production function', *Review of Economics and Statistics* Vol. 39, No. 3.

Wolfson, M. (1994), 'When inequalities diverge', *American Economic Review Papers and Proceedings*, May.

Gunnar Eliasson

Royal Technical Institute and Industrial Institute for Economic and Social Research
Sweden

# Chapter 18: Endogenous Economic Growth Through Selection

## Abstract

*Economic growth can be viewed as being propelled endogenously by the (1) entry, (2) reorganisation, (3) rationalisation and (4) the exit of firms, with the first two selection and innovation investments operating in the very long run, and the remaining two (rationalisation and exit) in the short and immediate-term. This growth process can only be represented by a micro organisation-based macro theory and is simulated here on the Swedish micro-to-macro model MOSES. It is found that the impact of the selection mechanisms operating through the entry and exit processes, while slow in the beginning, significantly affect economic growth in the very long run. A political economy that is unable to organise itself to wait for the very long-run benefits of significant immediate efforts to enhance firm exit and entry may never experience strong and sustained economic growth.*

## 1. Economic Growth as a Selection Process -The Problem

Adam Smith is the declared father of many basic ideas in modern economics (1937). He is, however, not as well-known for what he argued was most important of all for the wealth of nations, namely dynamic competition - particularly that maintained by *free innovative entry* to check the formation of monopolies detrimental to the creation of economic wealth. Such dynamic competition is made possible by the superior capacity of some firms to compete for customers and resources, and is driven by the rewards for that superiority. This superior capacity defines the organisational competence of the temporary monopoly that is called a firm (Eliasson, 1990). Such superior organisational competence is always temporary, since it is relative to the competence of all other firms, and firms are continuously generating new competence through organisational learning and innovation (Eliasson, 1992a).

One thesis of this paper is that without free innovative entry, rapid long-term growth in an economy cannot be supported. Innovative entry improves the technology level of firms, and forces competition on incumbent producers. The costs come in the form of increased exits by low performers, and transactions costs associated with the reallocation of resources.

Darwin suggested that the species adapted through selection to an exogenously changing environment. The economic question raised here is whether firms can overcome the natural filtering of winners and losers in markets through organisational learning. This paper is concerned with the growth dynamics of this game situation in what can be called the *experimentally organised economy* (Eliasson, 1987, 1991a). The experimental mode of behaviour is fundamental for entrepreneurial activity and economic growth. Individual agents are characterised by strongly bounded rationality and markets by free innovative entry. The latter drives the selection mechanisms in the model that also influence the economic environment, making it largely non-transparent to each agent. Only through curious experimentation will entrepreneurs be able to explore the enormous investment opportunity set of such an experimentally organised economy and discover the superior technologies that can be profitably invested in.

The nature of the entrepreneurial spirit and technological competition through innovation and imitation are therefore central to this paper. They are one side of free innovative entry. The other side is regulation, ie. the restrictions on competitive innovative entry imposed politically. Entry may therefore be insufficient for economic growth to occur for two reasons. First incentives for entrepreneurship may be lacking or, second, entrepreneurship may be held back by regulation, typically through restrictions on free competitive entry in markets (Eliasson, 1991b).

The source of entrepreneurial behaviour is partly to be found in the economic environment itself that compels actors (through competition) to be entrepreneurial, or fail. This can be established by demonstrating the existence, theoretically and through simulation, of the vastness of the business opportunity set. In doing that, standard economic theory is linked to the much richer but less rigorously formulated economics of the Austrian school.

The problem of the restriction of free competitive entry is much wider in scope than restrictive establishment rules and trade protection. The entire public sector of western market economies is characterised by almost complete prevention of innovative and competitive entry. In Sweden, for instance, only 20 to 30 per cent of total production can be said to be subjected to open competition (Carlsson, 1993). The extreme case is, of course, the formerly planned economies and most of the developing world. The institutions providing for more or less open selection mechanisms in markets are therefore not as typical of western market economics as one may have been led to believe. Paradoxically, it is difficult to imagine a true market economy without the widespread presence of such institutions. The institutions preventing the selection mechanisms in the market from being operative very much reflect the willingness of societies to accept change and their inability to cope with it. And the only way of modelling the macroeconomic importance of those mechanisms is through a micro-based simulation model.

This paper aims to demonstrate (through simulations on the Swedish micro-to-macro model, which is a quantitative approximation of the experimentally organised economy), that a very long time horizon is associated with a viable growth process. A rapid reorganisation of the model economy, in which competition is stimulated by more innovative entry, may create immediate negative effects in the form of exit and unemployment, while the benefit in the form of positive growth effects will come in the distant future. The trade off between the immediate future and the long term very much hinges on the flexibility of the labor market. Conversely, the negative effects at the macro level of less innovative entry will appear after decades, but then with a momentum that accelerates and requires enormous policy effort and patience to turn around. This means that a political economy unable to mobilise sufficient patience or political interest to give the very long-term benefits a fair chance against immediate negative effects may never experience growth and/or may never be capable of turning a declining economy around.

## 2. The Four Investment Mechanisms of Growth

Free innovative entry and technological competition expand the investment opportunity set to create an environment of local unpredictability for each agent. Agents not only choose what to do, but also choose modes of thinking about and deciding what to do. Competition through new entry exercises a competitive threat to all agents, which they have to counter by innovative competition of their own. This dynamic competition moves the entire economy and forces agents not capable of coping to exit. Exits, in retrospect, will then appear at the macroeconomic level as a standard cost of economic growth. If this cost is not incurred, it is synonymous with the absence of competition through innovative new establishment and of economic growth (Eliasson, 1994).

For viable, growth-promoting competition to occur, however, sufficient diversity has to be maintained in the economy (Eliasson, 1983, 1984) through entry, or through innovative behaviour among incumbent firms. For growth to occur, some superior or very competent firms will have to be constantly attempting to access the investment opportunity set. Such positive competition does not necessarily increase with the number of firms. The probability of some superior firms entering the opportunity set, however, normally increases with the number of firms attempting to enter.

This 'growth through selection' hypothesis of the experimentally organised economy (EOE) is the exact counterpoint to the static general equilibrium model. The Swedish micro-to-macro model MOSES is a model representation of the EOE. The model assumes that firms face such a varied menu of economic choices as to make even barely informed decisions at the micro level impossible. Under such circumstances, decisions must be characterised by *bounded rationality* (Eliasson, 1990). Bounded rationality in a complex economic environment also defines the *competence* of agents in markets to make successful business decisions. If this competence is accumulated through organisational learning in heterogeneous environments (Eliasson, 1992a; Ballot and Taymaz, 1993), the 'competence capital' becomes extremely heterogeneous. The diversity of environmental and firm competence characteristics needed to establish a model version of the experimentally organised economy can be introduced through a stylised so-called Salter curve analysis (Salter, 1960).

The Swedish micro-to-macro model MOSES is structured on technological competition in a Salter curve environment that constantly upgrades itself through the competitive process. The model emulates the competitive dynamics of the experimentally organised growth economy. Economic development is seen as an ongoing learning process through the immense investment opportunity set, creating new resources and new technologies. Free entry creates competition which drives the growth process, forcing reorganisation, rationalisation or exit on firms. This mechanism is moderated by physical (natural) barriers, culture and policy (Eliasson 1991b, 1994). A positive sum economic game is established. It is characterised by the investment/growth mechanisms of Table 1, all of which are represented in the Swedish micro-to-macro model (Eliasson, 1978, 1985).

Macroeconomic growth, therefore, occurs through (1) *entry*, (2) *reorganisation* and (3) *rationalisation* of existing firms and (4) through *exit* of low performing firms. Together these four mechanisms can explain all investment activities behind economic growth. The bulk of medium-term production growth occurs through the *reorganisation* and *rationalisation* of existing firms. The exit of low performers is a *divestment* that makes room for the expansion of high performers, not in the least through releasing labor and competence at reasonable costs. The productivity effects of exit are immediate, while reorganisation and rationalisation investments quickly affect growth. In fact, most of the growth inducements in Swedish industry of the 1980s came through the reorganisation and rationalisation of existing firms.

The *entry* mechanism is, however, critical in the long run. Without entry, existing industrial structures will not be upgraded and modernised effectively (Eliasson, 1991b) - and especially not through the injection of new technologies not related to investments in existing industrial facilities. Innovative entry works in the very long run. It is highly experimental, with the average performance of entrants being lower rather than higher than that of the average incumbent (Eliasson, 1991b; Granstrand, 1986) but the spread being much wider. Hence, the exit rate among new entrants is extremely high. The viable market economy has to learn to live with (1) the exit of incumbents forced by the successful entry of others, (2) the high failure rate of entrants, and (3) the long waiting time needed to see the few successes mature and grow sufficiently to influence macroeconomic behaviour.

While mainstream economic analysis focuses on efficient management (ie. rationalisation) among a given number of existing firms, the experimentally organised economy, as represented by the Swedish micro-to-macro model (see below) incorporates all four elements of change, and is fully capable of embodying development bloc formation (Carlsson, Eliasson and Taymaz, 1993), creating very different long-term macro dynamics. All mechanisms operate in different time dimensions and require different competence characteristics. They have to be in a certain balance over time to generate stable, sustained growth at the macro level. Three of the four mechanisms operate through selection. The macroeconomic growth process of the real world is composed of all four mechanisms and cannot be understood in terms of one of them only.

Table 1: The Four Fundamental Investment/Growth Mechanisms

| | |
|---|---|
| (1) | *entry*, new innovative establishments |
| (2) | *reorganisation* of existing firms |
| (3) | efficient *management* of existing firms (rationalisation) |
| (4) | bankruptcy, *exit* or creative destruction |

Classical and neoclassical theory assumes the market environment to be given, and populated by price-taking agents. Darwinian or Schumpeterian economic analysis (Winter, 1964; and Schumpeter, 1942) models agents that react to an exogenously (through technology) changing environment, by organisational adjustment, changing the character of firms and of industries. Micro-to-macro theory takes one final step further, by making the environment endogenous and dependent on the ongoing organisational adjustment of agents. Only through this final step will the properties of the experimentally organised economy become manifest: namely an economy whose structure an outsider observer - the firm - can only be partially informed about through a biased information system.

## 3. The MOSES Model – A Brief Presentation

The MOSES model has been well documented in many publications[1]. A few observations should, however, be made here. When seen from above, the MOSES economy is composed as an eleven sector Keynesian Leontief sector model, with endogenous investment, complete demand feedback through investment, and a non-linear Stone type expenditure system. Individual behavioural relations are typically rational and neoclassical, except for the extreme informational limitations on individual firms (bounded rationality) and the presence of important tacit and incommunicable competence in firms. As a result, markets never clear, the state of disequilibrium (the path) influences long-term growth, and traditional exogenous equilibria do not exist as possible operating domains of the MOSES economy.

In this model, the four manufacturing sectors (basic-, intermediate-, durable- and non-durable consumption goods) have been surgically removed and replaced by similarly structured real firm data, all being consolidated to exactly fit the aggregates of the removed sectors. To achieve that, a residual firm defining the difference between the consolidated real firms and the aggregate has been defined. This residual firm has then been split up into several smaller firms. Altogether the initial population in a MOSES simulation consists of some 250 firms (Taymaz, 1992). Endogenous entry and exit change the population of firms over time.

---

[1] Eliasson (1978, 1985) and Albrecht et al (1989) include a fairly complete presentation of the specification of the model, and Eliasson (1991a, 1992b) a brief version, including the main behavioural equations.

Table 2: Number of Initial, Entering and Exiting Firms in Simulations

| | Initial Year | New entries during 50 year simulation | Exits during 50 year simulation |
|---|---|---|---|
| Base run | 250 | 0 | 119 |
| Experiment I (entry) | 250 | 257 | 125 |
| Experiment II (low exit) | 250 | 0 | 87 |

Firms make long-term investment decisions based on expected long-term rents from (returns to) investments. These investments upgrade the production system of the firm.

While the individual firm's decisions about investment, entry and exit are based on long-term profitability expectations related to financing costs, the short-term production planning, recruitment and firing decisions are made on the basis of short-term profit improving decisions. There are explicitly modelled product, labour and financial markets. Through complete feedback during and between the quarterly planning period, quantities and prices are determined - not simultaneously but sequentially - within the same period. In this sense, the MOSES model represents a general game of monopolistic competition.

The entry and exit features are critical for the experiments to be reported on in the next section. *Exit* occurs in the model when a firm has reduced its net worth to zero, but it can also be made to occur earlier and on more rational grounds, when a firm repeatedly fails to satisfy its long term profitability targets. The consensus within the economics profession until the mid-1980s was that *entry* really did not matter (Eliasson, 1991b); but it does, and very significantly so. Entry in the current version of the model is based on a stochastic process moved by profitability conditions in the four industry markets.

Entrants are assumed (on the basis of empirical evidence, see Eliasson, 1991b; and Granstrand, 1986) to be on average equal to or somewhat inferior in performance to incumbents. However, the spread is very much larger. This means that the failure rate among entrants is very much higher than among the incumbents. To filter out enough success stories through competition to move long-term growth, therefore, resources are used up through more or less mistaken business decisions. In fact, at the macro level, *a significant rate of business failure* should be viewed as a standard cost for economic development (Eliasson, 1992a).

## 4. Simulation Results

Two experiments were run, which were designed to show the macroeconomic effects of typical entry, to illustrate the time lags involved before economic growth and other effects become apparent, to demonstrate the existence of indirect effects on competition and exit, and to document the non-linear character of the market machinery set in motion by a more active entry process. All experiments were run by quarter over 50 years, and were compared to a *base run*, on the parameter setting calibrated on data from the Swedish economy as described in Taymaz (1991, 1993).

In the base run, firm exit was allowed but new firm entry was shut off. In Experiment (I), *new firm entry* was allowed. In Experiment (II), new firm entry was turned off again, while the exit process was slowed down compared to the base run.

Allowing firm entry (Experiment I) makes little difference to manufacturing output for 10 to 15 years. After this, manufacturing output grows much more quickly than under the base scenario, and is more than 50 per cent higher after 50 years. The average difference in the growth rate in Experiment I is on average almost one per cent per year and slowly increasing[2], as surviving new high productivity firms grow while low performance firms contract or exit because of more intense competition (Table 2). In Experiment II, when exit is slowed down relative to the base run, some improvement in output growth is seen for the first 20 years or so (Figure 1), then development falls behind the base run. The explanation is simple. First, slack is used more efficiently, even though low productivity facilities are kept going that were forced to exit in Experiment I. After a while, diminished competition, due to the absence of competitive entry, makes factor costs increase and affect overall growth negatively. This negative development, however, takes some 20 years or more to become visible. After 20 years or so, the higher factor costs reduce investment incentives in expanding industries and the overall macro effect turns negative. The development of GNP is about parallel (Figure 2).

Unemployment responds in a similarly slow fashion (Figure 3). Significant reallocation of labor over the labour market occurs, but under both Experiments I and II, unemployment at the macro level is slightly lower than in the base run for about 20 years. In the entry experiment, strong expansion compared to the base run

---

[2] For practical reasons, experiments had to be terminated after 50 years. Judging from earlier experience with the model, this difference could be expected to keep expanding, since (see Eliasson, 1983) the base run with no entry will eventually stagnate and show negative growth.

Figure 1: Difference Between Manufacturing Output in Base Run and Experiments I and II (Expressed as Percentage of Manufacturing Output in Base Run)

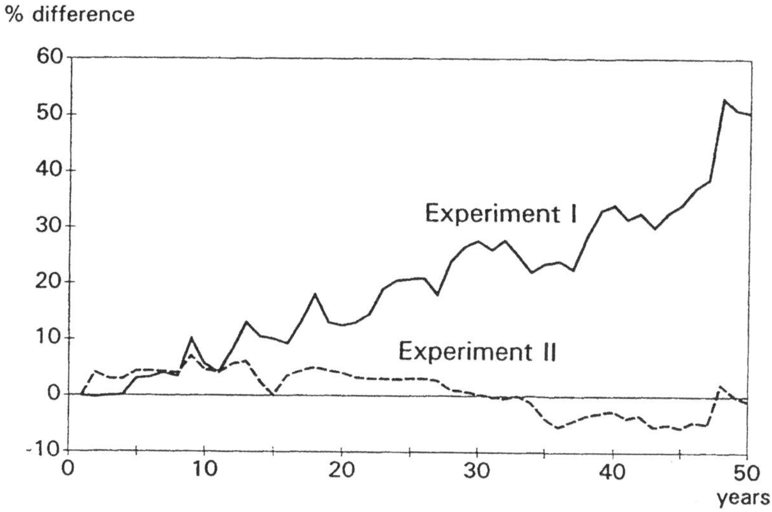


Figure 2: Difference in GNP Between Base Run and Experiments I and II (Expressed as Percentage of GNP in Base Run)

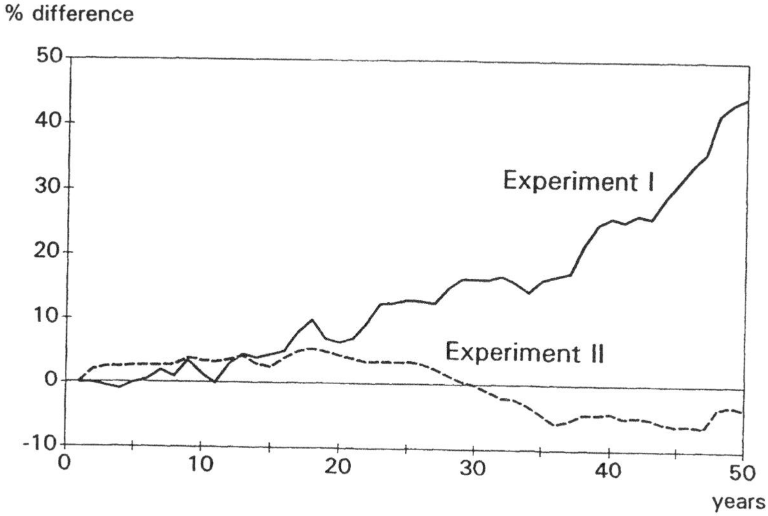

significantly lowers unemployment in the fourth decade, while an increase eventually is seen in the low exit case because of increasing wage costs. On the whole, however, wage adjustments eventually correct the unemployment situation under the labour market regime of the MOSES model, which assumes reasonably flexible wages.

Rates of return follow a time pattern similar to the unemployment cycle (Figure 4). The long expansion wave in the entry case after some 20 years is moved by high profitability, which is eventually competed away. In the low exit case, rates of return follow a different cycle but begin to exhibit instabilities towards the end of the 50 year run - a signal that a bad production structure with too little diversity is emerging (see below). It is known from earlier experiments (see for instance Eliasson, 1983, 1991a, 1994) that if competition is held back by limiting the exit of firms and increasing the presence of old obsolete structures, Salter structures eventually begin to be dominated by firms with bad performance. This result is also clear in the Salter distributions on the 50 year horizon (Figures 5 to 7).

Figures 5 to 7 illuminate the outcomes after 50 years of Experiment I. Rather well-behaved productivity and rate of return distributions are recorded on the 50 year horizon. It is interesting to observe that while old (initial) and new firms are fairly well distributed over the entire range, *new entrants dominate in the upper, high end of the rate of return distribution* (Figure 5). This should be the case (Eliasson, 1991b) but apparently, and very much in keeping with the hypothesis of the experimentally organised economy, high labor productivity is not necessarily the same thing as a high rate of return performance. Thus, while the most productive firms after 50 years are new entrants, some of the original firms also have high labour productivity (Figure 6). When rate of return distributions on the 50 year horizon are compared in Figure 7, the distorted distribution in the no entry and slow exit experiments come out clearly. As observed in Eliasson (1983) such shapes expose the industrial structure to changes in the competitive situation, or in the interest rate. A sudden jump in the interest rate can knock out the bulk of the industrial structure in the base and slow exit cases, if the increase is large (as it was so several times in the 1970s).[3]

The steep contrast between the best firms to the left and all the others in the low end tail imposes tension in, and disruptive behaviour of, the price system. When the

---

[3] The shapes of the labour productivity curves are similarly distorted. This has not been shown in this paper, because producer prices have changed very differently in the 50 year simulations and there is not enought space in this paper to go through the problem of how to deflate individual firm value added.

Figure 3: Difference in Unemployment Rate Between Base Run and Experiments I and II (Expressed as Percentage of Unemployment Rate in Base Run)

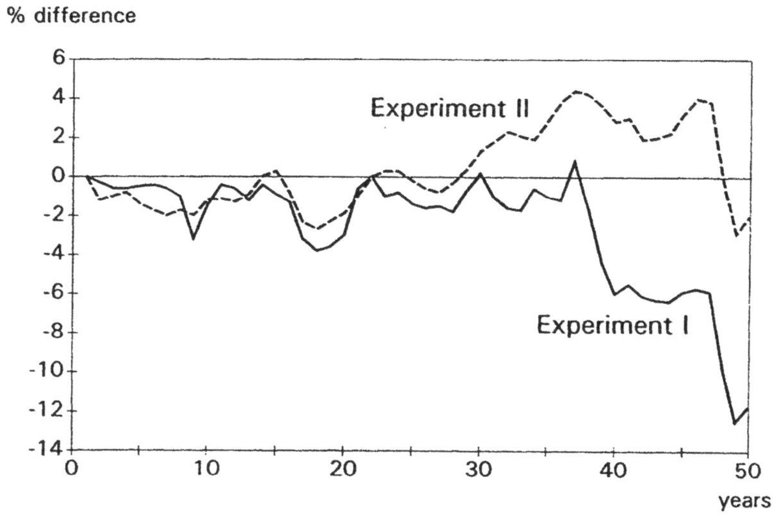


Figure 4: Difference in Rate of Return Between Base Run and Experiments I and II (Expressed as Percentage of Rate of Return in Base Run)

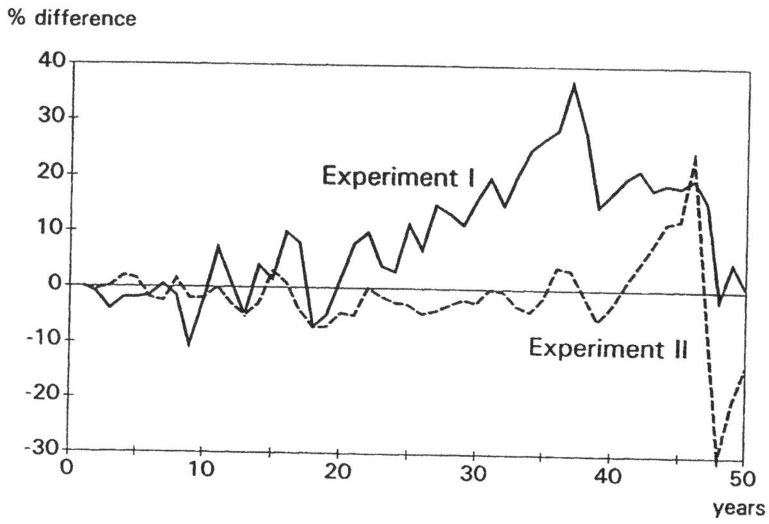

Figure 5: Rate of Return Distribution After 50 Years in Experiment I

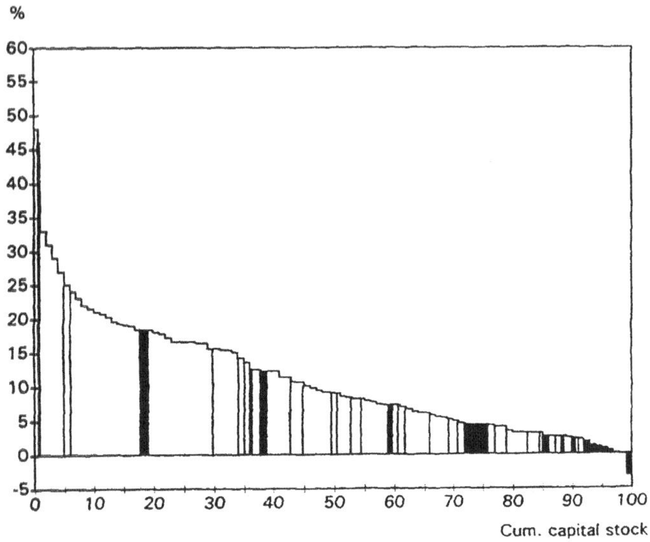


Figure 6: Labor Productivity (Salter) Distributions on 50 Year Horizon in Experiment I

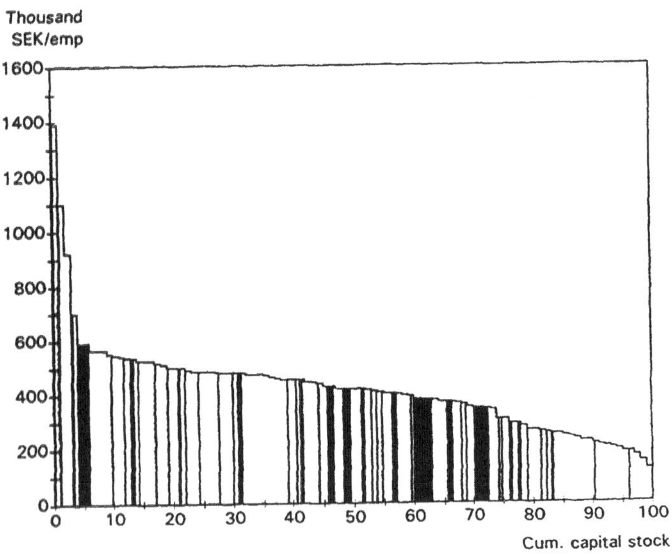

Figure 7: Rate of Return Distributions in Entry, Base and No Exit Experiments After 50 Years

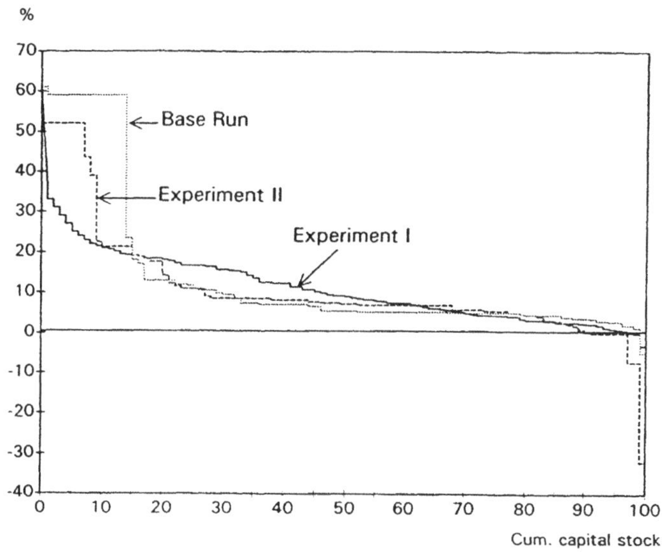


profitable firms want to expand they can easily outbid the large number of depressed firms in the low tail end in wages and interest rates, and suddenly force a large number of firms to exit. As can be seen, not much of an increase in interest rates is needed to make some three quarters of capital owners less concerned about staying business. The entry experiment I shows a much smoother performance distribution on the horizon and less cause for disruptive behaviour in the price system (Eliasson, 1983,1984).

## 5. Conclusions

Simulations with the MOSES Model suggest that encouraging new firm entry enhances future economic growth and manufacturing output in the long-term. In the short-run, however, the reallocation of labor caused by increased competition, reorganisation and exit may initiate short-sighted policy responses that prove detrimental to the necessary creation of new firms. The long-term decline in the number of new manufacturing firms being established in Sweden each year thus has worrying implications for future economic growth rates.

## References


Albrecht, J., F. Bergholm, G. Eliasson, K.A. Hanson, Chr. Hartlen, M. Heiman, Th. Lindberg and G. Olavi (1989), *MOSES Code*, Research Report No. 36, Industriens Utredningsinstitut, Stockholm.

Ballot, G. and E. Taymaz (1993), 'Firm sponsored training, technical progress and aggregate performance in a micro-macro model', Working Paper No. 402, Industriens Utredningsinstitut, Stockholm.

Carlsson, B. (1993), Den konkurrensskyddade och den oskyddade produktionen in *Den langa vagen (The Long Road)*, Industriens Utredningsinstitut, Stockholm.

Carlsson, B., G. Eliasson and E. Taymaz (1993), 'The macroeconomic effects of technological systems: micro-macro simulations', Paper presented to the 20th Annual Conference of the European Association for Research in Industrial Economics (EARIE), September 4-7, Tel Aviv, Israel.

Eliasson, G. (ed). (1978), 'A micro-to-macro model of the Swedish economy', *Conference Reports 1978:1*, Swedish paper from conference on microsimulation methods by IUI and IBU, Stockholm, Sweden.

Eliasson, G. (1983), 'On the optimal rate of structural adjustment', in: G. Eliasson, M. Sharefkin and B.C. Ysander (eds.), *Policy Making in a Disorderly World Economy*, Conference Reports 1983:1, Industriens Utredningsinstitut, Stockholm.

Eliasson, G. (1984), 'Microheterogeneity of firms and the stability of industrial growth', *Journal of Economic Behavior and Organization*, Vol. 5, Nos. 3-4.

Eliasson, G. (1985), *The Firm and Financial Market in the Swedish Micro-to-Macro Model - Theory, Model and Verification*, Industriens Utredningsinstitut, Stockholm.

Eliasson, G. (1987), *Technological Competition and Trade in the Experimentally Organised Economy*, Research Report No. 32, Industriens Utredningsinstitut, Stockholm.

Eliasson, G. (1990), 'The firm as a competent team', *Journal of Economic Behavior and Organisation*, Vol. 13, No. 3.

Eliasson, G. (1991a), 'Modelling the experimentally organized economy: Complex dynamics in an empirical micro-macro model of endogenous economic growth', *Journal of Economic Behavior and Organization*, Vol. 16, No. 1-2

Eliasson, G. (1991b), 'Deregulation, innovative entry and structural diversity as a source of stable and rapid economic growth', *Journal of Evolutionary Economics*, Vol. 1, No. 7.

Eliasson, G. (1992a), 'Business competence, organisational learning and economic growth: Establishing the Smith-Schumpeter-Wicksell connection', in: F.M. Scherer and M. Perlman (eds.), *Entrepreneurship, Technological Innovation, and Economic Growth*, , University of Michigan Press, Ann Arbor, Michigan.

Eliasson, G. (1992b), 'The MOSES model: Database and applications', in: J. Albrecht et al. (eds.).

Eliasson, G. (1994), 'The theory of the firm and the theory of economic growth', in: L. Magnusson (ed.), *Evolutionary and Neo-Schumpeterian Approaches to Economics*, Kluwer Academic Publishers, Dordrecht, Boston, London.

Granstrand, O. (1986), 'A note on measuring and modelling innovative new entry in Swedish industry', in: R.H. Day and G. Eliasson (eds.), *The Dynamics of Market Economies*, North-Holland, Amsterdam.

Salter, W.E.G. (1960), *Productivity and Technical Change*, Cambridge University Press, Cambridge.

Schumpeter, J.A. (1942), *Capitalism, Socialism, and Democracy*, Harper and Row, New York.

Smith, A. (1937), *An Inquiry into the Nature and Causes of the Wealth of Nations*, Modern Library, New York.

Taymaz, E. (1991), *MOSES on PC: Manual, Initialization, and Calibration*, Research Report No. 39, Industriens Utredningsinstitut, Stockholm.

Taymaz, E. (1992), 'A description of the initial 1982 and the synthetic 1990 database' (Ch II) and 'Initial state dependency: Sensitivity analyses on MOSES' (Ch. VI)', in: J. Albrecht et al. (eds.).

Taymaz, E. (1993), 'A calibration algorithm for microsimulation models', Working Paper No. 374, Industriens Utredningsinstitut, Stockholm.

Winter, S.G. (1964), 'Economic "natural selection" and the theory of the firm', *Yale Economic Essays*, Vol. 4, No. 4.

Gérard Ballot
Université Panthéon-Assas (Paris II) and ERMES-CNRS
France

Erol Taymaz
Middle East Technical University
Turkey

# Chapter 19: Firm-Sponsored Training, Technical Progress and Aggregate Performance in a Micro-Macro Model

## Abstract

*Microsimulation models have recently been used to study the effects of R&D decisions made by a population of competing firms on the aggregate economy. These models, which belong to the evolutionary school, yield new theoretical results in economics - results related to the importance of learning and variety in aggregate growth, and to the endogenous nature of technical progress. The model developed in this study goes further by introducing, for the first time, the role of training and human capital accumulation as a source of innovation and growth. Moreover, instead of remaining an academic exercise (as did previous models), it is an extension of the MOSES model, which is a complete micro- to macro-model of the Swedish economy with real data from a large sample of Swedish firms. The model deals with the many indirect effects of training and suggests that firms should be induced to increase their training effort, as such effort has major positive effects on growth and starts a 'virtuous cycle' of training and business success.*

## 1. Introduction

In the 1980s industrial countries increased their rate of intangible investment and decreased their rate of tangible investment. The measured intangible investment comprises expenditure on R&D, software and marketing, but generally not on training. At the same time, productivity growth and income growth declined. This productivity paradox, and the often heard complaint by firms that they have difficulties in hiring skilled workers, suggest that training may be a required complement to tangible and intangible investments if a high level of productivity and income growth is to be achieved. Training might be in insufficient supply. Even in countries like Sweden, where the government has funded an active training policy, deficiencies in skills appear (OECD, 1993). This argument for increased investment in training receives some support from the success of Japan, where firms devote considerable funds to training.

Some other explanations could be examined. However, the lack of skilled workers is clearly not caused by a global excess in labour demand, as unemployment is high and rising. Our theory is not easily integrated into the framework of standard theory. In a perfectly competitive world, it is not clear why the supply of education or training would be insufficient. According to standard human capital theory, workers would pay for investment in general human capital (as they take it with them when they quit a firm), while workers and firms should share some of the expenditures in specific human capital, as it is lost when they separate. In theory, the rates of return should be equal to the rate of interest, and profits should be zero.

These predictions are not confirmed in practice. The classical explanations are that workers face a moral hazard problem when they borrow - because of a lack of collateral, because they are risk averse or, more subtly, because they have difficulties in signalling to other firms the general character of the training acquired while working in the firm. These arguments may explain why workers invest too little. With the exception of the signalling theory (which cannot be the whole story) these arguments do not explain why firms invest in general human capital. That firms do invest in general training has been documented by various studies. These show that in France, the US, and Sweden, for example, firms pay much of the training, and do not lower the wage to recoup the cost (Stern and Ritzen, 1991).

This chapter offers another explanation of why firms invest in general training as well as in specific training and explains why some firms do not invest. This explanation takes us into another paradigm of the markets, the Schumpeterian paradigm. Microsimulation methods appear to be adequate tools to model theories within this paradigm. The justification of microsimulation relies not only on its technical complexity but also on its theoretical coherence. Microsimulation allows for the diversity of the firm's behaviour in areas such as human capital stock, training expenditures, learning and performance. MOSES (the Model Of the Swedish Economic System) is used for two reasons. First, it is one of very few microsimulation models based on decision algorithms (at the firm level), and this is a necessary feature in the study of the determinants of the training decisions. (For another example, see the ARTEMIS models developed by Ballot - 1991.) Second, it is a model of a complete economy, so indirect effects are accounted for.

The purpose of the chapter is to add a detailed training-innovation block to the MOSES model, in order to study the aggregate effects of training-innovation theory in a Schumpeterian competition framework. The paper is organised as follows. Part

2 presents the intellectual framework. Part 3 describes the model. Part 4 discusses the results of the simulations. Some conclusions complete the paper.

# 2. The Intellectual Framework

## 2.1 Training, innovation and schumpeterian competition

This chapter develops a novel idea to explain why firms might invest in human capital even if there is some loss through turnover (Ballot, 1992a; Ballot and Taymaz, 1993) - labelled the 'training for the rent' hypothesis. The basic argument is fairly straightforward: firms that innovate successfully obtain a quasi-rent. To obtain this rent firms need trained workers at the time of implementation of the new technology. It can then be rational for firms to pay for general training, even if some workers quit. Firms recoup the cost of the investment and the cost of higher wages or turnover with the quasi-rent.

The competition framework is Schumpeterian, and therefore admits the existence of rates of profits over the interest rate even in the long run. There is also a dispersion of these rates (which is partly related to the stochastic nature of R&D output, and partly to a 'race for competence', in which the winner gets all the market).[1] The framework allows for factors that influence the level of general investment and the rate of return, such as the rate of turnover or financial constraints. In pure competition such factors could have no influence, as the optimal investment in general training is zero and financial constraints do not exist. Deficiencies in human capital can occur and keep the firm inside the production possibility frontier in the short term and, simultaneously, slow down significantly the long-run outward movement in this frontier.

Ballot and Taymaz (1993) have made preliminary tests of the effects of firm-sponsored training and R&D expenditures on profitability. These support the hypothesis stated above. In France, the human capital stock, measured as cumulated (but discounted) training expenditures, has a very significant positive effect on profitability. At the same time, R&D has a negative effect and the interaction of the two variables has a positive effect. This result is only suggestive - but it points out the potential benefits of integrating the analysis of innovation and the analysis of training (general and specific), both at the theoretical and the empirical level.

---

[1] Ballot and Taymaz (1993) develop this notion, which is broader than, but similar to, the well known 'race for patents'.

While the analysis at the firm level reveals the importance of training for successful innovation, well-known diffusion effects of innovation imply that the social rates of return to investment in innovation are much higher than the private rates of return (Gomulka, 1990). Although some analytical work on R&D and diffusion at the market level exists, it does not encompass the multiple channels by which general training might influence the processes of innovation and diffusion among heterogeneous firms, and the effects of these processes on aggregate behaviour. The game-theoretic R&D literature does not discuss training. Numerical models are required and, among them, microsimulation is an appropriate way to model complexity. Moreover, it is not clear that the concept of equilibrium that is necessary to solve an analytical model is adequate in an environment characterised by innovation and Schumpeterian competition, involving learning by firms and continuous changes in the market structure.

## 2.2 General knowledge vs specific skills

Human capital theory bases its distinction between general and specific human capital on the loss of the capital when the worker leaves the firm and on the derived prediction that the firm will not sponsor general training. This chapter proposes a richer distinction, arguing that the derived prediction no longer holds.

Specific human capital is necessary to use a technology - specific skill. Such human capital is a direct factor of production. It is lost when the worker quits, so that hiring a worker means that he arrives with zero specific skills. It is acquired through training sponsored by the firm. The workers do not invest money but invest in the effort of learning.

Specific skills allow some learning-by-doing to take place. The productivity of equipment is improved without physical investment when the specific skill is available. The higher it is, the faster the average productivity in the firm converges to the productivity of the new equipment. It should be stressed that learning-by-doing, as specified in this model, does not correspond exactly to the phenomenon described by Arrow (1962). The latter corresponds to an automatic process by which the production cost is a decreasing function of cumulated past production. Such a function could be added in this study, but for this first version of the training block in MOSES, the many types of learning are restricted to some essential innovation and diffusion effects of human capital.

General human capital has a general nature in a double sense. First, it is transferable. Second, it is a competence which facilitates the learning of skills. Obvious examples are proficiency in mathematics or in foreign language. Some other competencies, such as strategic competence by managers, do not correspond to a precise curriculum and are acquired partly through training and partly through experience.

Since the purpose here is mainly theoretical and qualitative, it is assumed that there is only one general competence, called 'general knowledge' (or general human capital). It is not a direct factor of production. However, it has important effects on innovation and the diffusion of innovation. First, it allows firms to find new technologies and hence increase the likelihood of finding more profitable ones (not all new technologies in our model are more profitable than the existing ones). Such a competence corresponds, for example, to a higher scientific level of engineers and to better trained managers who seize opportunities faster.

Second, it enables the firm to search faster the technologies used by other firms and implement some of them (sometimes after modifying them). This competence has been given a prominent place in the analysis of the determinants of innovation under the name of 'absorptive capacity', coined by Cohen and Levinthal (1990). They state that this competence is acquired by prior R&D done by the firm. The concept is important because it introduces the novel idea that the diffusion of technology is not a public good and involves costs and time. We argue that absorptive capacity requires not only prior R&D, but also general knowledge.

## 2.3 Decision processes with bounded rationality, learning and training

In the model the economy is characterised by continuous innovation and by a process of creative destruction of heterogenous skills, jobs and firms.

This economy is very complex. It is important that modelling does not reduce this complexity too much because it is a fundamental determinant of the agents decision making processes - and of the macroeconomic coordination and evolution that result.

Firms cannot understand the workings of such an economy and build a model of it to compute their optimal decisions. The rhythm of innovation and the evolution of market structures generates too much uncertainty to assign probability distributions to key variables (market shares, net profit rates etc.) in the future. This uncertainty generates uncertainty in other firms. Uncertainty is endogenous and there is a kind of uncertainty multiplier.

Markets are often characterised by small numbers. The decisions of one firm can seriously affect the market shares and profits of other firms. However, the technique of optimisation under uncertainty over a long (or infinite) time horizon does not allow highly valued results to be obtained, because it cannot take into account variables with a high future variance and unstable and unpredictable interactions.

In the real world, agents compensate for the limitations of their information by learning continuously. They are adaptive agents. They know they have not taken the best decision, and try to improve the result next time. However, if they know *how* they have obtained a good result, they do not know necessarily *why* (by which mechanisms of the economy). A trial and error process allows the agents to learn the *how*. This contrasts with classical optimisation in which agents know the model of the economy (rational expectations) and do not make mistakes in a probabilistic sense (the error distribution has a zero mean). Under classical optimisation the agents do not learn.

Classical optimisation uses the tools of calculus. *Adaptive behaviour* (also called *bounded rationality* after H Simon) deserves the attention of the economist who wants to base the study of the workings of an economic system on microeconomic foundations only if adequate tools can be found. The version of the MOSES model that has served as a starting point contains some adaptive rules. The training-innovation block added here uses a more powerful tool - genetic algorithms.

The idea of this type of algorithm has been invented by Holland (1975) to identify an optimum in an environment in which one does not have any information and in which the objective function can be extremely complex.[2] These algorithms have proved their robustness in such environments in numerous disciplines but are only now starting to be used in economics. We are not aware of applications to a micro-macroeconomic model of a complete economy. If such applications do not exist the present model makes a qualitative step in the use of genetic algorithms.

---

[2] It can have discontinuities and local optima (see Goldberg, 1989 for a thorough analysis).

The genetic algorithm is used to search for more efficient technologies, with the objective of getting closer to the most efficient technology for a given state of fundamental science. This latter technology remains, however, unknown. The efficiency of search (according to several criteria) is determined by the level of the general knowledge of the firm and hence influenced by its expenditure on general training. This level also influences the capacity to adopt the technologies of other firms. These notions of efficiency of search and capacity can be summarised under the general term of 'rationality' of the firm (see Eliasson, 1990 and 1993).

This concept has a fundamental methodological implication for the modelling of decision processes. As the firms decide on their level of training the degree of rationality of a firm is endogenous.[3] Not only information but also rationality differs between firms. The heterogeneity of firms has a more fundamental nature than their endowments (physical capital, financial assets etc.) which, in the long run, are endogenous to their behaviour. One can then hope to account better for the diversity of firms' behaviour and for the fact that some meet success and others failure, generating a Schumpeterian dynamic process with its macroeconomic implications (external competitiveness, unemployment etc.) and, in the process, explaining innovation and technological progress.

## 2.4 Micro to macro links: externalities and aggregate outcomes

The previous section looked at microdecisions, while this section describes briefly how the microdecisions of a population of firms interact to yield interesting aggregate relations.

MOSES is a complete micro to macro model and one of the very few to contain some learning by firms. Firms in the model are inside their production possibility frontier and can improve their efficiency (see Appendix A for a short description of the model). The model also contains the possibility of failure and of reorganisation of the economy through exit and entry. This makes the rate of technical progress endogenous at the aggregate level under the constraint of the pace of scientific advance (the global technology).

The present training/innovation block makes both training and technical progress endogenous at the micro level. Interactions between firms, through diffusion and

---

[3] Lane (1993b, pp.178-180) develops a similar plea for a theory of rationality considered as an endogenous variable. We offer an operational model (limited to innovation).

competition/selection effects, allow for a much more realistic modelling of the determinants of technical progress and of its effects at the aggregate level.

The model includes many of the dynamic competition aspects of the recent simulation models of innovation initiated by Nelson and Winter (1982) and developed by Silverberg, Dosi and Orsenigo (1988) in a one industry framework and by Chiaramonte and Dosi (1992) in a two-sector environment.

MOSES, as a micro-macro model with firms learning through Genetic Algorithms (GAs), appears as a basic artificial world (or economy) of the type described by Lane (1993a, 1993b). It might also be labelled a complex adaptive system according to Holland and Miller (1991), as it has the three required characteristics. That is, it is a network of interacting agents; it exhibits a dynamic, aggregate behaviour; and its aggregate behaviour can be described without detailed knowledge of the behaviour of the individual agents.

The interesting property of these artificial worlds or complex adaptive systems is the emergence of coordination in the economy, which is self-organising. Such aggregate variables as the number of firms, the rate of growth, the rate of investment in training and the rate of technical progress are endogenous. The relations between these variables are not built into the model, but observed after the simulation. They emerge.[4]

However, these relations may be unstable. The extent of coordination may change. This open-ended dynamic behaviour, rather than bounded rationality/learning, might be the major departure from the standard modelling methodology, which imposes asymptotic equilibrium not only when analytic results are searched for but also in large computable equilibrium models which are solved numerically.

The model displays two important features of real economies that are not seriously integrated into the analytical models. The first is the set of externalities coming from diffusion and imitation. However, diffusion is assumed to be costless in endogenous growth models (knowledge is a public good) or, at another extreme, costs the price of the patent. In fact, it has a cost that varies with the knowledge of the imitating

---

[4] For a systematic exploration of the macro patterns that can emerge in a micro-macro simulation model as a result of different macro-conditions or institutions, see Ballot (1991, 1992b) in the context of a labour market.

firm, and this allows the innovating firm to keep a quasi-rent without patent. In the model, diffusion also improves the technologies because it increases the number of cross-overs. These occur between firms as well as within firms. Such inter-firm cross-overs emphasise the importance of variety if diffusion is to be fruitful (Jovanovic and Rob, 1989). Variety and diffusion then contribute to the rate of technical progress.

The second major feature is selection.[5] Diffusion is unequal and firms that spend more on R&D and general training will have an advantage in the long run. Such firms often even drive out of the market those that spend less - because of lack of funds or high turnover or because they give a priority to another allocation of funds (financial or physical investment). Selection among heterogeneous firms has important effects on the aggregate rhythm of technical progress and growth. However, it may not be the simple positive effect one would think of. If selection effects are so strong that the industry is dominated by a monopoly, there will be little diffusion and the rate of technical progress will be slower.

A real economy, and the model, is characterised by a complex set of interactions between training, knowledge, innovation, diffusion and market structure which determine (hopefully) some self-organising aggregate patterns. Through their mediation, patterns finally depend on the more stable cultural and institutional environment, and on the history of the economy (including stochastic shocks). The patterns are emerging properties, often not predictable from building blocks. The present version of MOSES should then provide a richer story about the determinants of growth than endogenous growth models that are aggregate (eg. Aghion and Howitt, 1992) because, for the first time, it adds the endogenous creation of competences to innovation and dynamic competition in markets.

## 2.5 Policy issues

The model has two important general features that make it suitable for policy experiments. First, it is a complete micro-to-macro model with feedbacks and micro-macro compatible accounts for a real economy (Sweden). As is the case with other microsimulation models, it allows the simulation of institutional policies (taxation, laws) at the proper level (which is the microeconomic level). Externalities and feedback effects occur and transform the micro effects by dampening or amplifying

---

[5] The standard version of MOSES, without endogenous R&D, already displays interesting selection effects through entry and exit (Eliasson, 1991).

them. Second, the model is rich and allows the study of effects of important and debated policies such as government intervention in education and training matters.

## 3. The Specification of MOSES

The Model of the Swedish Economic System (MOSES), which is summarised in Appendix A, is a micro-to-macro-simulation model of the Swedish economy. It has been constructed primarily to analyse industrial development. Therefore, manufacturing is modelled in greater detail than other sectors. The manufacturing sector is divided into four sectors (raw material processing, intermediate goods, durable and capital goods, and consumer non-durables). Each industry consists of a number of firms, some of which are *real* and some of which are *synthetic.*

225 firms or divisions are defined explicitly in the manufacturing sector in the 1982 database, which is the database used in simulation experiments summarised in this chapter. 154 of those firms are real - that is, data about those firms come from the Planning Survey conducted jointly by the Industrial Institute for Economic and Social Research (IUI) and the Federation of Swedish Industries. The rest of the firms (71 firms) are synthetic and they make up the difference between real firms and the industry totals in the national accounts. The model simulates the behaviour of firms in four manufacturing industries. Other sectors and the household sector are modelled at the macro level within a Leontieff-type I-O structure.

Learning and training (the accumulation of knowledge and skills) were not explicitly modelled in the original version of MOSES. In the following section the technology to explain how (general) knowledge and (specific) skills affect the performance of the firm is first described. Second, the learning and training modules are explained.

### 3.1 Technology

The technology of a firm can be represented by a set of 'techniques' as follows:

$$F = \{f_1, f_2,\ldots f_n\}$$

where F is the technology used by the firm, and $f_i$ is the $i^{th}$ technique, i=1,2,...,n. For simplicity (and without any lack of generality), a technique can be assumed to have only two values/alternatives, ie. $f_i \in \{f_i^1, f_i^2\}$.

The best practice technology (the 'global technology'), which is defined similarly, describes the best *combination* of techniques. The firm which uses *all* techniques in the set of global technology reaches the highest technological 'level'. The global technology can be interpreted in different ways. It may be the best combination of techniques for a certain technological paradigm (Freeman and Perez, 1988). A firm can know only *a part* of the global technology (or the opportunity set, see Eliasson, 1990) and its technological level is determined by the correspondence between the global technology and the technology applied by the firm.

In these experiments, a 40-element vector is used for the global technology. The technological level of the firm depends on the degree of correspondence (DC) between the global technology and the technology employed by the firm, as follows:

$$DC = \sum_{i=1}^{n} a_i w_i$$

$$a_i = \begin{cases} 0 & \text{if } t_i \neq f_i \\ 1 & \text{if } t_i = f_i \text{ and } t_j = f_j \\ -1 & \text{if } t_i = f_i \text{ and } t_j \neq f_j \end{cases}$$

$j=i+1$ if $i<n$, $j=1$ if $i=n$

where $w_i$ is the weight for the $i^{th}$ technique, T the global technology, and F the technology used by the firm. $t_i$ and $f_i$ denote $i^{th}$ techniques of T and F, respectively. The last line of the $a_i$ specification has a nonlinear DC function originated from the *interdependence* of techniques.

The technological level of the firm is computed by an exponential function of the DC value:

$$P = \alpha\, exp(\beta\, DC)$$

where $\alpha$ and $\beta$ are industry-specific parameters, P the technological level, and

*exp(..)* the exponential function. Although there may be many alternative specifications for the DC and P functions, this one is quite flexible and sufficient for our purpose.

The following simple example can be used to clarify the specification of technology in the model. Assume that the global technology for a hypothetical firm in the petrochemical industry consists of two techniques, $t_1$ and $t_2$. $t_1$ and $t_2$ refer to the wall thickness and the size of the storage tank, respectively. $t_1$ can be either 1 or 2 cm, and $t_2$ can take two values, $10m^3$ and $20m^3$. Economies of scale are important in building storage tanks and, therefore, the best practice technology requires large tanks with thick walls, T={2,20}. Also assume that $w_1$=1 and $w_2$=2 in the DC function and $\alpha$ =1 and $\beta$ =.5 in the P function. All alternative technologies with their corresponding 'pay-offs' (or 'productivity' levels) are shown in Table 1.

Table 1: Calculation of the Technological Level of a Hypothetical Firm

| $f_1$ | $f_2$ | $a_1$ | $a_2$ | DC | Pay off | Description |
|---|---|---|---|---|---|---|
| 2 | 20 | 1 | 1 | 3 | 4.5 | Large tank with thick walls |
| 1 | 10 | 0 | 0 | 0 | 1 | Small tank with thin walls |
| 2 | 10 | -1 | 0 | -1 | .6 | Small tank with (unnecessarily) thick walls |
| 1 | 20 | -1 | -1 | -2 | .4 | Large tank with thin walls (high risk of explosion) |

Note: DC=$a_1$+2$a_2$, Pay off=exp(DC/2)

When the size of the tank is 20 $m^3$ and the wall thickness is 2cm, the firm achieves its highest performance level (4.5 units). When $f_1$=1cm and $f_2$=10 $m^3$, the pay off is only 1 unit. But if the firm uses a thicker wall than is necessary for a small (10 $m^3$) tank, the pay off is lower because of the extra cost involved. Finally, if the firm builds a $20m^3$ large tank with 1cm thick wall, the pay off has the lowest value (0.4) which reflects the risks of explosion due to thin (weak) walls.

In the MOSES model, there are two critical firm-specific technology variables that determine the performance of the firm. These are the MEFF variable, which is, in a sense, (the inverse of) the incremental capital-output ratio, and the MTEC variable, which represents the level of labor productivity that can be achieved in new investment. Thus, two global technology vectors are used, one for the MEFF variable, and the other for the MTEC variable. The MEFF variable of a firm depends on the correspondence of that firm's technology vector and the global technology for the MEFF variable (as explained above). The MTEC variable is computed similarly.

The firm then tries to 'discover' global technologies by learning to improve its MEFF and MTEC variables. In the experiments, global technologies are represented by 40-element binary arrays where 0's and 1's denote alternative techniques. The elements of global technologies were generated before simulation experiments by a pseudo-random number generator.

## 3.2 Learning

Firms use 'genetic algorithms' to discover the global technology. A firm has a memory to retain k-number of alternative technology sets at a time (in these experiments, three sets) and actually uses the set that has the highest degree of correspondence with global technologies. Firms 'learn' about the global technology by recombining their own technology sets (*experimentation*), recombining their sets with other firms' sets (*imitation*), or by mutations (*innovation*). One of the most important processes of evolutionary dynamics, *selection*, takes place at the sectoral level through the selection of firms. Badly performing firms (and technologies used by them) will be nullified by the competition process in the market. This is the way learning takes place at the national economic system level (see Eliasson, 1992).

The learning process which takes place at the beginning of each year is executed in four steps for all technologies in each firm's memory. First, the firm decides if it will try experimentation or imitation. The firm decides to try experimentation with INSEARCH/(INSEARCH+1) probability. For example, if the INSEARCH variable of the firm is equal to 1, the probability that the firm will try experimentation is 50 per cent.

Second, if the firm decides to try experimentation, it will select a technology from its memory for recombination with the set under consideration. The probability of selection depends on the relative degree of correspondence with the global technology. If the firm decides to try imitation, then it will select a firm in the same market for recombination. As may be expected, the probability of selection depends on firms' technological levels (the MEFF or MTEC variables).

Third, the firm selects randomly NTRIAL number of elements of the technology to be used in recombination. Then, the values of those elements (ie. techniques) are replaced by the corresponding elements from the selected technology vector. If the degree of correspondence improves, the firm keeps the modified technology in its memory. Otherwise, the existing technology remains in the memory.

Finally, if the firm cannot find an improved technology in a NOTECH number of trials, then it will try a mutation. The firm can achieve a mutation with PMUTAT probability. In the case of mutation, a randomly selected NTRIAL number of elements of the technology vector are replaced by their opposites (0 →1, and 1 →0)

The learning specification has three critical variables - INSEARCH, NTRIAL and PMUTAT. (The fourth variable, NOTECH is constant and equal to three.) That is, if the firm cannot improve any one of technologies in its memory in a year, it will try a mutation. A decrease in INSEARCH means the firm will have a stronger tendency for out-search (imitation). Intuitively, out-search is usually better than in-search as the set of available technologies is broader in the case of out-search (because the number of firms in the sector is higher than the number of technologies that reside in the firm's memory). Moreover, except for the most advanced firm, at least one firm's technological level is higher than that of the imitating firm. Simulation experiments with the learning module support this intuition. When the INSEARCH variable is reduced, the learning process goes faster - that is, firms quickly discover the elements of global technology.

The second variable, NTRIAL, is another critical variable for the performance of the learning process. A low value for NTRIAL means the firm can change only a few elements (techniques) at a time. This implies a slow learning process. Experiments with the learning module show that increasing the NTRIAL variable improves learning performance. Finally, the PMUTAT variable, which determines the probability of mutation after a certain number of unsuccessful experiments, has a positive impact on learning.

The learning module based on genetic algorithms as specified here creates 'standard' results. For example, the diffusion of new innovations follows the usual S-curve. The learning process has also decreasing returns if the global technology is constant in the sense that, when firms get closer to the (constant) limits of their technological frontier, it becomes increasingly difficult to improve firms' technological levels.

## 3.3 Training

Training in a firm generates two types of assets - general knowledge and firm-specific skills. It is assumed in the model that both types of assets are created by training and embodied in personnel employed by firms. General knowledge, once created, is applicable in all firms and, therefore, transferable. If employees with a high level of general human capital move to another firm, they will increase the stock of general knowledge of the new firm. The firm-specific skills, as the name implies, cannot be transferred from one firm to another.

In the model, it is assumed that the stock of knowledge of a firm affects its problem-solving capabilities. To be precise, the INSEARCH, NTRIAL and PMUTAT variables explained in the preceding section depend on the knowledge stock of the firm. Accumulation of knowledge in a firm either by general training or by hiring highly qualified employees from other firms will improve those variables, so that the firm will be able to learn rapidly the global technology. In other words, firms with a large stock of general knowledge, on average, will rapidly learn the global technology and will improve their technology variables, MEFF and MTEC.

Firm-specific skills play a critical role in the application of what is learnt about the global technology. There are two aspects of the application process. First, a firm learning more about the global technology can update/improve its current stock of productive equipment. The improvement of existing fixed capital stock as a result of learning global technology depends on the stock of firm-specific skills. Second, the actual use of existing fixed capital depends on the stock of firm-specific skills. A firm endowed with the most productive equipment cannot produce any output at all if its employees are not trained to use that machinery. Thus, firm-specific skills can be used both in updating existing equipment (of the old vintage) and in effectively operating the (updated) equipment.

For the specification of the accumulation of knowledge, it is assumed that the knowledge and skills generated by a unit of training expenditure is constant in real terms and equal for all firms. In other words, there is no assumption of 'learning by learning' or 'learning to learn'. The 'productivity of training' does not change by time or by firm.

## 4. Experiments

In order to gain an understanding of the impact of general and specific training and of different institutional and behavioural settings, a set of simulation experiments on the Swedish micro-to-macro model were designed. In each case the impact at the macro level was analysed - that is, the impact on macro-performance. The modelling of training and learning modules is not yet complete. More work is planned, especially for the calibration of parameters. Therefore, the results should be considered as a first, exploratory phase of the modelling effort.

### Experiment BASE

The first experiment is the base experiment. In this experiment, firms have simple decision rules to determine the level of training expenditures. They desire to spend on training 0.4 per cent of their total stock of human capital and one per cent of the wage bill every quarter if they fully utilise their productive equipment. Otherwise, those numbers are multiplied by the inverse of utilisation ratio. The separation rate (quits and lay offs) by employees does not have any effect on training expenditures. Unemployed people are trained by the government to the extent that the growth rate of the stock of general knowledge of unemployed is equal to the industry average.

In the BASE experiment, firms cannot properly monitor the stock of general knowledge in each other - so that when they try to hire employees from each other (when they 'attack' each other to get the people they need), they can neither target firms with high human capital stock or hire employees with above-average knowledge. Employees in the attacked firm do not accept the offer from other firms unless the average wage bill in the attacking firm is at least 30 per cent higher.

### Experiment TRANS

The second experiment, TRANS, differs from the BASE experiment in the specifications of the labor search and of the training decision functions, which are more sophisticated than those in the BASE run. In the TRANS experiment, firms can, at least to some extent, observe the level of human capital in each other. Thus, when a firm wants to hire new employees, it tries to attack those firms with high human capital stock. If the wage level of the attacked firm is lower, then the attacking firm gets employees with above-average human capital. Thus, hiring skilled employees from other firms may be more profitable for a firm than training its own employees. Moreover, firms' training decisions depend on the separation rate as well as the variables mentioned before. If a firm has a high separation rate, it will spend less on training. As may be expected, firms spend less on training in this experiment than in the BASE experiment.

### Experiment GOVTR

The third experiment is similar to the TRANS experiment. However, in this experiment, government tries to solve the problem of inadequate training in firms by

Table 2: Experiment Specifications

| Experiment | Training expenditures | Inter-firm relations | Training of the unemployed |
|---|---|---|---|
| BASE | 1 per cent of the wage fund plus 0.4 per cent of the value of the firms' human capital | No observation on firms' human capital | Equal to the industry average |
| TRANS | Same as the BASE run but decreases with the separation rate of employees | Observes other firms' human capital, attacks firms with high level of human capital to hire new employees | Same as the BASE run |
| GOVTR | Same as the TRANS experiment | Same as the TRANS experiment | Equal to the level of the firm which trains the most |

training unemployed people. Government trains the unemployed to the extent that the human capital stock of the unemployed is higher than the stock in all firms. Thus, firms can increase their average stock by hiring from the pool of unemployed.

Experiment specifications are summarised in Table 2. A group of selected macro-performance indicators are shown in Table 3. This Table reveals that stocks of both skills and general knowledge become much smaller if the externality problem is significant for training activities. A low level of training expenditures has considerably slowed down the learning process, as shown in the very low levels of technology variables (the MTEC and MEFF variables) for the TRANS experiment. As the accumulated value of the specific skills is low, actual productivity gains are lower in the TRANS experiment during the whole simulation period. Note that the stock of skills is less than the stock of knowledge in both the BASE and TRANS experiments, although firms spend more for specific training. In the BASE experiment, for example, firms spent 15 per cent more on average on specific training than on general training. The stock of skills is lower because it is lost during inter-firm transfers, whereas general knowledge is not lost if workers leave their firms.

As may be expected, the growth rates of real GNP and manufacturing output are considerably higher in the BASE experiment than in the TRANS experiment. A paradoxical result seems to be generated for the unemployment variable. In spite of the high real growth rates achieved in the BASE experiment, the average unemployment rate is much lower in the TRANS experiment. The high growth rate of labor productivity attained in the BASE experiment, however, can, at least partially, explain this seemingly abnormal result. The labor saving character of technological progress may cause relatively high unemployment rates.[6]

Table 3: Training and Macroeconomic Performance: Experiment Results

| | BASE | TRANS | GOVTR |
|---|---|---|---|
| *First 15 years* | | | |
| GNP growth (period average, %) | 4.47 | 3.32 | 3.46 |
| Manufacturing growth (period average, %) | 6.56 | 5.50 | 5.53 |
| Rate of unemployment (period average, %) | 6.51 | 5.02 | 4.04 |
| Wage growth (period average, %) | 7.74 | 4.56 | 4.70 |
| Productivity growth (period average, %) | 4.51 | 2.95 | 3.12 |
| MTEC (end-of-period value) | 15.21 | 13.07 | 15.26 |
| MEFF (end-of-period value) | 2.37 | 2.10 | 2.33 |
| SPECTR (end-of-period value) | 121.14 | 39.82 | 35.87 |
| GENTR (end-of-period value) | 148.88 | 88.88 | 138.71 |
| *Last 15 years* | | | |
| GNP Growth (period average, %) | 4.07 | 3.42 | 3.45 |
| Manufacturing growth (period average, %) | 4.36 | 3.90 | 4.16 |
| Rate of unemployment (period average, %) | 5.72 | 2.95 | 4.27 |
| Wage growth (period average, %) | 7.73 | 3.80 | 3.67 |
| Productivity growth (period average, %) | 5.56 | 4.38 | 4.43 |
| MTEC (end-of-period value) | 26.22 | 19.64 | 26.63 |
| MEFF (end-of-period value) | 3.30 | 2.62 | 3.21 |
| SPECTR (end-of-period value) | 155.79 | 64.01 | 83.41 |
| GENTR (end-of-period value) | 196.93 | 83.67 | 178.32 |

Notes: BASE is the base experiment. In the TRANS experiment, the separation rate negatively affects firms' training expenditures. GOVTR is similar to the TRANS experiment but now the government trains unemployed people to a very significant extent. Simulation period is 30 years for all experiments. MTEC and MEFF are model variables that represent labor productivity of the latest equipment and (the inverse of) the incremental capital-output ratio, respectively. SPECTR and GENTR are average values of the stocks of skills and general knowledge, respectively.

[6] There are only process innovations in the current version of the model. Firms do not innovate new products. Therefore, technological progress has a strong labor saving bias.

The macro-performance of the economy in terms of real growth rates of GNP and manufacturing output is marginally better when the government trains unemployed people (the GOVTR experiment). The growth rate of labor productivity of this run is also marginally better than that of the TRANS experiment. The training of unemployed people does not increase macro-performance, as firms do not spend much on specific-training because of high separation rates. Although firms receive highly qualified employees from the pool of unemployed and become quite successful in learning about the global technology (high MTEC and MEFF variables), the employees are not able to apply what they have learnt. A comparison of the values of the SPECTR and GENTR variables for the BASE and GOVTR experiments shows that general training of unemployed people by the government can be a method to increase the stock of general knowledge. But the wrong mix of specific and general skills may impede the use of advanced technologies. It must be emphasised here that the assumption about the level of training the unemployed receive in the GOVTR experiment may not be realistic.

## 5. Conclusions

The MOSES model with the training-innovation block is a laboratory for the study of training policies. The previous paragraph offers the very first results. The essential contribution of the block in its present state corresponds to a renewal of the theory of the links between training, technical progress and growth. The model contains endogenous growth phenomena with virtuous and vicious cycles where continued training plays an important role, both endogenous and exogenous. The results can be summarised in three propositions.

First, investment in training, if high at the start, creates a virtuous cycle with strong innovation and growth. In return, growth creates the monetary resources necessary to finance the growth of training. A complete dynamic micro-macro model like MOSES allows the study of those cycles and of the role of the various factors in economic growth.

Second, non-cooperation by firms yields general and specific training expenditures inferior to the optimum. (This proposition may not be new as far as specific training is concerned. However, it is new for general training, as in human capital theory the optimal expenditure of firms in general training is zero. The externality justifies the intervention of the government that must respect the following proposition.)

Third, the type of training is important. A certain balance between general and specific training is necessary.

Experiments presented here should be considered as a first attempt to model complex activities like learning and training, which are certainly very important for the long-run performance of all economies. The model needs further work in two major directions. First, the specification of various functions (for example, training expenditures, effects of separations, etc.) should be based on detailed and, if possible, econometric studies. A parallel study on the relationship between micro-performance and training sheds light on the subject (Ballot and Taymaz, 1993). Second, there remain a number of functions/relationships that cannot be observed in reality (learning the global technology, etc.). Therefore, the model needs to be calibrated to produce reasonable results before being used in policy experiments.

APPENDIX A

# THE MOSES MODEL

## Overview of the Model

Firms in the model constitute short and long-run planning systems for production and investment. Each quarter, each firm begins by forming price, wage, and sales expectations and a profit margin target. These expectations and targets are then used as inputs into the production planning process, in which each firm sets a preliminary production/employment plan.

The firm's initial (*ex ante*) production and employment plans need not be consistent with those of other firms in the model. If, for example, the aggregated employment plans for all the firms exceed the number of workers available at the wage levels the firms intend to offer, an adjustment mechanism is invoked to ensure *ex post* consistency. In the case of labor, the adjustment takes place in a stylised labor market, in which the firm's employment plans confront those of other firms as well as the labor supply. This process determines the wage level, which is thus endogenous in the model. In a similar manner, firms' production plans are revised after a market confrontation in the domestic product market, and domestic prices are set.

There is also a capital market where firms compete each quarter for investment resources and where the rate of interest is determined. Given this interest rate, firms invest as much as they find it profitable to invest in the light of their profit targets.

The model is quite complex (see Eliasson, 1977 and 1985; Albrecht et al., 1989 and 1992; Bergholm, 1989; Taymaz, 1991). The parts of the model most pertinent for present purposes are presented below. In the following presentation, each variable is calculated for each firm in each year or quarter (although time subscripts have been dropped when so doing does not cause any confusion).

### Expectations, targets

Expectations are generated on an annual basis with quarterly modifications concerning percentage changes in sales, prices, and wages for each firm according to a simple 'adaptive expectations' formula. The profit target, TARGM, is defined as the share of non-wage value added in total value added. This is also generated by adaptive expectations.

### The production function

The production function for each firm in MOSES is of the following form:

$$Q_t = QTOP_t*(1\text{-}exp(\text{-}TEC_t*L_t/QTOP_t)) \qquad (1])$$

where Q is the potential output (in physical units) for a given employment level in number of hours ($L_t$); QTOP the maximum level of output which is approached asymptotically when infinite amounts of labor are used, given a certain level of capital stock; TEC the productivity of the first unit of labor (to be more precise, the slope of the production function at the origin), and *exp*(.) the exponential function. Subscript t denotes time.

The maximum output, QTOP, depends on the (real) stock of physical capital, the stock of specific skills, and the efficiency of the stock of physical capital as follows.

$$QTOP_t = QTOPFR_t*\{MINRT+[(1\text{-}MINRT)*ST_1*(1\text{-}exp(\text{-}SPECTR_t/ST_2))]\} \qquad (2)$$

where QTOPFR is the level of productive capacity (the asymptotic limit of the QTOP variable) and SPECTR the stock of specific skills. $ST_1$ and $ST_2$ are industry-specific parameters. MINRT is the minimum (percentage) level of output that can be produced with no stock of specific skills. It shows the productivity level of completely unskilled workers. Thus, if SPECTR=0, QTOP=QTOPFR*MINRT. On the other hand, as SPECTR $\rightarrow \infty$, QTOP $\rightarrow$ QTOPFR. The productive capacity, QTOPFR, is determined by the efficiency and amount of physical capital (QTOPFR = EFF*PK, where EFF is the output/capital ratio, and PK the stock of physical capital).

### Upgrading technological level

As shown in the specification of production functions (equations 1 and 2), there are two critical *technology* variables that determine the performance of the firm: EFF and TEC. There are two methods to upgrade the EFF and TEC variables. In the case of *embodied* change, the technological level is upgraded by investment as new equipment embodies the stock of knowledge possessed by the firm. In the *disembodied* case, the EFF and TEC variables are increased by implementing what is known by the firm (organisational changes, rationalisation, etc.).

Investment is the first mechanism to increase the EFF and TEC variables since newly installed capital embodies MEFF and MTEC levels of technology. The technological level of the capital stock is upgraded by investment as follows.

$$TEC_t^{\tau}=[(TEC_{t-1}*QTOPFR_{t-1})+(MTEC_t*\Delta QTOPFR_{t-1})]/(QTOPFR_{t-1}+\Delta QTOPFR_{t-1}) \quad (3)$$

$$EFF_t^{\tau}=[(EFF_{t-1}*PK_{t-1})+(MEFF_t*INV_{t-1})]/(PK_{t-1}+INV_{t-1}) \quad (4)$$

where INV is the amount of investment and $\Delta$QTOPFR is the additions to the productive capacity by investment ($\Delta QTOPFR_{t-1}=INV_{t-1}*MEFF_{t-1}$). Superscript $\tau$ indicates that the $TEC_t^{\tau}$ and $EFF_t^{\tau}$ variables have now temporary values. Their final values for the period t are calculated in equations 5 and 6 below.

MEFF is the efficiency of newly installed capital and MTEC is the level of labor productivity associated with new capital. As shown in equations 3 and 4, the TEC and EFF variables are calculated as weighted averages of technological levels of different vintages of capital. In a sense, the MEFF and MTEC variables reflect the stock of knowledge possessed by the firm. The technological level of the productive equipment actually used (measured by the TEC and EFF variables) is lower than the level known by the firm because of the vintage effect.

It is assumed that the firm can also gradually update its existing equipment without any investment in physical capital stock by applying what is learnt as follows.

$$TEC_t=TEC_t^{\tau}+(MTEC_t-TEC_t^{\tau})*ST_3*[1-exp(-SPECTR_t/ST_4)] \quad (5)$$

$$EFF_t=EFF_t^{\tau}+(MEFF_t-EFF_t^{\tau})*ST_5*[1-exp(-SPECTR_t/ST_6)] \quad (6)$$

where $ST_3$, $ST_4$, $ST_5$ and $ST_6$ are industry-specific parameters. Thus, as SPECTR $\rightarrow \infty$ the TEC and EFF variables are updated to the amount equal to $ST_3$ and $ST_5$ per cent of the difference between known and applied knowledge. Thus, we assume that the stock of *specific* skills (those skills specific to the physical capital used in the firm) determines the pace of disembodied change for a *given* level of the MTEC and MEFF variables. The values of the MTEC and MEFF variables depend on the efficiency of learning which is determined by the stock of *general* knowledge.

Equations 1 and 2 define the production function. Equations 3 to 6 show how technology variables are changed in the model. The following capital accumulation equation closes the system of production:

$$QTOPFR_t = [QTOPFR_{t-1}*(1-\delta_q)] + \Delta QTOPFR_{t-1} \quad (7)$$

where $\delta_q$ = the (constant) rate of depreciation,

## Learning

The critical technology variables, MEFF and MTEC, are determined at the beginning of each year by using the following equations.

$$MEFF_t = LT_1*[1\text{-}exp(\text{-}DCEFF_t/LT_2)] \quad (8)$$

$$MTEC_t = LT_3*[1\text{-}exp(\text{-}DCTEC_t/LT_4)] \quad (9)$$

where DCEFF is the degree of correspondence between GLOBEFF and $FIRMEFF^m$, DCTEC the degree of correspondence between GLOBTEC and $FIRMTEC^m$, and $LT_1$, $LT_2$, $LT_3$ and $LT_4$ industry-specific parameters. The calculation of the degree of correspondence is explained in the text.

GLOBEFF and GLOBTEC are 40-element global technology vectors for the MEFF and MTEC variables, respectively. FIRMEFF and FIRMTEC are the *sets* of corresponding firm-technology vectors. There are three (40-element) vectors in each set and the superscript $^m$ denotes the vector that has the highest degree of correspondence to the relevant global technology vector.

Firms try to 'discover' the GLOBEFF and GLOBTEC vectors by modifying their FIRMEFF and FIRMTEC variables (using genetic algorithms as explained in the text). The critical variables of the learning process, INSEARCH, NTRIAL and PMUTAT variables are exponential functions of the stock of general knowledge.

$$INSEARCH_t = GT_1*[exp(\text{-}GENTR_t/GT_2)] \quad (10)$$

$$NTRIAL_t = GT_3*[1\text{-}exp(\text{-}GENTR_t/GT_4)] \quad (11)$$

$$PMUTAT_t = GT_5*[1\text{-}exp(\text{-}GENTR/GT_6)] \quad (12)$$

where $GT_i$'s are industry-specific parameters. The firm decides to try experimentation (in-house search) with INSEARCH/(INSEARCH+1) probability. Therefore the probability of experimentation will decrease by GENTR: as GENTR $\to \infty$, INSEARCH $\to 0$. The alternative to experimentation is imitation of other firms. As explained in the text, imitation is usually more efficient than experimentation. The number of elements to be changed in each trial (NTRIAL) and the probability of mutation (PMUTAT) increases by GENTR. $GT_3$ and $GT_4$ parameters are asymptotic values of the NTRIAL and PMUTAT variables, respectively.

### Training and the stock of human capital

Stocks of specific skills and general knowledge are accumulated by investment in training:

$$SPECTR_t = (SPECTR_{t-1}*(1-\rho_s)) + INVST_{t-1} \tag{13}$$

$$GENTR_t = (GENTR_{t-1}*(1-\rho_g)) + INVGT_{t-1} \tag{14}$$

where $\rho_S$ and $\rho_g$ are depreciation parameters, and INVST and INVGT are (real) specific and general training expenditures per employee, respectively.

Firms can increase the stock of specific skills only by training, whereas the stock of general knowledge can also be increased by hiring highly educated workers (as explained in the section on the labor market).

### Short-run production planning

The (quarterly) production planning in the firm starts with the profit margin target and expectations on changes in sales, prices, and wages. The firm tries to find a production plan that is both satisfactory (satisfies the profit target) and feasible (under the production function). A simple iterative search algorithm is used for this purpose. The production plan gives the level of (planned) output and employment. If the firm needs more employees, it tries to hire people from other firms or from the pool of unemployed. If it has more employees than it needs, it starts the firing process.

### Labor market

In the labor market, firms are first ranked by their relative labor requirements. Then, starting with the firm that wants to hire relatively more people, they attack each other or the pool of unemployed to get people they want. The probability of being attacked depends on the size and (the inverse of) the wage level of the firm in the BASE experiment and the stock of general knowledge in the TRANS and GOVTR experiments.

In the BASE experiment, if the wage level in an attacking firm is at least 30 per cent higher than the level in the attacked firm, then CHL number of workers are transferred from the attacked to attacking firm (where CHL = min{DEMAING, KSI*LATTACKED}, where DEMANING is the number of workers demanded by the attacking firm, LATTACKED is the number of employees in the attacked firm, and KSI is a parameter). The average stock of general knowledge of transferred workers is equal to the average stock in the attacked firm. The new stock of general knowledge in the attacking firm is computed as the weighted stocks of existing and transferred workers.

In the TRANS and GOVTR experiments, if the wage level in the attacking firm is higher than the level in the attacked firm, then CHL number of worker are transferred (where CHL is as defined above). The stock of general knowledge in the attacked firm is assumed to be uniformly distributed within the range {(1-DIST)*GENTR, (1+DIST)*GENTR} where DIST is an industry-specific parameter. The average stock of knowledge of transferred workers is equal to the average stock of the top CHL/LATTACKED percentage of workers in the attacked firm. The new stock of knowledge in the attacking firm is computed as the weighted average of existing and transferred workers. This specification implies that the attacking firm is, at least partially, able to observe stocks of general knowledge in other firms.

The separation rate is calculated after the adjustments in the labor market as follows:

$$SEPRT_t = SUMSEP_t / L_{t-1} \tag{15}$$

where $SEPRT_t$ is the separation rate, $SUMSEP_t$ is the number of employees who left the firm in the current quarter, and $L_{t-1}$ is the number of employees in the last quarter. Then the separation intensity is computed as follows.

$$SEPINT_t = (QTPR*SEPRT_t)+[(1-QTPR)*SEPINT_{t-1}] \tag{16}$$

where SEPINT is the separation intensity variable and QTPR is a parameter.

### Determination of investment

There are four kinds of assets in MOSES: fixed assets (PK), liquid and other current assets (LK), inventories (IK), and educational assets (SPECTR and GENTR). The demand for funds and borrowing is met from investment demand and cash flow:

$$DEMFUND_t = DESINVPK_t + DESCHLK_t + DESINVTR_t \quad (17)$$

$$DESCHBW_t = DEMFUND_t - CASH_{t-1} \quad (18)$$

where DEMFUND is the demand for funds, DESINVPK the desired investment in fixed assets, DESCHLK the desired change in liquid assets, DESINVTR the desired investment in training, DESCHBW the desired change in debt (or borrowing), and CASH the cash flow.

The desired investment in training (training expenditures) is computed as follows.

$$DESINVTR_t = (QTOPFR_t/QTOP_t) * \{[ALFATR*PSER_t*(GENTR_t+SPECTR_t)] + [BETATR*W_t*L_t] + [GAMMATR*SEPINT_t]\} \quad (19)$$

where $PSER_t$ is the price of training services, and ALFATR, BETATR and GAMMATR are parameters.

Depending on the resources of the bank and total demand for borrowing ($\Sigma DESCHBW_i$), the firm borrows CHBW from the bank and allocates its funds proportionately between DESINVPK, DESCHLK and DESINVTR. The investment in training, INVTR, is divided between specific and general training as follows:

$$INVGT_t = [1-(DELTR*QTOPFR_t/QTOP_t)]*INVTR_t/PSER_t \quad (20)$$

$$INVST_t = INVTR_t - INVGT_t \quad (21)$$

where DELTR is an industry-specific parameter.

Note that if the difference between QTOPFR and QTOP increases, the firm will spend more on specific training to exploit what is learnt.

## References

Aghion, P. and P. Howitt (1992), 'A model of growth through creative destruction', *Econometrica* Vol. 60, No. 2.

Albrecht, J.W., F. Bergholm, G. Eliasson, K.A. Hanson, C. Hartler, M. Heiman, T. Lindberg and G. Olavi (1989), *MOSES Code*, The Industrial Institute for Economic and Social Research, Stockholm.

Albrecht, J.W., P. Braunerhjelm, G. Eliasson, J. Nilson, T Nordström and E. Taymaz (1992), *MOSES Database*, The Industrial Institute for Economic and Social Research, Stockholm.

Arrow, K.J. (1962), 'The economic implications of learning by doing', *Review of Economic Studies*, Vol. 29. No. 3.

Ballot, G. (1991), 'Modelling the labor market as an institution: model ARTEMIS, ERMES Working Paper No. 91-09, Université Panthéon-Assas, Paris.

Ballot, G. (1992a), 'Continuing education and Schumpeterian competition: Elements for a theoretical framework', Paper presented to the Nordic Seminar on Human Capital Creation, The Research Institute of the Finnish Economy (ETLA), Helsinki, 14-15 May.

Ballot, G. (1992b), 'Labor market policy and the distribution of labor incomes, a dynamic analysis', ERMES Working Paper No. 94-02, Université Panthéon-Assas, Paris.

Ballot, G. and E. Taymaz (1993), 'Firms' sponsored training and performance: A comparison between France and Sweden based on firms' data', ERMES Working Paper No. 93-09, Université Panthéon-Assas, Paris.

Bergholm, F. (1989), *MOSES Handbook*, The Industrial Institute for Economic and Social Research, Stockholm.

Chiaramonte F. and G. Dosi (1992), 'Microfoundations of competitiveness and their macroeconomic consequences', in: D. Foray and C. Freeman (eds.), *Technology and the Wealth of Nations*, Pinter Publishers, London.

Cohen, W.M. and D.A. Levinthal, (1990) 'Absorptive capacity: A new perspective on learning and innovation', *Administrative Science Quarterly*,Vol. 35, No. 1.

Eliasson, G. (1977), 'Competition and market processes in a simulation model of the Swedish economy', *American Economic Review*, Vol. 67, No. 1.

Eliasson, G. (1985), *The Firm and Financial Market in the Swedish Micro-to-Macro Model*, The Industrial Institute for Economic and Social Research, Alongvist and Wicksell, Stockholm.

Eliasson, G. (1990), 'The firm as a competent team', *Journal of Economic Behavior and Organization*, Vol. 13.

Eliasson, G. (1991), Deregulation, innovative entry and structural diversity as a source of stable and rapid economic growth, *Journal of Evolutionary Economics*, Vol. 1, No. 1.

Eliasson, G. (1992), The MOSES Model - Database and Applications, in: J.W. Albrecht et al. (eds.).

Eliasson, G. (1993), 'Technology, economic competence and the theory of the firm', Working Paper No. 385, The Industrial Institute for Economic and Social Research, Stockholm.

Freeman, C. and C. Perez (1988), 'Structural crises of adjustment: Business cycles and investment behaviour', in: C. Freeman, R. Nelson, G. Silverberg and L. Soete (eds.), *Technical Change and Economic Theory*, Pinter, London.

Gomulka, S. (1990), *The Theory of Technological Change and Economic Growth*, Routledge, London.

Holland, J. (1975), *Adaptation in Natural and Artificial Systems*, University of Michigan, Michigan.

Holland, J. and J.H. Miller (1991), 'Artificial adaptative agents in economic theory', *American Economic Review*, Vol. 81, No. 2.

Jovanovic, B. and B. Rob (1989), 'The growth and diffusion of knowledge', *Review of Economic Studies*, Vol. 56.

Lane, D.A. (1993a), 'Artificial worlds and economics', Part I, *Journal of Evolutionary Economics*, Vol. 3, No. 2.

Lane, D.A. (1993b), 'Artificial worlds and economics', Part II, *Journal of Evolutionary Economics*, Vol. 3, No. 3.

Nelson, R. and S. Winter (1982), *An Evolutionary Theory of Economic Change*, The Belkup Press of Harvard University, Cambridge, Mass.

OECD (1993), *Industry Training in Australia, Sweden and the United States*, OECD, Paris.

Silverberg, G., G. Dosi and L. Orsenigo (1988), 'Innovation diversity and diffusion: A self-organization model', *Economic Journal*, Vol. 98, December.

Stern, D. and J.M.M. Ritzen (eds.) (1991), *Market failure in training?*, Springer-Verlag, Berlin.

Taymaz E. (1991), *MOSES on PC. Manual, initialization and calibration*, The Industrial Institute for Economic and Social Research, Stockholm.

PART FIVE

# Models and Data

Keith Blackburn
Warren Richter

Australian Bureau of Statistics
Australia

# Chapter 20: Centralised Data Management: Can It Help Micro-Modellers?

'... the policy world needs "a second revolution". The first revolution of the past two decades institutionalised the use of detailed estimates of cost and population effects of alternative proposals as part of the legislative process, and contributed to the development and widespread application of large computerised models as estimation tools. The second revolution requires significant investments in data, research knowledge, and computing to improve the quality of these models and the estimates they produce' (Citro and Hanushek, 1991, p. 27).

## Abstract

*Data management, the management of data through all stages of its life cycle, involves statistical integration, transformational integration and physical systems integration. To provide context, in this chapter the role of data management in facilitating macro-economic policy is contrasted with its role in facilitating social policy analysis, based on micro-data. Historical and current institutional arrangements, the states of development of overarching conceptual frameworks and resourcing stand out as the areas of greatest difference. To achieve the 'second revolution advocated by the 1991 report of the US Panel to Evaluate Microsimulation Models, new institutional arrangements at both national and international levels are needed to provide a basis for more effective investment in the data needed for social policy.*

## 1. Introduction

The focus of this chapter is the role of and organisational arrangements for data management in the social policy process. It is addressed to a wide audience: to those concerned with national and international statistical policy; collection and survey statisticians; statisticians involved in the construction of data sets for use in micro-modelling; micro-model builders and social policy advisers and decision makers.

* The views expressed in this paper are those of the authors and do not necessarily reflect the policies of the Australian Bureau of Statistics.

To structure the subsequent discussion, micro-models are considered in terms of two major elements:

- the database used to describe the entities to be modelled (the information which is to provide the objective foundation of an analysis); and
- the meta-data description of government programs (the formalised algebraic description of government programs expressed as a function of database variables).

Central to microsimulation modelling is the task of data management. A number of steps are necessary: assemble the source data sets to describe the entities to be modelled; formally describe the government programs to be modelled; and establish common naming conventions, data concepts, definitions and classifications appropriate to the formal program description and derivable from available source data. Both the policy program and the data must be described in the same meta-data language. These common data concepts, relationships, and structures are needed in order to link the database and the program meta-data system. The source data sets can then be transformed to produce a data set conforming to the needs of decision makers and modellers.

The transformed data set, ideally, will also possess a number of other important qualities described later in this chapter. To achieve a data set conforming to user requirements and possessing these special qualities requires substantial financial and technical resources and high levels of communication between all the relevant stakeholders: collection and survey statisticians, statisticians responsible for creating the transformed data sets, microsimulation modellers, and policy analysts. The resulting integrated data set, the meta-data description of that data set, and the government program meta-data system are the outcome of communications between the statistical, modelling and policy communities. They also provide the basis for ongoing communications. Publicly accessible, integrated and documented national data sets are envisaged which would become the starting point for individual modelling groups.

## 2. Data Needs of Modellers

Public policy decisions involve a major part of government outlays and affect everyone in the community, through either resulting taxes or benefits. Both the public policy analyst and the modeller are obliged to ensure that the data used for such decisions are suitable for the purpose.

## 2.1 Authoritative, reliable and reproducible

Data used for policy analysis must be authoritative and have the confidence of all involved, including the legislators, the media and informed public opinion. To be *authoritative,* the data must be seen to have been created by an organisation or organisations whose statistical integrity and competence is accepted without reservation. The process by which the data are created must be open and well documented.

To be *reliable,* the data would possess predictable and acceptable error characteristics. Sampling errors would be commensurate with the intended use. Non-sampling errors would be researched, documented and, where possible, quantified. Discrepancies between data sets would be documented and, within reasonable bounds, given the known limitations of measurement. If there were no discrepancies, the process by which they had been removed would be publicly documented and its statistical veracity established. Revisions, if any, would be minimised and of acceptable size.

In order to be *reproducible,* use of the data would be from properly archived data sets. Documentation of model outputs used to provide policy advice would be able to identify the transformed data set which had underpinned the advice (for example, by archive edition number) and would enable independent reproduction of model outcomes – given access to the model.

## 2.2 Conceptually relevant, structured, consistent and comprehensive

The data needs to be *conceptually relevant*. Individual concepts and classifications used must be consistent with the policy context to be modelled. If a preferred concept were not directly measurable, a derived measure would be required. Ideally the data would be *structured* in accordance with internationally agreed conceptual frameworks and the frameworks would be *consistent*. The data would be internally consistent and, in aggregate, would be consistent with relevant macro-data sources. The data set underlying a particular model would be *comprehensive*, describing the required concepts without gaps and with appropriate classification detail. The unit scope of the data set would also be comprehensive, with careful attention paid to ensuring proper measurement of institutional boundaries.

### 2.3 Relatable/linkable, available and accessible

Data needs to be *relatable* across data sets. To facilitate this, concordances would relate the data items and classifications used in different data collections. Ideally it would be possible to link data from one data set to data in another. *Linkage* would be assisted by duplication of data items across surveys and by provision of appropriate data management software.

A major area of difficulty for the public policy use of micro-data is the "increased emphasis placed by statistical agencies on restricting data access in order to guard against possible breaches of confidentiality" (Citro and Hanushek, 1991, p. 7). In Australia the ABS releases micro-data in the form of unidentifiable unit records, subject to a careful review of the steps taken to ensure against possible disclosure and a legally binding undertaking by the using organisation that use will be restricted to 'statistical purposes only'. However, there are limits to the extent that policy uses for micro-data can be satisfied given the legal obligations under which the ABS operates. An important task of official statisticians is to maximise this *availability* and use of statistics within the constraints set by law.

A number of factors are important in achieving *accessibility*. Users must be able to readily discover the existence of data, they must be able to relate the data to their needs, and electronic access must be in accordance with international standards for data structures and systems interconnection, etc. Awareness of data availability and the ability to relate it to needs would be facilitated by the provision of a data catalogue drawing on data and meta-data conforming to a defined information model.

## 3. The Nature of Data Management

Data management can be defined as the successful management of data through all stages of its life cycle. Effective data management will address each of the data needs identified above. For the purpose of this paper, it will be considered in terms of three primary elements:

- statistical integration;

- transformational integration; and

- physical systems integration.

*Statistical integration* relates to the concepts and methodologies underlying source data (data 'as collected') and can be regarded as the *collection or survey dimension* of data management. It is concerned with the management of statistical concepts (data item and unit definitions, scope and coverage definitions, etc.), coding, collection, estimation and imputation methodologies, classifications and primary source data derivation. The extent to which statistical integration is achieved in practice is quite variable. Within some relatively limited subject areas, high levels of integration have been achieved. Between subject area (and in some cases, between time) integration is generally somewhat weaker. Between countries, integration is particularly weak.

*Transformational integration* relates to the process of transforming collected data so that it conforms more closely with the conceptual constructs of end users. It can be regarded as the *primary data analysis dimension* of data management. Required conceptual constructs may not be directly measurable. Measurement error may mean that required identities do not hold in practice. Particular data collections may provide only a partial picture, leaving gaps that have to be filled. Such integration involves modifying primary source data in order to ameliorate such problems. It is also required to adjust for the effects of methodological and conceptual changes to individual data sources over time, and the effects of methodological and conceptual differences between data sources at each point of time.

*Physical systems integration* relates to the way in which information management processes and systems facilitate the physical linking, indexing and accessing of data and can be regarded as the *information science dimension* of data management. The primary focus of the 'systems' aspect of data management is the assembly, documentation and meta-data description of statistical output.

These three elements of data management may be undertaken in relative independence or with tight coordination. The effective use of information management processes (physical systems integration) can greatly assist the processes of statistical integration. Information science can assist the management of both in the development and implementation of statistical classifications and standards (statistical integration) and in the design and delivery of data and meta-data access methods. The need for transformational integration is diminished the further statistical integration is taken. An important aim of data management is to minimise the extent to which data must be transformed further in order to meet end user data requirements, including the need to fit their conceptual constructs, by meeting such needs through sound collection design and execution. Despite this long-term goal, an important theme of this paper is the need to substantially increase the effort put into transformational integration. It is argued that only after very substantial investments in this area and in the other two areas of data management will the data needs of microsimulation modellers and social policy advisers be met adequately.

# 4. Situation Analysis

The situation confronting microsimulation modellers with respect to data management practices varies considerably from country to country. However, many common elements are apparent, particularly when the situation with respect to micro-social data is contrasted with the situation that typically exists in respect of macro-economic data. The comparison reveals deficiencies that must be addressed if the data needs of micro-modellers previously outlined are to be more adequately met. One of the needs noted above was compliance with international standards for data structures and system interconnection. These standards are still evolving, although most users seem reasonably satisfied with this aspect of data management. As a consequence, this issue is not discussed further.

## 4.1 Macroeconomic data situation

In February 1993, as the culmination of more than 13 years of international deliberation, the third edition of the UN System of National Accounts (SNA) was ratified by the UN Statistical Commission. The 1968 edition of the SNA, before that the 1953 edition and various other international standards, have provided a blue print for the collection and transformation of economic data for four decades.

The SNA (and related standards) has had a profound impact on the *statistical integration* of economic statistics and has provided the standard against which national progress in the collection of economic data has been measured. SNA has assisted national statistical organisations to achieve appropriate levels of resourcing for their economic statistics. It has had a profound impact on the design of data collections influencing the definition of units, the definition of data items, the design and development of classifications, and the collection methodologies adopted. It has also influenced the design and construction of basic statistical infrastructure, such as integrated business registers from which collection samples could be drawn.

The process of ensuring the authority, conceptual relevance, consistency, and comprehensiveness of economic statistics has also been profoundly influenced by the SNA. The *transformational integration* of economic data is typically done within the context of such economic accounting frameworks as the national accounts, balance of payments, flow of funds, and input output. Central economic policy agencies have recognised the need for a single agreed publicly available primary data analysis that could form the basis for all subsequent economic analysis and macro-economic modelling, and have committed substantial resources to the task. In Australia, for example, more than 25 per cent of the total resources the Australian Bureau of Statistics (ABS) commits to business data is used in primary data analysis of this kind.

*Physical systems integration*, the third element of data management with which this paper is concerned, is less developed, but many countries make all their major economic data series available from a single (usually a time series) database, using common data structures and a variety of media. The extent of on-line meta-data support is limited, although hard-copy documentation of classifications, concepts, sources, and methods is now generally well developed.

## Institutional situation

Responsibility for data management is shared. International standards are determined jointly by national statistical agencies and the relevant international agencies. In the case of the SNA, the UN Statistical Commission provides final ratification of the agreed standards. The origins of the SNA lie in the conceptual and empirical work of academic economists and in the pioneering institutionalisation of that work by government economic policy agencies during World War II. Responsibility for the primary data analysis (compilation of conceptually compliant data sets using the available – but incomplete and inconsistent – source data) moved first from academia to government Treasuries and then to national statistical or economic analysis agencies. As this occurred, the ways in which the national accounts, balance of payments, flow of funds and similar statistical overviews were constructed became increasingly formalised, ensuring the resulting data possessed the many properties required for efficient use as the basis of economic policy. Typically, international standards form the basis of national standards, modified to reflect national circumstance. Formal statistical advisory bodies now exist in most countries to advise statistical agencies about economic data requirements and priorities. This advice feeds directly into statistical outcomes and into international deliberations concerning future directions of the standards.

The process of improving the synthesis of economic source data is not static. Substantial resources continue to go into the development of new data structures or the improvement of existing structures, further improving the scope and coherency of the available macro-economic view.

Responsibility for data collection may be vested in a central statistical agency or a group of agencies with statistical responsibilities. Responsibility for primary data analysis (compilation of national accounts, balance of payments, flow of funds and similar analyses) may be vested in a central statistical agency or a group of agencies. Successful statistical integration (ie. getting data as correct as possible at source) depends on a high level and quality of communication between areas responsible for data collection and for primary data analysis. Where both groups are within the same agency, the statistical integration role has generally been enhanced significantly, which in turn has enhanced the role of transformational integration as a means of improving the integrity and useability of data for the purposes of macro-economic analysis.

## Implications for economic analysis

A single authoritative, comprehensive, conceptually relevant and consistent view (both structural and dynamic) is now available for all major market economies.[1] That view is broadly consistent over time and between countries, facilitating international comparative analysis. Because the synthesis of primary source data is done only once to mutually agreed standards, economic policy debate is able to focus on policy rather than on how inconsistencies in view of the primary source data should be reconciled. Inconsistent primary source data views co-exist with the integrated synthetic view, but are of relatively limited value for the purposes of economic policy analysis, because of their inconsistency and the special characteristics designed into the integrated views. However, because transformation of primary data sources is the responsibility of the statistical agency, and is undertaken in accordance with agreed and documented standards, economic modellers and policy analysts can focus on modelling and analysis rather on data management issues.

---

[1] By the end of the 1940s, integrated statistical reporting systems were in place in Australia, the United States, the United Kingdom, Canada, the Scandinavian countries, the Netherlands and France.

## 4.2 Microsocial data situation

International standards are available for most areas of social data collection. These standards have assisted the process of statistical integration. However, no internationally agreed and overarching framework has been endorsed, although attempts have been made to develop such a framework. Data management has tended to be approached in a fragmentary way.

The resources committed to social data integration are significantly smaller than those committed to economic data integration and are generally less focussed. For example, in Australia, the ABS allocates less than one per cent of its social data resources to data integration and primary data analysis.

International standards and international consultations have both contributed to the *statistical integration* of social statistics. They have influenced the design of data collections, the definition of units, the definition of data items, the design and development of classifications, and the choice of collection methodologies. In the absence of an overarching framework, statistical integration has been more collection-specific. Integration across collections is generally less developed than is the case for economic statistics.

The task of achieving unit record or micro-data integrity is particularly daunting. Techniques such as uprating, re-weighting, re-referencing and imputation have been developed to assist the analyst achieve consistency when working with more than one primary data source. The sensitivity of analytical outcomes to the technique(s) used to achieve superficial consistency is an area still requiring a substantial amount of basic research and codification. Work on *transformational integration* of micro-social data is generally not ongoing nor publicly available. Such work is generally undertaken by individual modelling groups in conjunction with their other work. An exception is the pioneering work of Statistics Canada in the development of publicly accessible microsimulation databases.[2] Another exception, which adopts a different approach, is the work of the Netherlands Central Bureau of Statistics, directed to compiling consistent and ongoing mezo or intermediate level socio-economic accounts.

---

[2] Statistics Canada has devoted considerable resources to developing microsimulation databases such as the Social Policy Simulation Database and Model (SPSDM). The work uses techniques such as imputation and statistical matching, which permit public access to unit records but is informed by exact matches of administrative and survey data carried out within the agency.

The systems supporting *physical systems integration* are generally quite weak. Facilities for referencing available and potentially available data (for example, new data items derivable from existing micro-data to a client's specifications) are generally poor. One of the main reasons for this is the multiplicity of sources of information about statistical collections and statistical output, and the necessarily generalised nature of the descriptions of the statistics themselves.

In Australia, for example, the data and descriptions of that data (ie. meta-data) required by micro-data users come from a large number of collections and are held in a variety of data structures. It is difficult, time consuming and expensive to prepare unit record files from ABS social surveys. Although there are national or corporate standards for classifications and for the administration of units and data item definitions etc. with the exception of classifications, there are no national or corporate standards for storing descriptions of data (including data item and unit definitions) and statistical processes. Standard data item names and definitions are only partly implemented. Standardisation is weakest between different subject fields and over time. Meta-data standards are needed in order to address the problem of inconsistent and misleading data item names and definitions and non-standard classifications.

The identification of potential and actual data item and classificatory inconsistencies within and across collections and processing systems would be facilitated if current, previous and intended data items and classifications were captured and described in a standard way. There is no standard framework for describing and presenting the full range of ABS social statistics, nor are there standards for detailing the processes and methods used in ABS operations (ie. the concepts, sources and methods). The available meta-data relating to concepts, sources and methods are not readily accessible by microsimulation modellers. Such data are stored in a variety of locations and in a variety of formats – usually on paper – and are often incomplete. The authors believe this unsatisfactory situation is paralleled in most other countries.

### Institutional situation

International data standards are determined using similar mechanisms to those applying to economic data. Although the SNA is relevant to many household data items, there is no overarching framework providing a focus for the resolution of cross-cutting standards issues. Consequently there is no clear international forum to focus on cross-cutting statistical integration of social micro-data.

At a national level, the proportion of statistical budgets committed to the primary data analysis of social micro-data is a very small fraction of commitments to economic macro-data. A significant proportion of the work that is done on preparing social data sets for input into microsimulation models is not done within the data collection agency, but is done as a part of the modelling work. Few shared standards exist for defining how meta-data descriptions of government programs should be generated. This reduces the scope for sharing such meta-data between different modelling groups and results in duplication and potential inconsistency between models.

### Implications for social analysis

For most countries, no single authoritative, comprehensive, conceptually relevant, and consistent micro-data set is currently available. The data sets that are in use in microsimulation models generally fail to comply with the preferred characteristics for such data established earlier in this paper. Despite the valuable progress that has been made in international comparative analysis as a result of the Luxembourg Income Study initiative, the scope of available internationally comparable data is very limited.

Because there are few mutually agreed standards and generally no publicly available syntheses of primary source data, modelling groups frequently need to commit a significant part of their research resources to the resolution of data issues. Even with such commitments, the total investment in the primary data analysis of social micro-data is insignificant compared to the investments made in respect of economic data. One consequence is that data quality is compromised and the analyses that use the data are less robust. Another consequence is that data archiving and documentation are often inadequate, resulting in weak public accountability. There are also two efficiency consequences. Because primary data analysis processes are fragmented, both the primary data analysis task and the total data management task are done less efficiently. This leads to duplication of effort and the need for additional data reconciliation.

## 5. Constraints and Opportunities

Resources, institutional arrangements (in terms of responsibilities and coordination), software and hardware capabilities, and limitations imposed by measurement processes all place constraints on improving data management for social policy.

However, substantial gains in data management capabilities and growing recognition of the importance of social micro-data for the purposes of public policy have created an opportunity to realise the 'second revolution' advocated by the report of the US Panel to Evaluate Micro-simulation Models. Constraints and opportunities are considered below, mainly from the perspective of recent ABS experience.

There has been an increasing awareness in the ABS in recent years of the need to adopt an active data management strategy. The strategy adopted aims to better meet client needs (including those of the microsimulation modelling community) by making data more available, accessible, linkable and reproducible, and by improving cost-effectiveness. An increased investment has been made in the development and management of social data standards and classifications.

In the late 1970s, the ABS attempted centralised data management, but it was not cost-effective. However, given the rapid improvements in information technology and the developments in computing science, the ABS decided a new data management initiative was appropriate. A data management functional area has been formed which has the broad aim of ensuring successful management of data through all stages of its life cycle. A small investment has also been made to create a unit responsible for the development of an integrated publicly accessible database suitable for selected microsimulation modelling purposes. Subject to successful pilot work, further investment is planned.

These initiatives have required strong management commitment. This has been essential because of the resources required, but also because significant changes are often required to the work practices and culture of statistical processing areas. Reasons can readily be found for continuing to do things the way they have always been done; reaching agreement on new directions requires considerable negotiation and resolve. In this regard, coordination and decision process mechanisms become extremely important and have played an important role in progressing the work to its current stage.

Improved data management within a statistical agency will provide benefits to microsimulation modellers and the public policy community in two ways. First, the quality of data used in their research can be improved as a result of data management, such that the data used in model building are more statistically integrated and have undergone more primary data analysis. This should result in data that are more authoritative, reliable, conceptually relevant, structured, consistent, and comprehensive. Second, access to the data should be enhanced. In addition to being more linkable, available, accessible and reproducible, the data would be better described. Although not currently an opportunity, as discussed below, transfer of meta-data could open the possibility of standardisation of aspects of work relating to program meta-data.

## 5.1 Statistical integration

Achieving statistical integration requires negotiation and compromise to balance the needs of users for comparability against their need for specificity, and to balance long-term improvements in relatability against short-term losses. A major constraint to improved statistical integration is the considerable inertia of existing data collection and processing operations. The large body of established practices, the multiplicity of lines of responsibility, the variety of systems, and the variety of clients make it difficult to effect change. Each of these constraints must be balanced against the very large potential gains in the information value of data which will flow from improved statistical integration.

The task of improving data comparability and relatability requires the eventual standardisation of all areas of the statistical process, including definitions and naming conventions for concepts and variables, classifications, collections methodologies, processing procedures, and output conventions. To facilitate across-collection linkage, core variables such as age, sex, marital status, occupation, and labour force status have been selected for initial standardisation. The initial standards are regarded as interim. National standards are developed later.

The interim standards are developed using a thorough process of internal consultation and limited external consultation. The process involves collecting information about all internal uses of a particular concept or group of related concepts (for example, labour force variables) and identifying the degree of variation. After consultation, a draft standard is developed which attempts to identify existing best practice and satisfy the needs of all important variants. A standard format is used to ensure that all significant aspects of the standard are addressed and documented in a consistent way. After further exposure, a draft standard is endorsed. This process allows relatively rapid development. Edited versions of the standards are then published to inform the data-using community and to seek views prior to the next step, which is endorsement as national standards.

The process outlined is expensive in its use of dedicated resources and in the opportunity cost of the time required of collection and survey statisticians. Developing and updating major classifications is particularly expensive, involving intensive consultation with users and a substantial research effort.

## 5.2 Physical systems integration

Statistical integration provides a necessary foundation for physical systems integration by providing unique naming conventions and by generating the meta-data required to describe the entities recorded by the database system.

It will soon be possible to access household unit record meta-data across the full range of ABS social surveys. The access will be in terms of a common statistical language (standard data item names and definition, and standard classifications) and common physical access arrangements (a single system with details of all ABS data), using a consistent and controlled interface for the presentation of meta-data.

# 6. Information Model

Set out below is a brief description of the attributes of the key entities within the informational model which lie at the heart of ABS data management strategies. To date, the focus of the model has been on macro- rather than micro-data, so the comments in parentheses regarding 'micro-data' entities should be regarded as tentative.

A substantial part of the design of the model reflects recommendations made by Professor Bo Sungren of Statistics Sweden and the University of Stockholm following a consultancy to the ABS (1992)

The *source entity* details the processes followed in producing the statistical output or in obtaining the micro-data about the survey/collection/census/etc. which the data were obtained. These textual meta-data detail the processes followed in producing the statistical output, the design of the statistical collection (frame, relationships between answers and questions asked, etc.), how the collection was undertaken (how non-response was handled, response rates, etc.) and how micro-data were processed (editing and derivation rules, confidentiality rules and processes, aggregation processes, etc).

Figure 1: Information Model

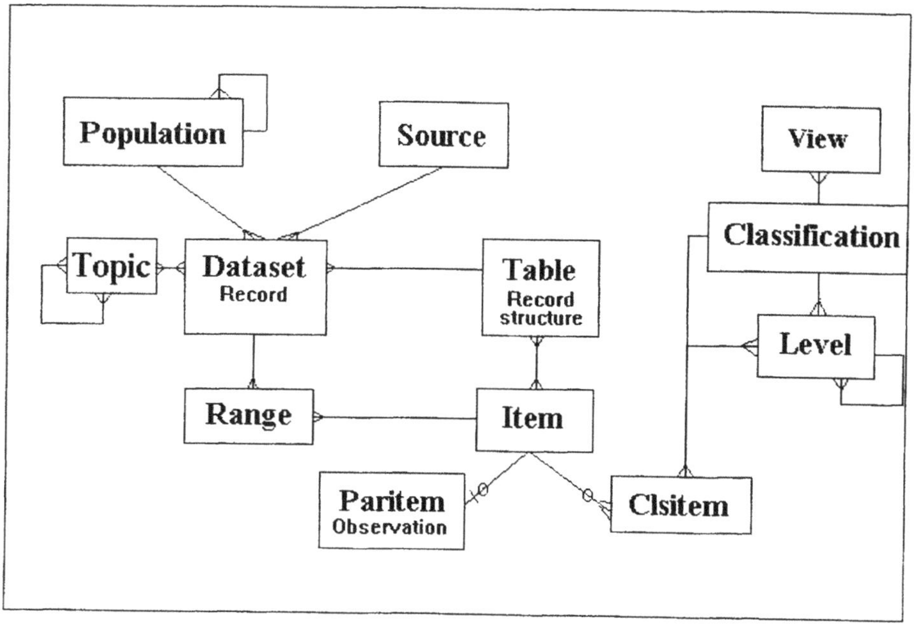


The *population entity* holds information about the population being measured. When a population is part of another, this relationship can be recorded.

The *topic entity* holds indexing terms. Topics can be associated with other related topics forming a kind of 'thesaurus'. The system automatically maintains topics and associations matching the sources, populations and part-of-population information above. Each *data set* describes a specific table of macro-data. If macro-data are available, they can be obtained and presented to a client. Source data sets have an associated population, source, and various associated topics. (For micro-data, a *record entity* would be defined.)

The *table entity* gives the collection of data items present in a data set, and their macro-data name. It also identifies the prime reference period dimension for a data set. (For micro-data, a *record structure entity* would be defined.)

The *item entity* relates to the data items in the table. There are two main kinds of item - parametric and classificatory. The *paritem entity* relates to parametric data items. These contain the estimates of statistical characteristics concerning the population and (sub)domains of objects of interest. They are numeric values with an associated unit of quantity, and sometimes have other associated information such as variance. (For micro-data, the *observation entity* would be stored.) The *clsitem entity* relates to classificatory data items. These identify sub-domains of the population for which the parametric estimates are available. Text descriptions of survey forms and questions, data items and relationships between data items, and classifications and concordances are all stored in other entities which are linked to the *item* entity (and eventually to any *observation* entities). The *range entity* entries are used to define the range of values present in an item in a data set.

The *classification and level entities* form the classification meta-data sub-system. These record the legal category codes in each level, and the way they form hierarchies. Multiple alternate hierarchies can be represented. Loose 'family' relationships between classifications can also be recorded. The *view entity* links a classification to various presentation options. Different labels and different presentation orders can be selected.

The benefits from investing resources in developing and capturing data and meta-data in accordance with this model are substantial. Clients will be able to explore the full range of ABS data and meta-data using a variety of search mechanisms (for example, by data item and data set names, classifications and associated descriptions and values, concepts and topics, text strings, and populations and collections).

## 6.1 Data set and model integration

Improved communications with major data users is expected to be a side benefit of the development of a common data catalogue, common meta-data and information model. Provision of a common statistical language is expected to significantly enhance ready physical access and a consistent interface across the complete range of ABS data, the basis for discussing data priorities, data concepts and survey designs.

## 6.2 Transformational data integration

There is a growing recognition within the ABS of the need to do more primary data analysis in respect of unit record social data. A similar view was expressed by the US Panel to Evaluate Micro-simulation Models, which recommended that the relevant US statistical agency adopt 'a more active role in adding value to databases for modelling and research purposes and for generating published data series' and attempt to produce 'best estimates of the income distribution and related variables' in order to remove source data inconsistencies.

Several areas of constraint are relevant to the development of publicly available synthetic unit record level structures. National statistical agencies have seen the transformation of social unit record files as the responsibility of the microsimulation modelling community. International statistical agencies have given little attention to the implications of the needs of microsimulation modelling for the development of international standards with respect to relevant social data sets. There is a need for ongoing work to research and document inconsistencies between relevant macro- and micro-data sources and to research ways to incorporate such information into any reconciliation process.

## Role of statistical agencies

Because of the statistical complexity of the task, because of the specific nature of user needs, because of the limited demand for such analysis and because of an emphasis on economic macro-data analysis in funding provisions, data collection agencies have tended to resile from attempts to create unit record level data structures that are internally consistent and consistent with major external data sources. For macro-economic analysis, statistical agency involvement has resulted in authoritative data sets compiled according to internationally agreed standards. Such involvement has also ensured a national focus for resource allocation for the purpose of primary data analysis, and has resulted in a large total commitment (relative to the commitment with respect to social data). More widespread involvement by national statistical agencies in constructing databases suitable for microsimulation modelling would provide a basis on which international agencies could expand their own involvement in microsimulation modelling to provide a forum covering all aspects of data management, including the development of relevant standards and the use of appropriate techniques for constructing integrated unit record databases.

## Data reconciliation

Even if full statistical integration were achievable, the need for primary data analysis would remain in order to achieve consistency. Data reconciliation is necessary if measurement error is present or if the required data structure is over-determined (ie. if the information content of the structure is duplicated for particular data concepts).

Although data inconsistency can also occur in fully-determined and under-determined systems (due to weak statistical integration or due to measurement error), it is most apparent in over-determined systems because data conflicts cannot be hidden in such systems. Redundant information not only implies more than one estimate for the same concept, but also results in the failure of identities defined by the structure.

Irrespective of the approach taken to resolve differences, the over-determined system will result in problems for macro/micro linkage. If data aggregates are retained in unadjusted form, two types of linkage problem occur. If micro-data from a duplicated source not utilised within a structure are to be linked to macro-data within the structure, the micro-data will be inconsistent with the duplicated macro-measure and with other macro-measures within the structure of which the macro-measure forms a part. Similarly the micro-data from the macro-aggregate within the structure will be inconsistent with the macro-aggregate duplicated outside the structure. Alternatively, if the duplicated macro-measures are adjusted to ensure agreement, the problem may appear to be exacerbated, because three distinct measures of the concept will then exist: the two measures that are inconsistent at source *plus* the synthetic measure intended to reconcile the two source estimates.

With respect to macro-economic data, two approaches to data integration are commonly used: one stresses the integrity of individual sources, and allows the use of explicit statistical discrepancies. The other stresses the consistency of the synthetic database and eliminates any discrepancies.

Using the 'integrity of sources' approach, individual estimates are compiled on a 'best estimate' basis. In the absence of knowledge of detailed error structures (all errors, including sample and non-sample error), inconsistencies between aggregates will remain. Even after careful investigation, the evidence provided by inconsistencies is usually insufficient to justify the arbitrary adjustment of a survey-based estimate of an aggregate. With this approach, the user of the data gets both individual best estimates and, because the statistical discrepancy is explicit, some indication of the extent of inconsistency of the underlying data sources.

The 'balanced relationships' approach eliminates the use of any explicit statistical discrepancy. Usually this is done using some form of mathematical procedure. Because there is no explicit statistical discrepancy, the error structure of the data is hidden. Offsetting the ease of use of the balanced data set is the problem that, unless the error structure of the data is known, knowledge of the sensitivity of analysis outcomes to data error is restricted to analysts involved in constructing the balanced data set.

The explicit discrepancy approach is more amenable to macro/micro linkage. The transformations involved are deliberately of limited scope so that the macro-aggregates will more accurately reflect the micro-database. The balanced approach is less amenable to linkage because within-unit-record structural relationships will often be distorted by the balancing procedures.

# 7. Conclusion

There is a strong need for centralised or highly coordinated data management in the areas of *statistical and physical systems data integration*. The work requires a high level of management commitment. It also requires the commitment of significant financial and specialised human resources if it is to be successful.

The benefits to the policy research and microsimulation modelling communities from these two areas of data management are likely to be substantial. The benefits also accrue to the data producers and other users of social data.

Much of this work is breaking new ground, particularly in the area of the application of data management to unit records. Consequently, there would be substantial benefits from establishing and maintaining close international communications between all of the groups involved in this work.

The development of a publicly documented common statistical language with ready physical access opens the prospect of extending use of the language into the meta-data description of government programs used in micro-simulation models. Although this language would necessarily be specific to each country, achieving a common language at the national level would greatly facilitate international comparative studies.

The substantial commitment by central agencies to primary macro-economic data analysis has resulted in significant benefits to the macro-economic policy community. To achieve the 'second revolution' in the area of social policy analysis requires a similar commitment of resources. This is most likely to happen if the work is focussed on a central agency with strong data responsibilities.

An international focus is also required to provide a forum within which it is possible to progress cross-cutting standards issues, greater integration and improved articulation of social data frameworks, and techniques for primary data analysis and *transformational integration*.

## References

Citro, C. F. and E.A. Hanushek (eds.) (1991), *Improving Information for Social Policy Decisions: The Uses of Microsimulation Modelling, Volume 1, Review and Recommendations*, National Academy Press, Washington DC.

Sungren, B. (1993), 'Organising the meta information systems of a statistical office', Report presented to the Australian Bureau of Statistics, Canberra.

Stephen Pudney

University of Leicester
United Kingdom

Holly Sutherland

University of Cambridge
United Kingdom

# Chapter 21: Statistical Reliability in Microsimulation Models With Econometrically-Estimated Behavioural Responses

## Abstract

*The basis for this paper is the construction of asymptotic confidence intervals for a representative set of microsimulation results produced by the UK tax-benefit model, POLIMOD, which has been extended to incorporate a multinomial logit model of female labour supply. The confidence intervals allow for errors associated with sampling variability, parameter estimation, and stochastic simulation. The study finds that sampling error is the dominant source of uncertainty for most summary statistics. However, measures of policy effects on female participation are very uncertain, mainly because of parameter estimation error. There is little potential for efficiency gain through replicated simulations or through conditioning on observed behaviour in the basic simulation.*

## 1. Introduction

Static microsimulation models use cross-sectional survey data to analyse the impact of public policy on the distribution of living standards. Whether they are used to evaluate policy change or as a tool for policy design, model outputs are typically quoted as though they were precise and reliable indications of the actual impact of policy on the population. It is often difficult to form a clear idea of the degree to which results might be subject to error - or the extent to which the size of the error might differ with the type of output from the model.

* We are grateful to Gerry Redmond for his advice and assistance. A longer and more technical version of this paper was presented at the International Association for Research in Income and Wealth Special Conference on Microsimulation and Public Policy in Canberra, December 1993. We are grateful to conference participants and particularly our discussant, Anders Klevmarken, for helpful comments. The Economic and Social Research Council, the Department of Social Security and the Newton Trust provided financial support. We are also grateful to the Central Statistical Office and ESRC Data Archive for providing us with access to Family Expenditure Survey data.

Error may arise in a number of ways. *Bias* in model estimates may be introduced because the *sampling* is unrepresentative, either because of the sample design - for example, where the non-household population is excluded - or because of differential response by groups in the population. A second source of bias arises from the presence of systematic *measurement* errors. This is a common area of concern in models based on survey data where, for example, there is evidence of under-reporting of certain sorts of income or where respondents behave differently during the reporting period (for example, spending less on alcohol and tobacco products than normally occurs). Another common type of measurement error is that introduced by *uprating*. Typically, models use survey data collected in some past year to simulate policy change in the current year. Adjustments are made to the data to bring them into line with current population characteristics but, inevitably it is not possible to adjust the data in all relevant respects or at a sufficiently disaggregated level to represent precisely the current situation. The crude adjustments that *are* possible - and the lack of adjustment where none is made - will both introduce bias in the model outputs. A third source of bias arises by failing to adjust model outputs for the changes in *individual behaviour* that may result from a change in policy. Furthermore, if behavioural effects are predicted by an econometric model, then specification and estimation problems may also introduce bias.

In parallel with these sources of bias, there are analogous sources of *variance* in the model estimates. First, even if the household survey yields a sample which is representative of the population on average, a simulation using these data will inevitably be subject to sampling error in the same way as any other calculation from survey data. Second, unsystematic measurement error generates additional variability. Third, sampling error introduces uncertainty into the estimation of any behavioural effects, which in turn increases the variance of the simulation model outputs.

A set of microsimulation results should be seen as a statistical estimate of a particular type of population characteristic: a summary of the effect of a perturbation in some policy instrument (Pudney and Sutherland, 1994a). Like any other statistical estimate, the set should be accompanied by a standard error or confidence interval. This chapter extends our earlier analysis to incorporate the effect on statistical reliability of estimating the impact of behavioural response. The chapter also investigates the magnitude of the additional element of uncertainty introduced by behavioural estimation, focussing on a model of female labour supply which assumes that a discrete choice between participation states exists. Other important sources of statistical error (for instance, measurement error) are not considered here.

A second focus of the chapter is on the role of stochastic simulation in the calculations. The effect of the number of replications that are performed on the efficiency of the simulation is examined. How to relate the participation decision 'after' the policy change to that `before' the policy change is also considered - since it is clear that there should be a relationship but less clear how it should best be specified. Whether or not the initial participation state should be conditioned on the actual state observed in the data is examined, as is the extent to which conditioning on the observed state increases the reliability of the simulation results.

The chapter is organised as follows. Section 2 describes the econometric models used in this exercise, in the first place, to generate earnings in alternative participation states and, in the second place, to build a labour supply model to predict the probabilities of participation in each state. In both cases the intention is to construct models which are typical of methods which are in widespread use, rather than to concentrate on refining these models. Section 3 discusses the issues involved in the choice of method of stochastic simulation. Section 4 spells out the assumptions made about the nature of errors in the simulation process and the nature of the sampling error which underlies the survey data. Section 5 gives a brief description of the model we use to carry out the simulations and Section 6 describes the policy change used to demonstrate the model. Section 7 describes our results and Section 8 presents the conclusions.

## 2. Econometric Models

The aim of this chapter is to investigate the importance of errors which stem from the estimation of parameters in econometric models of the behavioural response to policy changes. In this context, the behaviour that has been studied most thoroughly by applied econometricians is that involving labour supply. Particularly important is the labour supply of women, for whom there is perhaps the strongest evidence of 'second-round' behavioural responses. This chapter is not concerned with specifying and estimating a labour supply model that is in any sense a correct description of behaviour: merely one that is representative (in terms of estimation variability) of models that might be used in practice to support tax-benefit simulations. Econometric labour supply models fall into two broad classes. The first views labour supply as a discrete choice between alternative forms of labour force participation (see Zabalza et al., 1980), and Ilmakunnas and Pudney (1990) for examples). The second views labour supply as a continuous hours variable, with some allowance for a discrete state of non-participation (eg. Burtless and Hausman (1978) and Blundell et al. (1988)).

In general, we strongly favour the former type of model. Continuous hours models imply that a large number of women will make small adjustments to their hours of work in response to any policy change. It seems extremely implausible to suggest, for example, that millions of women will increase their hours of work by a few minutes per week in response to a tax cut. In contrast, discrete choice models imply the more reasonable prediction that response will take the form of a relatively small number of women making large changes to their hours of work, with the majority making no change at all. The implications of this distinction for income distribution may be very important.

It is important to realise that our behavioural modelling is only partial. In practice, responses to a major policy change will be very widespread: the labour supply of men and the job search behaviour of unemployed men and women will both be affected, for instance - and, through the demand side of the labour market, there will be further effects on wages. Thus our analysis is likely to understate both the extent of behavioural response and the accompanying level of parameter estimation uncertainty that would result from a full-scale modelling effort.

The model is a multinomial logit (MNL) model with three states (non-participation (N), part-time work (P) and full-time work (F)) as discrete alternatives. The explanatory variables include demographic variables as well as three simulated income variables which represent the levels of net family income corresponding to the three possible participation choices of the adult female member of the family. The construction of these variables is as follows.

(i) Gross part-time and full-time earnings (not hourly wage rates) are predicted from separate earnings regressions in which demographic and educational variables are used as regressors. These equations are fitted to the subsamples of women observed to be in part-time and full-time employment respectively, and are used to predict potential earnings for every woman in each of the two possible employment states.

(ii) For women who are observed to be in work, and for whom a gross earnings variable is therefore observed, this observed level is used in place of the prediction. Furthermore, the prediction of earnings for the other (unobserved) participation state is scaled up or down in proportion to the actual and predicted earnings in the observed state.

(iii) POLIMOD (a tax-benefit model designed for the analysis of the distributional effects of changes in public policy) is then used to simulate net family income corresponding to the three possible levels of gross earnings of the woman. In these simulations, the gross earnings of other family members are held fixed, but direct taxes and income-related social security benefits are simulated in full.

The three net income variables are denoted $Y_N$, $Y_P$ and $Y_F$, and are treated in the same way as directly observed variables. In particular, the fact that estimation of the underlying earnings functions introduces further error is ignored and, in this respect, the results can be regarded as conservative. Note that participation is only modelled for women who are out of full-time education, are aged under 60 and who do not describe themselves as retired. Participation is also not modelled for families where predicted net income is non-positive in the base simulation.

After some experimentation with alternative forms, a model specification implying the following form for the probability of participation state j (where j = N, P, F) was used:

$$P_j = \frac{\exp\left(\beta'_j z + \gamma \ ln Y_j + \delta\left(ln Y_j\right)^2\right)}{\sum \exp\left(\beta'_i z + \gamma \ ln Y_i + \delta\left(ln Y_i\right)^2\right)} \qquad (1)$$

where z is a vector of demographic variables and the vectors $\beta_j$ contain the coefficients to be estimated. We adopt the normalisation $B_n$=0 to permit identification. Note that the coefficients of income variables are restricted to be equal across states.

Probabilities like (1) can be interpreted in many different ways. One common interpretation is in terms of stochastic utilities. The specification (1) implies that the utility derived from alternative j has the following log-quadratic form:

$$U_j = \beta'_j z + \gamma \ ln Y_j + \delta\left(ln Y_j\right)^2 + \varepsilon_j \qquad (2)$$

where the $\varepsilon_j$ are mutually uncorrelated unobserved variables which are distributed conditionally on z and $Y_j$, as type I extreme value variates. Under these assumptions, the conditional selection probabilities are of the logit form (1). Simulated participation states can be generated either stochastically through the underlying utilities (2) or directly from the selection probabilities. As described below, the method of generation can have important implications for the way policy simulations are done and the results they produce. The estimated parameters are presented in Appendix A.

These estimates cannot be interpreted directly as the model involved is nonlinear. However, Table 1 illustrates their economic implications in terms of the predicted impacts of variations in the characteristics of a single representative individual on labour force participation frequencies. The base case individual is aged 35, married, has one school-age child and left school at age 16. The potential net family incomes for the three possible labour force states are £163, £209 and £268 per week. For this individual, the probabilities of non-participation, part-time work and full-time work are 18 per cent, 39 per cent and 43 per cent respectively, implying an expected labour supply of just under 22 hours per week (assuming 18 and 40 hours for part-time and full-time jobs).

In this model, labour supply is not very responsive to economic incentives, in comparison with its responsiveness to the large predicted effects of demographic variations. An equivalised £10 increase in any of the potential net incomes (with the other two held fixed) produces an increase in expected labour supply of no more than two per cent. This is in sharp contrast to the fall in expected labour supply of 13 hours as the predicted effect of adding a pre-school child to the family. This is quite a typical finding. For example Blundell et al., (1988) estimated that two alternative and quite radical reforms of the taxation of married women's income would produce average changes in labour supply of only 6-25 minutes per week from a base of around 15 hours. Note also that the estimated behavioural responses to economic incentives embodied in the net income terms, despite being small, are statistically significant at the conventional five per cent level in terms of the underlying estimated income coefficients. Thus, the model is not unusually poor in relation to estimator precision.

The primary objective here is not to estimate the size of policy effects, and no strong claims are made for this participation model as a 'correct' description of behaviour. Instead, the aim is to estimate the degree of uncertainty introduced into microsimulation results by using an econometrically estimated labour supply model. For this purpose, it is sufficient that the model be representative of econometric models generally, in terms of its structure and the sampling properties of its estimated parameters.

Table 1: Predicted Effects of Variations in Explanatory Variables

| Change in explanatory variables | Participation probabilities (%) | | | Hours |
|---|---|---|---|---|
| | $P_N$ | $P_P$ | $P_F$ | |
| | 18.01 | 39.21 | 42.78 | 24.17 |
| | Change (percentage points) | | | Change (hours) |
| Age + 5 years | 1.19 | 5.24 | -6.43 | -1.63 |
| Add 1 pre-school child | 33.39 | 0.78 | -34.17 | -13.53 |
| Lone parent | 3.83 | -23.72 | 19.90 | 3.69 |
| Left school after 16 | -6.74 | -11.99 | 18.73 | 5.33 |
| £10 increase in $y_N$ | 1.82 | -0.87 | -0.95 | -0.54 |
| £10 increase in $y_P$ | -0.55 | 1.84 | -1.30 | -0.19 |
| £10 increase in $y_F$ | -0.44 | -0.95 | 1.39 | 0.34 |

## 3. Simulation Methods

The econometric model provides estimated forms for the three participation probabilities for each woman in the sample. Because of the complexity of both these probabilities and the joint distribution of the variables that appear in them, it is usual practice to resort to stochastic simulation in order to estimate a policy outcome. Stochastic simulation is a simple process that involves the following steps:

(i) identify the 'random' elements of behavioural response (the 'error terms') and make suitable assumptions about their distribution (for example, normal, extreme value, etc.);

(ii) for each individual in the sample, generate a pseudo-random number(s) drawn from the assumed distribution; and

(iii) calculate for that individual the implied behavioural response.

To estimate the impact of a policy change, this process must be implemented both for the pre- and post-change policy environments and a comparison made.

Used in this way, stochastic simulation is not straightforward. There are four major questions to be addressed in setting up a simulation.

*(1) What is the relationship between the random elements underlying pre- and post-change behaviour?*

Some assumption must be made about the relationship between the random elements of behaviour in the pre- and post-change simulations. If the simulation is regarded as a picture of the change that takes place over time after the new policy is implemented, then this assumption is essentially one about the autocorrelation of behaviour during the period of change. Viewed in that way, it seems appropriate to assume a high (but possibly not perfect) degree of correlation between the pre- and post-change participation decisions. Alternatively, if the policy simulation is viewed as an exercise in applied comparative statics, the most natural assumption is that the unobserved random factors in the participation model are held fixed across the policy change. In our view, the latter interpretation is the more realistic (and certainly the easiest to implement), as it does not require estimation or assumption of a particular autocorrelation structure.

*(2) In what form should the random elements be embedded in the behavioural model?*

The problem here is that an estimated econometric model may be consistent with more than one stochastic structure. To illustrate this, consider the MNL model of participation. The empirical content of this model is the set of three participation probabilities, $P_N$, $P_P$ and $P_F$, given by equation (1). But these probabilities are consistent with a number of different stochastic mechanisms, each of which implies a different method of carrying out the stochastic simulation. In the MNL case, there are the following two obvious ways of simulating the model.

The *interval method*:

- divide the [0,1] interval into the sub-intervals:

$$S_N = [0, P_N),$$

$$S_P = [P_N, P_N + P_P), \text{ and}$$

$$S_F = [P_N + P_P, 1];$$

- generate a pseudo-random number v, chosen randomly from the interval [0,1]; and

- generate the participation variable as state i, where i is the state for which $\upsilon \in S_i$.

The *structural method:*

- generate three pseudo-random numbers, 1, 2 and 3, drawn from the type I extreme value distribution (this can be done by constructing them as:

  $\varepsilon_i = -\log(-\log(\upsilon_i))$, where the $\upsilon_i$ are three independent uniform random numbers);

- calculate the stochastic utilities $U_i = \beta_i' z + \gamma Y_i + \delta Y_i^2 + \varepsilon_i$ and

- generate the participation state as the one corresponding to the largest of the $U_i$.

These two methods deliver (on average) the same correct frequencies of the three states, given the correct specification of the econometric model. Moreover, in the case of the interval method, this is true however the unit interval is partitioned. Ordering the three states as (say) F, P, N would produce a different actual outcome, but one still drawn from the same correct distribution. The method of performing the stochastic simulations is thus of little consequence for simulating the outcome of a single policy environment.

However, it may be much more important when examining the effects of a policy change if we are interested in any summary measure that depends on the impact at the level of the individual family (eg. the numbers of gainers and losers, or the number of transitions into employment). In that case, different approaches to stochastic simulation will generate systematically different results. To appreciate this, consider the interval method, which (for any one observation) rests on a single pseudo-random number. Assume that we are taking the comparative-statics view and holding fixed across the policy change. Suppose that in the pre-change case this value lies slightly below the value of $P_N$ implied by the econometric model, so that non-participation is generated as the base outcome. Now assume that the policy change reduces $P_N$ to the lower value $P^*_N$ which is less than $\upsilon$. It is apparent that, in the post-change case, there will be a switch of state to whichever one comes after non-participation in the partition of the unit interval. The simulated result is thus clearly sensitive to ordering: if we choose to order the states N, P, F, the policy change will be simulated to produce many switches from N to P; however, if we choose to order the states N, F, P, the policy change will produce switches from N to

F. Thus, the outcome of the simulation depends in a critical way on what appears at first sight to be a rather arcane technical decision about the method of performing the stochastic simulation. The structural method is less arbitrary in this respect, as no ordering of the states is required, but it does rest on a strong interpretation of the response model as a stochastic utility maximisation process. Such an interpretation is entirely *a priori*: it is not inherent in the econometric modelling procedure.

*(3) Should we make use of the survey observations on actual behaviour in making the pre-change simulation ?*

Macro-economic forecasters who use econometric models normally adopt the practice of making *ad hoc* 'constant-term adjustments' to keep the model on track - in other words, to make it consistent with the most recent actual data. An analogous possibility arises in the static microsimulation case. We could attempt to bring the pre-policy change predictions in line with the observed outcomes by generating the underlying random numbers in such a way that the pre-change simulation of labour force participation coincides with that observed in the survey data. In other words, the simulation of policy responses could be conditioned on the full set of observed data. If conditional simulations are to be made, there are some technical issues to be faced, as the method of implementing this conditioning is dependent on the nature of the econometric model being used. We are able to derive an appropriate conditional simulation algorithm for the MNL model used in this study (see Pudney and Sutherland (1994b) for details), but for more complex models this may not be possible. In general, conditioning the simulation on the actual pattern of labour supply does make a difference to statistical reliability, but it does not necessarily improve matters; it is even possible that the results could become less reliable. This is investigated empirically below.

*(4) How many replications should be performed ?*

The statistical reliability of stochastic simulation results can be improved by performing multiple replications or, in other words, by generating independently a number of predicted behavioural responses for each family in the survey. The final output summaries are then calculated from the expanded set of simulation results. Although increasing the number of replications must increase reliability, it entails a significant increase in computing effort, and the size of the gain in reliability may not be large enough to justify the additional effort.

It is the purpose of this chapter to shed light on these four issues and give an empirical illustration of the influence that econometric estimation has on the statistical reliability of microsimulation.

# 4. Statistical Assumptions

The calculation of confidence intervals rests on a number of assumptions about the sampling process used to generate the underlying survey data, the pattern of non-response, and the nature of errors inherent in the simulation process. For the sake of simplicity, we work with a set of assumptions that abstract from much of this complexity.

## 4.1 The sampling process

We assume that the underlying population is infinite so that finite population corrections can be avoided. It is composed of a set of M strata denoted $S_1 \ldots S_M$. The population frequencies of these strata are $P_1 \ldots P_M$. When a sampling unit is drawn at random from stratum $S_m$ there is a probability $Q_m$ that a successful response will be obtained. The sampling process used by the survey designers is simple random sampling, and a fixed set of N units is contacted. Of these N contacts only $N^*$ provide acceptable responses. Note that the achieved sample size $N^*$ is a random variable the distribution of which is binomial (N, $Q^*$) where $Q^*$ is the weighted mean of the $Q_m$. The observations on some variable of interest, x, are denoted $x_n$ (n = 1...N), with an arbitrary 'missing' code allocated to non-responses. To correct biases originating from differential response, a set of survey weights is available. This weighting system may take different forms, depending on the external information that is available. We assume that population frequencies are unknown but response rates are known, allowing re-weighting in such a way that expected sample deficiencies are corrected. For the sake of simplicity, we assume weighting based on known response rates, where the weight for observation n is constructed as follows:

$$w_n = (Q_{m_n})^{-1} \tag{3}$$

and $m_n$ identifies the stratum to which observation n belongs. Note that a formally identical sampling distribution arises if there is also two-stage probability sampling based on the strata $S_1 \ldots S_M$. In this more general case, the weight must be multiplied by the ratio of the population frequency to the sampling probability for stratum $m_n$.

Note that the weighting scheme corrects for expected biases in the sample rather than actual 'deformities'. It thus requires knowledge (or estimates) of the response rates $Q_m$, but not the population frequencies, $P_m$, of the strata. A more common approach to weighting uses information on the population frequencies, and constructs the weights in such a way that sample and population frequencies coincide. This leads to weights of the form:

$$w_{2n} = P_{m_n} / \hat{P}_{m_n} \tag{4}$$

where $\hat{P}_{m_n}$ is the sample frequency of stratum $m_n$. In general, the weights generate more efficient estimates of population characteristics. However, they are awkward in the sense that they lead to variance formulae that are more complicated and require separate estimation of stratum means and variances. This is a drawback, because practical applications often involve complex weighting schemes with many different weighting classes. For example, the optimisation-based methods of Merz (1986) and Atkinson et al. (1988) can result in weighting variables which are continuously-variable. Pudney and Sutherland (1994a) discuss these two approaches, and present an illustrative calculation that suggests a relatively small difference between (3) and (4) in practice.

A second important simplification used in this paper is the assumption that the sampling unit coincides with the basic unit of analysis. In fact, this is not the case. Our simulations are based on the family unit, defined as a single adult or married (or cohabiting) couple, together with any dependent children. The Family Expenditure Survey (FES), from which our data are taken, is household based, with each household supplying an average of 1.35 family units. The failure of any household member to respond causes the whole household to be treated as non-respondent. The clustering of family units within households raises difficult statistical problems which we neglect in this paper, and we proceed as if the FES sampling process were based on family units rather than households. If there is a positive correlation between the incomes of different family units within the same household, the precision of estimation is reduced by clustering. In other words, the effective sample size lies somewhere between the number of households and the number of families, rather than being equal to the latter as we assume. On these grounds, we expect our confidence intervals to understate the true sampling variability to some degree.

## 4.2. Formulae for confidence intervals

All our confidence intervals are of the usual normal form. For a $(1-\alpha)100$ per cent significance level the interval is:

$$\hat{\mu} \pm C_{\alpha/2} \sqrt{V} \qquad (5)$$

where $\hat{\mu}$ is the estimated quantity, $C_{\alpha/2}$ is the appropriate two-sided critical value for the normal distribution and V is the asymptotic approximation to the sampling variance of $\hat{\theta}$. Appendix B gives formulae for V for all the types of statistic used in our analysis. All are simple expressions, capable of routine calculation in standard statistical software.

By using (5), we have avoided re-sampling methods such as the bootstrap, which is the obvious alternative approach. In the present context, the bootstrap has serious computational disadvantages, as it requires the re-calculation of *both* the policy simulations and the econometric estimates a large number of times. We rejected this as not feasible.

In its most basic form, microsimulation is a mechanical procedure that generates a variable $x_n$ from other more basic sampled variables. This procedure rests only on knowledge of the administrative rules of the tax-benefit system and entails only non-stochastic relationships. It is discussed extensively in a companion paper, Pudney and Sutherland (1994a). The more ambitious simulations considered in that paper involve behavioural modelling and the estimation of unknown parameters: a process that introduces a further source of error. The reader is referred to Pudney and Sutherland (1994b) for a more detailed treatment of this case.

In the current paper we assume the use of stochastic simulation, working under the assumption that parameter estimation is done by maximum likelihood (or some asymptotically equivalent method), using the same data set as is used for the policy simulation itself. Assume there is a behavioural model relating a discrete variable $y_n$ to a vector of explanatory variables $z_n$. They are related through a conditional distribution with frequency function $f(y_n|z_n;\theta)$, where $\theta$ is a vector of parameters. Assume that we are interested in (say) the mean of a derived variable x constructed as follows:

$$x_n = \tilde{x}(y_n, z_n) \qquad (6)$$

This has conditional mean function:

$$\mu(z;\theta) = E(x|z;\theta) \qquad (7)$$

which is relevant to the case where the simulation is *not* conditioned on the observed base value of y, or:

$$\mu(z, y;\theta) = E(x|z, y;\theta) \qquad (8)$$

which is relevant to the case where the stochastic simulation *is* conditioned on the observed base participation state.

The variable x has unconditional mean function:

$$\mu(\theta) = E\mu(z;\theta) \qquad (9)$$

We assume, for the purposes of illustration, that the aim of the simulation is to estimate $\mu$. Note that many typical summaries of simulation results do have the form of sample means, including quantities like the proportions of gainers and losers or the average change in some variable resulting from a policy change.

Under stochastic simulation, the sample mean is:

$$\hat{\mu} = \frac{\sum_{n=1}^{N}\sum_{r=1}^{R} w_n x(z_n, \varepsilon_{rn};\hat{\theta})}{R\sum_{n=1}^{N} w_n} \qquad (10)$$

where the summation with respect to n is over the set of successful survey responses and the summation with respect to r is over the set of R replications. Define also the following mean across replications:

$$\bar{x}_n(\hat{\theta}) = \frac{1}{R}\sum_{r=1}^{R} x(z_n, \varepsilon_{rn};\hat{\theta}) \qquad (11)$$

The most convenient variance estimate to use in the confidence interval is the following, which is valid both for conditional and unconditional simulations.

$$V = \sum\left[\frac{w_n(\bar{x}_n(\hat{\theta}) - \hat{\mu})}{\sum w} + \hat{\mu}'_\theta\left(\sum \frac{\partial \ln f_n}{\partial\hat{\theta}}\frac{\partial \ln f_n}{\partial\hat{\theta}'}\right)^{-1}\frac{\partial \ln f_n}{\partial\hat{\theta}}\right]^2 \qquad (12)$$

This variance expression is seen to be a sum of squares of a constructed variable with two components. The first is the usual (weighted) mean deviation of the average across all replications of the simulation for family n. The second component results from the error involved in estimating behavioural parameters. The latter term is itself composed of two multiplicative parts: $\hat{\mu}_\theta$ is the estimated derivative of the mean of $x$ with respect to $\theta$ ; and the second part is the contribution of observation n to the sampling variability of the maximum likelihood estimator of $\theta$ . The derivative vector $\mu_\theta$ can be estimated most conveniently by difference approximation:

$$\hat{\mu}_\theta = \frac{\sum w_n \left( \bar{x}_n(\hat{\theta}+\Delta_N) - \hat{\mu}(\hat{\theta}+\Delta_N) - \bar{x}_n(\hat{\theta}) + \hat{\mu}(\hat{\theta}) \right)}{\Delta_N \sum w_n} \qquad (13)$$

where $\{\Delta_N\}$ is a sequence of vectors converging to zero with N. (For instance, $\Delta_N$ could be set equal to a scale factor multiplied by the vector of coefficient standard errors).

Many of the statistics commonly used to describe simulation results are weighted sample means of the kind discussed above. However, some are not: examples include quantiles, many inequality and poverty indices, and means within sub-populations. Formulae for many summary statistics besides the mean are given in Appendix B. All are of the general form (12), with the mean deviation $\bar{x}(\theta) - \hat{\mu}$ replaced by some other appropriately-constructed variable, with $\mu_\theta$ the relevant derivative vector (for detailed derivations see Pudney and Sutherland, 1994a).

# 5. The Simulation Model: POLIMOD

POLIMOD is a tax-benefit model designed for the analysis of the distributional effects of changes in public policy. The model uses the survey data on individual gross incomes, individual and family circumstances and housing costs to simulate the main components of the personal tax and social security systems as they stood in 1988. These components are income tax and employee and self employed National Insurance Contributions (NICs); contributory National Insurance benefits and pensions; sickness and maternity pay; and non-contributory benefits such as child benefit, one parent benefit and disability benefits, together with the system of means-tested benefits. Significant changes in both the income tax system and the system of means-tested benefits took place in April of 1988. Interviews with households in the FES take place throughout the year and we model the policy as it existed for each family when they were interviewed. Details of the 1988 British social security, income tax and National Insurance Contribution rules can be found in Matthewman and Sloss (1988) and Harvey and Saunders (1988).

The version used here is based on 1988 data from the UK Family Expenditure Survey. We confine our analysis to the year in which the survey data were collected. There is therefore no uprating process required to bring earnings, other sources of income, imputed capital, prices and so on in to line with a later year. Nor is there any need to adjust for structural shifts such as changes in housing tenure or unemployment. In practice, in a typical use of POLIMOD or other similar model, uprating is an extremely important source of error, so our confidence intervals are likely to be conservative indications of those that would apply in practical applications.

In general, the model uses recorded information about gross incomes to simulate the tax and benefit system before and after a policy change. In this application the gross earnings of a sample of the adult women are changed according to the procedures described above. Our analysis focuses principally on net equivalised family income and on changes to it. Net income is defined to include normal current earnings; self-employment income uprated from the reported date to the date of interview; current social security benefits, including housing benefit; the value of mortgage interest tax relief; income from investments; and other regular payments made to the family, such as maintenance payments or student grants. Income tax and National Insurance Contributions are deducted. We exclude irregular elements such as insurance receipts, inheritance, redundancy or pension lump sums. The value of fringe benefits is not included and housing costs, local taxes[1], travel to work costs and child-care costs are not deducted. Indirect taxes are not modelled in this exercise.

In simulating tax expenditures and social security benefits associated with housing costs we assume that it is the head of household who pays all housing costs and hence receives these benefits. We assume that there is no income sharing between the family unit for whom we measure income, and the rest of the household. On the other hand, we assume that family income is fully shared and that the income of any member of the family is available for use by the rest of the family. These are the conventional assumptions and are thus appropriate to use in the current exercise, but they may not be valid in all cases. The resulting net income figure is equivalised by dividing by a simple scale based on the one used by the OECD. This scale

---

1 Domestic rates in 1988.

consists of a factor equal to 1 for a single adult or 1.7 for a married or cohabiting couple plus 0.5 for each dependent child.[2]

All our simulation results are based on weighted data. The weighting scheme we used is designed to correct the demographic composition of the sample and is based on a 17-category classification of the population. An optimisation-based algorithm was used to produce weights which vary as little as possible, whilst producing a corrected demographic profile both at the household and individual level.

## 6. Policy Simulations

The need to consider the effects of individual behaviour on estimates of the impact of policy changes is usually considered to be most acute for radical changes in policy. In some cases these may be specifically designed to affect labour supply decisions. In other cases they are intended to improve the system of income maintenance but on a scale sufficiently large to provoke significant labour supply responses. An example of this sort of change is the introduction of a Basic Income (BI) scheme to replace personal tax allowances and some social security benefits.[3] This would provide an income for all adults, independent of employment status, marital status or income level. It is often argued that such an income would deter married women from taking paid jobs, particularly those who face low wage rates and currently are not entitled to any income of their own from social security benefits (except child benefit in the case of mothers).

We look at the effect of a BI scheme as an example of a reform that might be expected to induce relatively large labour supply changes by women. For clarity, we consider a simplified version of the scheme which ignores many of the issues that would have to be tackled in practice were this policy to be implemented, or indeed were we to be modelling it as a serious policy proposal. The results of the simulations should not therefore be taken as indicative of what we believe the full effects of the detailed proposals would be. It is certainly the case that the precise behavioural effects would be highly dependent on the way in which the changes were implemented and administered.

---

2 The OECD scale defines children as being under the age of 14. Here we have assumed that all dependent children (aged under 16 or up to 19 if still in full-time non-advanced education) count as children rather than adults.

3 See Parker (1989) and Parker and Sutherland (1991).

Details of the BI scheme modelled are shown below. Compared with the tax and benefit system as it existed in 1988, there is a small revenue cost (before taking behavioural effects into account). It should be noted that we are not constraining the policy package to be revenue-neutral, either before or after behavioural effects have taken place. Taking account of a revenue neutrality condition in a simulation without estimation of behavioural effects is addressed in Pudney and Sutherland (1994a).

Figure 1: A Basic Income Policy Package, 1988

| | January-March | April-December |
|---|---|---|
| Adult B1 | £39.50 | £41.15 |
| Child B1 | £15.30 | £15.65 |

The payment rates are equal to the current state basic pension (adult) and the sum of current child benefit and NI benefit child additions (child). BIs are non-taxable and are paid to all citizens, regardless of status.

All non means-tested benefits are abolished. Means-tested benefits remain as a safety net but the BIs are included in the income test.

National Insurance contributions (NICs) are abolished.

All personal tax allowances are abolished, as is tax relief on mortgage interest.

Income tax rates are all increased by 20 percentage points (giving rates of 45 per cent and 60 per cent after April 1988).

Essentially, what this policy reform does is to tax income on as wide a base as possible under a unified tax structure; abolish all social security benefits except residual means-tested benefits; and pay an income, approximately at a level sufficient for subsistence, to all individuals. The implication is that incomes are increased for most women who are not employed while, for most people, incomes are reduced through higher taxes. A few employed people are better off than before because their increased tax net of BI is smaller than their pre-reform tax. As we assume it is the income of the whole family that affects the labour supply decisions of its members, in the case of married women it is the net effect of these changes on the combined incomes of themselves and their husbands that is of significance here.

# 7. Results

In this Section we apply the confidence interval formulae discussed above and in Appendix B. For the base simulation (the system as it existed in 1988), the results are given in Table 2. These have moderately good precision in most cases. However a margin of error of, for example, £2.4bn (or 10.4 per cent of the estimate) associated with the estimated level of net government transfers to families, or of 79,000 in the number of working lone mothers (18.2 per cent of the estimate), may give grounds for concern.

At first sight it is encouraging to the econometrician that parameter estimation adds relatively little to the width of the confidence intervals in most cases. However, there are exceptions in the case of summary statistics focussing particularly on women's labour supply. For instance, the margin of error associated with the estimated numbers of working women or of working lone mothers is increased by up to 50 per cent when estimation error is taken into account. For other statistics, the increase is generally much less.

Table 2 also allows us to explore the efficiency gain achievable through an increase in the number of replications. These efficiency gains turn out to be surprisingly small. Increasing the number of replications from one to four gives a decrease in the width of confidence intervals of 12 per cent or less in the cases we consider. This is encouraging, given the considerable increase in computing effort required for replication of policy simulations. However, note that we have only demonstrated this for a static simulation model: complex dynamic models produce output with a much larger stochastic component, and, for such models, replication may be worthwhile (Hancock et al., 1992).

Tables 3, 4 and 5 examine simulated changes produced by the introduction of the hypothetical BI scheme. The simulation in Table 3 abstracts from labour supply responses by holding fixed women's participation states at their predicted outcomes under the base simulation. Note that these participation states are not restricted to be the same as those observed in the original survey data. (The predictions of participation state produced by the behavioural model are correct *on average*, but inevitably will not describe correctly the behaviour of each individual woman.) The simulation in Table 4 allows a second-round effect through women's labour supply response; again, the simulations are not conditioned on the observed participation state. Table 5 then explores the effect of conditioning on observed participation in the base simulation. (The model forces the pre-reform participation state of each

Table 2: Simulation Estimates - Base Policy

| Summary measure | Simulated (R=1) | 90% Confidence interval | Simulated (R=4) | 90% Confidence interval |
|---|---|---|---|---|
| Mean income | £105.62 | ± £1.62 (± £1.58) | £105.73 | ± £1.60 (± £1.56) |
| Bottom quintile point | £53.20 | ± £0.83 (± £0.73) | £53.32 | ± £0.76 (± £0.65 |
| Number of families on means-tested benefits | 7.251m | ± 0.235m (± 0.223m) | 7.253m | ± 0.221m (± 0.209m) |
| Number of individuals below half mean income | 10.560m | ± 0.523m (± 0.447) | 10.497m | ± 0.492m (± 0.411) |
| Number of individuals below £75 | 22.581m | ± 0.619m (± 0.584) | 22.557m | ± 0.581m (± 0.543) |
| Atkinson inequality index ( = 1.5) | 0.3356 | ± 0.0262 (± 0.0256) | 0.3394 | ± 0.0260 (± 0.0254) |
| Aggregate receipts of means-tested benefits | £9.840bn | ± £0.477bn (± £0.437bn) | £9.893bn | ± £0.447bn (± £0.403bn) |
| Aggregate income tax payments | £47.751bn | ± £1.900bn (± £1.884bn) | £47.980bn | ± £1.897bn (± £1.880bn) |
| Aggregate net government transfers to families | -£23.642bn | ± £2.449bn (± £2.413bn) | -£23.926bn | ± £2.434bn (± £2.397bn) |
| Number of working women | 9.555m | ± 0.303m (± 0.240m) | 9.585m | ± 0.271m (± 0.198m) |
| Number of working lone mothers | 0.434m | ± 0.079m (± 0.058m) | 0.445m | ± 0.071m (± 0.047m) |

Notes: Income is net equivalised family income throughout; all derivatives are evaluated using four replications. Formulae ignoring parameter estimation error appear in parentheses.

woman to correspond to that observed in the data.) Tables 4 and 5 both use the structural method of stochastic simulation, and each is based on the same set of underlying pseudo-random numbers.

Table 3 assumes that there are no changes in behaviour as a result of the reform, but assumes the pre-reform participation state is that predicted by the behavioural model, rather than the state observed in the data. The baseline situation is thus the same as that shown in Table 2. The most important point to emerge from these results is the very large increase in the margin of error *in relation to the simulated change*. For example, although the absolute margin of error for net government transfers falls from ± £2.4bn for the base level see Table 2 to ± £1.2bn for the effect of the policy change, this margin of error increases from 10 per cent of the estimate for the level to 160 per cent of the estimate for the change. This is an extreme example because of the relatively small size of the change, but the same tendency is observed in other measures. For example, the number of individuals living in families with less than half mean equivalised net income has a confidence interval of ± 5 per cent for the estimated level, but ± 12 per cent for the (very large) estimated change.

Table 3: Changes Produced by BI Policy - Participation State Held Fixed at Predictions Under Base Policy - Simulation Not Conditioned on Observed Base State

| Summary measure | Level under base policy (R=4) | Simulated change (R=1) | 90% Confidence interval | Simulated change (R=4) | 90% Confidence interval |
|---|---|---|---|---|---|
| Mean income | £105.73 | + £0.26 | ± £0.44 (± £0.44) | + £0.26 | ± £0.44 (± £0.43) |
| Bottom quintile point | £53.32 | + £2.92 | ± £0.70 (± £0.56) | + £2.85 | ± £0.67 (± £0.52) |
| Number of families on means-tested benefits | 7.253m | - 2.256m | ± 0.173m (± 0.165m) | - 2.243m | ± 0.157m (± 0.147m) |
| Number of individuals below half mean income | 10.497m | - 2.712m | ± 0.351m (± 0.322m) | - 2.726m | ± 0.315m (± 0.282m) |
| Number of individuals below £75 | 22.557m | - 4.628m | ± 0.441m (± 0.394m) | - 4.658m | ± 0.386m (± 0.331m) |
| Atkinson inequality index ($\varepsilon = 1.5$) | 0.3394 | -0.1527 | ± 0.0247 (± 0.0243) | -0.1559 | ± 0.0246 (± 0.0243) |
| Aggregate receipts of means-tested benefits | £9.893bn | - £5.744bn | ± £0.374bn (± £0.342bn) | - £5.776bn | ± £0.344bn (± £0.308bn) |
| Aggregate income tax payments | £47.980bn | + £83.943bn | ± £1.572bn (± £1.542bn) | + £84.191bn | ± £1.557bn (± £1.527bn) |
| Aggregate net government transfers to families | -£23.926bn | + £0.763bn | ± £1.218bn (± £1.200bn) | ± £0.592bn | ± £1.203bn (± £1.185bn) |

Notes: Income is net equivalised family income throughout; all derivatives are evaluated using four replications. Formulae ignoring parameter estimation error are in parentheses.

Again, the confidence intervals are dominated by sampling error rather than estimation error for macro-level statistics such as budgetary aggregates or inequality and poverty measures, but not for statistics focused closely on female labour supply. As before, increasing the number of replications has a small impact on efficiency, with the potential gains being even smaller for simulated changes than for levels.

The simulations reported in Table 4 use the same starting point as those in Table 3 but allow behavioural change as a result of the reform. A comparison of Tables 3 and 4 reveals an increase in uncertainty as we allow for participation responses to the policy change. For the estimated increase in numbers of working women and lone mothers, the estimated response is completely insignificant in the statistical sense. Parameter estimation error is an extremely important source of uncertainty for these two statistics; at least as large as sampling error in most cases. The contribution of parameter estimation error is critical in the sense that (at least in the case of $R = 4$ replications) the simulated policy impact would otherwise be significantly different from zero. It is hard to justify the claim that econometric

Table 4: Changes Produced by BI Policy - Participation State Generated by Econometric Model Using Structural Method - Simulation Not Conditioned on Observed Base State.

| Summary measure | Level under base policy (R=4) | Simulated change (R=1) | 90% Confidence interval | Simulated change (R=4) | 90% Confidence interval |
|---|---|---|---|---|---|
| Mean income | £105.73 | + £0.18 | ± £0.47 (± £0.45) | + £0.18 | ± £0.46 (± £0.43) |
| Bottom quintile point | £53.32 | + £2.67 | ± £0.81 (± £0.56) | + £2.60 | ± £0.78 (± £0.52) |
| Number of families on means-tested benefits | 7.253m | - 2.249m | ± 0.174 (± 0.165) | - 2.239 | ± 0.158 (± 0.147) |
| Number of individuals below half mean income | 10.497m | - 2.686m | ± 0.351m (± 0.324m) | - 2.666m | ± 0.312 (± 0.282) |
| Number of individuals below £75 | 22.557m | - 4.568m | ± 0.455m (± 0.394m) | - 4.600 | ± 0.401 (± 0.330) |
| Atkinson inequality index ($\varepsilon = 1.5$) | 0.3394 | -0.1519 | ± 0.0248 (± 0.0243) | - 0.1549 | ± 0.0248 (± 0.0243) |
| Aggregate receipts of means-tested benefits | £9.893bn | - £5.743bn | ± £0.374bn (± £0.342bn) | - £5.778bn | ± £0.344bn (± £0.308bn) |
| Aggregate income tax payments | £47.980bn | + £83.988bn | ± £1.593bn (± £1.545bn) | + £84.270bn | ± £1.578bn (± 1.529bn) |
| Aggregate net government transfers to families | - £23.926bn | + £0.719bn | ± £1.232bn (± £1.202bn) | + £0.510bn | ± £1.216bn (± £1.186bn) |
| Number of working women | 9.585m | - 0.045m | ± 0.070m (± 0.030m) | - 0.043m | ± 0.065m (± 0.018m) |
| Number of working lone mothers | 0.445m | -0.006m | ± 0.011m (± 0.007m) | - 0.004m | ± 0.010m (± 0.004m) |

Notes: Income is net equivalised family income throughout; all derivatives are evaluated using four replications. Formulae ignoring parameter estimation error are in parentheses.

modelling contributes much to microsimulation in the face of results like these. In contrast, for the statistics not closely dependent on female labour supply, such as income tax receipts or inequality measures, the changes in the estimates of the impact of the reform between Tables 3 and 4 are small. In every case the differences between the tables are within the tolerance of the 90 per cent confidence intervals for the simulations (computed either with or without changes in behaviour).

A similar comparison of Tables 4 and 5 indicates that conditioning the simulations on the observed participation data (in other words matching the predicted pre-reform state to the actual state) does not improve matters significantly. Indeed, for most of the statistics in Table 5, the confidence intervals widen as a consequence of conditioning in this way, although not by a substantial amount. The main effect of conditioning the simulations on the observed initial state is seen in the estimates themselves, rather than the confidence intervals associated with them. This underlines its role mainly as a device for keeping the simulations 'on track' - in other words, reducing the simulation bias introduced by inevitable shortcomings in the specification of the econometric behavioural relationships, rather than improving statistical reliability.

Table 5: Changes Produced by BI Policy - Participation State Generated by Econometric Model Using Structural Method - Simulation Conditioned on Observed Base State

| Summary measure | Level under base policy (R=4) | Simulated change (R=1) | 90% Confidence interval | Simulated change (R=4) | 90% Confidence interval |
|---|---|---|---|---|---|
| Mean income | £109.46 | - £0.70 | ± £0.48 (± £0.44) | - £0.73 | ± £0.48 (± £0.44) |
| Bottom quintile point | £53.75 | + £1.44 | ± £0.79 (± £0.54) | + £1.41 | ± £0.79 (± £0.54) |
| Number of families on means-tested benefits | 7.111m | - 2.016m | ± 0.176m (± 0.159m) | - 2.013m | ± 0.176m (± 0.158m) |
| Number of individuals below half mean income | 11.213m | - 1.979 | ± 0.333m (± 0.305m) | - 1.987m | ± 0.333m (± 0.305m) |
| Number of individuals below £75 | 21.262m | - 3.743m | ± 0.439m (± 0.363m) | - 3.726m | ± 0.438m (± 0.362m) |
| Atkinson inequality index ($\varepsilon$ = 1.5) | 0.3475 | - 0.1541 | ± 0.0245 (± 0.0245) | - 0.1541 | ± 0.0245 (± 0.0245) |
| Aggregate receipts of means-tested benefits | £10.215bn | - £6.066bn | ± £0.425bn (± £0.362bn) | - £6.068bn | ± £0.425bn (± £0.362bn) |
| Aggregate income tax payments | £51.489bn | + £86.874bn | ± £1.678bn (± £1.606bn) | + £86.851bn | ± £1.678bn (± 1.606bn) |
| Aggregate net government transfers to families | -£27.678bn | - £1.817bn | ± £1.225bn (± £1.203bn) | - £1.795bn | ± £1.225bn (± £1.202bn) |
| Number of working women | 10.000m | - 0.030m | ± 0.069m (± 0.028m) | - 0.036m | ± 0.065 (± 0.016) |
| Number of working lone mothers | 0.391m | - 0.000m | ± 0.011 (± 0.006) | -0.002 | ± 0.009 (± 0.003) |

Notes: Income is net equivalised family income throughout; all derivatives are evaluated using four replications. Formulae ignoring parameter estimation error are in parentheses.

## 8. Conclusions

This chapter has examined the statistical uncertainty associated with a representative set of static tax-benefit simulations by constructing asymptotically-valid confidence intervals for a widely-used UK tax-benefit model, POLIMOD. This analysis extends earlier work by incorporating a multinomial logit model of female labour supply. The resulting confidence intervals allow for errors associated with sampling variability, parameter estimation and stochastic simulation. We have found that sampling error is the dominant source of uncertainty for most summary statistics, but that measures describing the impact of policy on female participation are very uncertain, mainly because of parameter estimation error. We have found little potential for efficiency gain through replicated simulations or conditioning on observed behaviour in the base simulation.

Our results should be regarded as very conservative. Our assumptions about the basic sampling process probably understate the significance of sampling error, and our model of female labour supply is only a partial model of behavioural response. In practice, there may be some response in male labour supply, in job search behaviour and, taking account of the demand side of the labour market, in the distribution of rates of pay. A full consideration of all these issues would require the construction of a huge survey-based general equilibrium model and is almost certainly not feasible at present. However, with an expansion in the scale of econometric modelling, it seems inevitable that there would be a major increase in the contribution of parameter estimation error to the level of statistical uncertainty.

It may well be the case that the additional uncertainty introduced by econometric modelling is often so large that econometric estimates of behavioural impacts are of no practical use to economists attempting to simulate the effects of policy changes. Our computed confidence intervals for the impact of a radical Basic Income scheme on female labour supply are certainly consistent with this pessimistic view. It remains to be seen whether or not this sort of finding is typical of other econometric models, other types of behaviour change or other types of policy change. This is an issue that should be an urgent priority in the research agenda in policy simulation.

# Appendix A

## Econometric Estimates of a Labour Supply Model

| Explanatory variable | Part-time | Full-time |
|---|---|---|
| Constant | -4.201 | -.841 |
| | (0.542) | (0.464) |
| Age/10 | 1.900 | 1.609 |
| | (0.290) | (0.258) |
| $Age^2/100$ | -0.237 | -0.275 |
| | (0.037) | (0.034) |
| Married | 0.677 | 0.127 |
| | (0.132) | (0.120) |
| Children | -0.188 | -0.684 |
| | (0.065) | (0.078) |
| Completed full-time education after 16 | 0.104 | 0.832 |
| | (0.089) | (0.086) |
| Household size | -0.021 | -0.128 |
| | (0.043) | (0.039) |
| Married (children > 0) | 0.465 | -0.189 |
| | (0.141) | (0.154) |
| Pre-school child | -0.820 | -1.840 |
| | (0.107) | (0.130) |
| Log equivalised net family income | 0.729 | |
| | (0.226) | |
| Squared log equivalised net family income | -0.098 | |
| | (0.034) | |

# Appendix B

## Asymptotic Properties of Sample Statistics

Define the following notation. $x_n$ is the variable of interest. $y_{nj}$ is a binary indicator of the event that the nth sampled individual belongs to stratum j. The weighting factors, $w_n$, are defined as the reciprocal response rates. The summary statistics presented in the paper are examples of common types of sample statistics. We now give asymptotic approximations to the sampling variances of these generic statistics, without derivations. Details can be found in Pudney and Sutherland (1994a, 1994b). Note that these variance estimates require no direct knowledge of the population frequencies, within-stratum moments, or the design sample size.

All our formulae are based on asymptotic approximations to the sampling variances of the relevant summary statistic. The approximations all have the following generic form:

$$\sum_n \left[ \frac{w_n \lambda_n}{\sum w} + \hat{\mu}'_\theta \left( \sum \frac{\partial \ln f_n}{\partial \hat{\theta}} \frac{\partial \ln f_n}{\partial \hat{\theta}'} \right)^{-1} \frac{\partial \ln f_n}{\partial \hat{\theta}} \right]^2 \qquad \text{(A1)}$$

where $\lambda_n$ is some constructed variable specific to the relevant summary measure; $\mu_\theta$ is the vector of derivatives of the corresponding population measure with respect to each of the parameters of the labour supply model; $f_n$ is the component of the log-likelihood function contributed by observation n. Summations are over the set of successful responses. It suffices for our purpose to give formulae for the estimate of $\mu$, and the corresponding $\lambda_n$ in each case. This Appendix deals only with three of the four types of statistics used in the empirical section of the paper - endogenously-truncated headcounts, sample quantiles and Atkinson inequality statistics.

The fourth class of statistics is the weighted sample mean, which was discussed in detail in Section 4 of this paper. It is relevant to many of the summary statistics commonly used to represent the results generated by microsimulation models. With $x_n$ defined appropriately, the weighted sample mean encompasses the calculations made in the paper for mean income, the number of families on means-tested benefits, the number of individuals in families with equivalised income below £75, all financial aggregates, and the numbers of working women and lone mothers.

## An endogenously-truncated headcount

A commonly-used poverty measure is the proportion of individuals or family units below half mean income. Consider first the case of a family-based measure. For R replications, the statistic is:

$$\hat{H} = \hat{F}(\hat{\mu} / 2) \tag{A2}$$

where $\hat{\mu}$ is the weighted mean of $x_{rn}$ and $\hat{F}$ is some estimate of the c.d.f. of x, usually the (weighted) empirical c.d.f.:

$$\hat{F}(x) = \frac{\sum w_n R^{-1} \sum_{r=1}^{R} \xi_{rn}(x)}{\sum w_n} \tag{A3}$$

where $\xi_{rn}(x) = I(x_{rn}<x)$ and I(.) is the indicator function. In this case, the appropriate form for $\lambda_n$ is:

$$\lambda_n = \left(\xi_n\left(\frac{\hat{\mu}}{2}\right) - \hat{F}\left(\frac{\hat{\mu}}{2}\right)\right) - \frac{\hat{f}(\hat{\mu}/2)}{2}\left(R^{-1}\sum_{r=1}^{R} x_{rn} - \hat{\mu}\right) \tag{A4}$$

where $\hat{f}$ is the estimated p.d.f. Summations are over all respondents. A weighted kernel estimator can be used for $\hat{f}$.

In the case of a simulated change, the estimate is:

$$\Delta\hat{H} = \hat{F}_2\left(\frac{\hat{\mu}_2}{2}\right) - \hat{F}_1\left(\frac{\hat{\mu}_1}{2}\right) \tag{A5}$$

and $\lambda_n$ has the form:

$$\lambda_n = (\hat{\xi}_{2n} - \hat{F}_2) - (\hat{\xi}_{1n} - \hat{F}_1) - \frac{\hat{f}_2}{2}(x_{2n} - \hat{\mu}_2) + \frac{\hat{f}_1}{2}(x_{1n} - \hat{\mu}_1) \tag{A6}$$

where $x_1$ and $x_2$ are the values of x before and after the policy change and $\hat{F}_1, \hat{F}_2, \hat{f}_1$ and $\hat{f}_2$ are their estimated c.d.f.s. and p.d.f.s, evaluated at half the relevant mean.

The estimates (A2) and (A5) relate to the proportion of family units below half mean income. However, we may be more interested in the proportion of individuals who are members of such family units. In this case the relevant population characteristic is:

$$F(\mu/2)=\frac{E(v|x<\mu/2)\Pr(x<\mu/2)}{E(v)} \tag{A7}$$

where $v$ is the number of family members. The natural weighted estimate of F is:

$$\hat{F}(\hat{\mu}/2)=\frac{\sum w_n v_n \xi_n(\hat{\mu}/2)}{\sum w_n v_n} \tag{A8}$$

In this case, $\lambda_n$ takes the following form:

$$\lambda_n=\frac{v_n}{\bar{v}}(\hat{\xi}_n-\hat{F})-\frac{\hat{f}}{2}(x_n-\hat{\mu}) \tag{A9}$$

Note that the estimated density, $\hat{f}$, must now be calculated differently, using weights equal to $w_n v_n$. If we multiply by the (known) total number of families in the population, we have an estimate of the absolute number of individuals living in families which are below half mean income. It is in that form that the statistic is used in this paper. The extension of (A8) and (A9) to the case of a policy change is obvious.

### An inequality index

We consider only one representative of the class of inequality measures - the Atkinson index, which can be expressed as a simple function of moments:

$$A_\varepsilon=1-\frac{(E(x^{1-\varepsilon}))^{\frac{1}{1-\varepsilon}}}{\mu} \tag{A10}$$

$A_\varepsilon$ is estimated by replacing population expectations by their weighted estimates averaged over replications. This is equivalent to computing the statistic from a single enlarged dataset constructed by merging all R replications. Note that, for $\varepsilon \geq 1, A_\varepsilon$ is only computable if $x_n>0$ for all n. For this statistic, $\lambda_n$ is as follows:

$$\lambda_n=-\left(\frac{(1-\hat{A}_\varepsilon)^\varepsilon}{(1-\varepsilon)\hat{\mu}^{1-\varepsilon}}(x_n^{1-\varepsilon}-\hat{\eta})-\frac{(1-\hat{A}_\varepsilon)}{\hat{\mu}}(x_n-\hat{\mu})\right) \tag{A11}$$

where $\hat{\eta}$ is the weighted mean of $x^{1-\varepsilon}$.

For the change in the Atkinson statistic induced by a policy change, this expression becomes:

$$\lambda_n = -\left[\begin{array}{l} \dfrac{(1-\hat{A}_{\varepsilon 2})^{\varepsilon}}{(1-\varepsilon)\hat{\mu}_2^{1-\varepsilon}}\left(R^{-1}\sum_{r=1}^{R} x_{rn2}^{1-\varepsilon}-\hat{\eta}_2\right)-\dfrac{(1-\hat{A}_{\varepsilon 2})}{\hat{\mu}_2}\left(R^{-1}\sum_{r=1}^{R} x_{rn2}-\hat{\mu}_2\right) \\ -\dfrac{(1-\hat{A}_{\varepsilon 1})^{\varepsilon}}{(1-\varepsilon)\hat{\mu}_1^{1-\varepsilon}}\left(R^{-1}\sum_{r=1}^{R} x_{rn1}^{1-\varepsilon}-\hat{\eta}_1\right)+\dfrac{(1-\hat{A}_{\varepsilon 1})}{\hat{\mu}_1}\left(R^{-1}\sum_{r=1}^{R} x_{rn1}-\hat{\mu}_1\right) \end{array}\right] \qquad \text{(A12)}$$

### A sample quantile

The p-level quantile of x in the population is the value $X_p$ satisfying:

$$p = F(X_p) \qquad \text{(A13)}$$

where F(.) is the c.d.f. of x. The weighted estimate of $X_p$ is the solution to the following equation:

$$\hat{F}(\hat{X}_p) = p \qquad \text{(A14)}$$

where $\hat{F}$ is an estimate of the c.d.f. of x. The simplest weighted quantile is based on the empirical c.d.f. defined by (A3) above. For that estimate of F(.), there is no exact solution to the equation (A14), which only defines an interval within which the quantile must lie. An arbitrary rule is then used to produce a unique estimate. The variance approximation in this case is based on the following definition of $\lambda_n$:

$$\lambda_n = \frac{1}{\hat{\mathrm{f}}(\hat{X}_p)}\left(R^{-1}\sum_{r=1}^{R} \xi_{rn}(\hat{X}_p) - p\right) \qquad \text{(A15)}$$

where $\varepsilon_{rn}(\hat{X}_p) = I(x_n < \hat{x}_p)$ and $\hat{\mathrm{f}}$ is an estimate of the p.d.f.

We are also interested in the change in a quantile produced by a policy change. This statistic is defined as follows:

$$\Delta\hat{X}_p = \hat{F}_2^{-1}(p) - \hat{F}_1^{-1}(p) \tag{A16}$$

where $\hat{F}_1$ and $\hat{F}_2$ are the estimated c.d.f.s of $x_1$ and $x_2$. The definition of $\lambda_n$ is now extended to the following:

$$\lambda_n = \frac{\left(R^{-1}\sum_{r=1}^{R}\xi_{2rn}(\hat{X}_{2p}) - p\right)}{\hat{f}_2(\hat{X}_{2p})} - \frac{\left(R^{-1}\sum_{r=1}^{R}\xi_{1rn}(\hat{X}_{1p}) - p\right)}{\hat{f}_1(\hat{X}_{1p})} \tag{A17}$$

$\xi_1$ and $\xi_2$ are dummy variables indicating respectively respondents who are below the pre-reform quantile point and respondents who are below the (changed) quantile point post-reform.

## References

Atkinson, A.B., J. Gomulka and H. Sutherland (1988), 'Grossing-up FES data for tax-benefit models, in: A.B. Atkinson and H. Sutherland (eds.), *Tax-Benefit Models*, STICERD Occasional Paper 10, London School of Economics, London.

Blundell, R.W., C. Meghir, E. Symons and I. Walker (1988), 'Labour supply specification and the evaluation of tax reforms', *Journal of Public Economics*, Vol. 36.

Burtless, G. and J.A. Hausman (1978), 'The effects of taxation on labour supply', *Journal of Political Economy*, Vol. 86, No. 6.

Hancock, R.M., J. Mallender and S.E. Pudney (1992), 'Constructing a computer model for simulating the future distribution of pensioners' incomes for Great Britain', in: R.M. Hancock and H. Sutherland (eds.), *Microsimulation Models for Public Policy Analysis: New Frontiers*, STICERD Occasional Paper 17, London School of Economics, London.

Harvey, E.L. and G. Saunders (1988), *Income Tax 1988-89*, Tolley, London.

Ilmakunnas, S. and S.E. Pudney (1990), 'A model of female labour supply in the presence of hours restrictions', *Journal of Public Economics*, Vol. 41.

Matthewman, J. and E. Sloss (1988), *Social Security and State Benefits 1988-89*, Tolley, London.

Merz, J. (1986), Structural adjustment in static and dynamic microsimulation models, in: G.H. Orcutt, J. Merz and H. Quinke (eds.), *Microanalytic Simulation Models to Support Social and Financial Policy*, Elsevier, New York.

Parker, H. (1989), *Instead of the Dole*, Routledge, London.

Parker, H. and H. Sutherland (1991), *Child Tax Allowances? A Comparison of Child Benefit, Child Tax Reliefs and Basic Incomes as Instruments of Family Policy*, STICERD Occasional Paper No. 16, London School of Economics, London.

Pudney, S.E. and H. Sutherland (1994a), 'How reliable are microsimulation results? An analysis of the role of sampling error in a UK tax-benefit model', *Journal of Public Economics*, Vol. 53.

Pudney, S.E. and H. Sutherland (1994b), *'Statistical reliability and microsimulation: The role of sampling, simulation and estimation error*, Discussion Paper MU 9402, Microsimulation Unit, University of Cambridge, Cambridge.

Zabalza A., C. Pissarides and M. Barton (1980), 'Social security and the choice between full-time work, part-time work and retirement', *Journal of Public Economics*, Vol.14.

Steven Caldwell

Cornell University
United States of America

# Chapter 22: Health, Wealth, Pensions and Life Paths: The CORSIM Dynamic Microsimulation Model

## Abstract

*This chapter describes the development of CORSIM, which is probably the world's first PC-based dynamic microsimulation model. Following the declining use of dynamic microsimulation models in the 1980s in the US, a key goal in CORSIM's development was to reduce costs and increase benefits. Apart from an overview of CORSIM, this chapter discusses the various types of validation which may be undertaken on dynamic models and assesses why dynamic microsimulation model results may differ from benchmark data.*

## 1. Introduction

Under development since 1986, CORSIM is a full-featured dynamic microsimulation model (DMM) of families and individuals in the United States. The model has been used to dynamically age a sample representing the full US population of individuals and families and to conduct experiments on the dynamically aged population over seven decades of historical and future time (namely the 1960-1995 historical period and the 1995-2030 future period). Among the policy concerns to which CORSIM has been addressed are health care financing, welfare reform, family wealth holding and pension reform. The model is being used as well to address basic social science questions. What changes are occurring in kinship networks and why? What are the patterns and sources of intergenerational mobility? How do individual life paths depend upon aggregate societal conditions?

This dual focus on policy and basic science is fully intended. From its inception CORSIM has been constructed with the aim of providing an intellectual tool versatile enough to explore a broad array of both policy issues *and* basic social science questions. The twin aims for CORSIM are seen as complementary. Strengthening the model's capacity to address fundamental social science questions (eg. understanding life path dynamics) is seen as likely to enhance its effectiveness in treating policy issues (eg. policy impacts on individual health and pensions and family wealth holding), and vice versa. This paper discusses several key issues surrounding the construction, alignment and validation of CORSIM, its use as a platform for historical and future experiments, and its application to important policy and basic science questions.

The major components of the DMM paradigm are well-known. The first realisation of the DMM paradigm was constructed by Orcutt and a team of three graduate students at Harvard University (Orcutt et al., 1961). The second realisation - the DYNASIM model constructed at The Urban Institute in the early 1970's (Orcutt et al., 1976) - appeared fifteen years later. Because DYNASIM was expensive to construct and operate, the model was little used for actual policy exploration. The amount of validation done on the overall model, relative to the amount required for achieving policy-level credibility, was also inadequate.

Nevertheless, the DYNASIM model did justify its cost in another sense, in that it demonstrated the potential of DMM in a way that made a permanent difference to the intellectual and policy community. Virtually all dynamic microsimulation models built since DYNASIM can be said to trace their lineage at least partly to the original DYNASIM (Merz, 1991).

Why did microsimulation not grow more vigorously in the 1980's, given that its costs were declining? A simple 'consumption' hypothesis emphasises three sets of factors (Caldwell, 1987):

- costs relative to alternative tools;
- benefits as perceived by potential users; and
- resources of potential users.

In terms of this model, one interpretation of events in the US goes as follows. During the 1970's the US federal government's surprisingly high support of microsimulation was driven by an optimistic perception of benefits and generous resources. These factors overcame the downward pull of very high costs. But the dynamics changed in the 1980's. Perceived benefits and policy research resources both declined markedly, and this dragged microsimulation 'consumption' down in the 1980's, even as declining cost was pushing use upward.

## 2. Early CORSIM Goal: Reduce Costs, Increase Benefits

A primary motivation for building CORSIM was to develop a model which would break new ground by delivering more 'bang for the buck' - more benefits for less cost. This would show by example that DMM was not just a *potentially* useful tool for policy purposes in the distant future, but rather was an *actually* cost-effective tool for both policy and basic science purposes in the present. In pursuit of this goal attempts were made to:

Reduce the costs of *constructing and validating components* of CORSIM by:

- utilising microdata files in electronic form (CD-ROM, disks, tapes);
- analysing microdata with user-friendly software on inexpensive hardware; and
- including debugging and alignment procedures as integral elements.

Reduce the costs of *programming and running* CORSIM by:

- programming modular code in a powerful language (C);
- using efficient variance reduction and replication techniques; and
- porting the model across multiple platforms from PC's to supercomputers.

To *increase the benefits* of CORSIM by:

- deepening and broadening the theoretical and empirical content;
- validating extensively against substantial historical data of multiple types;
- increasing the sample size for higher subgroup accuracy;
- applying the model to basic social science as well as policy analysis; and
- presenting results more effectively (eg. visual output).

In 1986-88 two undergraduate engineering student programmers and I constructed what was--to the best of our knowledge--the first PC-based DMM (Caldwell: 1988a;

1988b; 1988c; 1991). The completed model, much influenced by DYNASIM in its design and content, was named CORSIM.

The desire for portability across platforms was a primary force motivating the design and construction of CORSIM. PC's are inexpensive and omnipresent and there is a large amount of software available for them; the limitations include relatively low processing power, small memories, and small disks. From the start, a model was needed which would run on multiple platforms. Among the general purpose programming languages (BASIC, PASCAL, FORTRAN, C), the C language stood out in portability. That potential has been realised; typically the process of porting CORSIM is mostly a matter of uploading the code from one platform to another and compiling. Incompatibilities, though inevitable, are usually trivial to fix. The costs of portability are that a sophisticated user interface has so far been sacrificed and programmers have the added burden - a small one - of writing portable code. But the gain is the ability to use standard PC's for writing, testing and debugging code and for carrying out validation runs. One is then able to move the model to more powerful platforms for large production runs aimed at exploring the model's response surface in a substantial way and with very large samples. Portability in effect allows CORSIM to be run on the least expensive platform available for the task at hand.

The impact of DMM should grow as declining costs draw new classes of customers into the arena. In that context, portable models which can migrate among desktop PC's, mainframes and supercomputers are an important development.

## 3. Development of CORSIM Versions 2.0 and 3.0

From 1990-93, funding from the National Institutes of Health made possible a project to greatly modify and expand CORSIM, as well as to develop a health module. Each equation in CORSIM was completely re-specified and re-estimated. Entire new modules were added. The code was improved in a variety of ways. Major debugging and validation features were added. The resulting model, new from the ground up, was labelled CORSIM 2.0. Since that time, substantial new improvements have been made and the model is currently in its third version.

The two major entities in CORSIM are the *Person* and the *Family*. For every person in the simulation, there is an associated *Person* record and, for every family entity, a *Family* record. Person records are grouped into families. *Primary* attributes in

CORSIM are those derived directly from the functioning of a process. A far larger class of CORSIM attributes is derived as simple functions of primary attributes. Examples are identities, life history (lagged) attributes and attributes derived via relationships.

CORSIM retains links among all persons who ever lived together, and hence attributes for one person can be related to attributes of other persons linked to the first. For example, children who do not live with their fathers nevertheless can be tabulated by the income of fathers. Or the achievements of adults can be compared to those of their siblings.

As it now stands, the model consists of about 900 equations representing 28 equation-based processes; numerous rule-like algorithms representing about 30 rule-based processes; and over 10,000 parameters in all. Data used to test and estimate the equation-based modules are drawn from nine large national microdata files. Data used to construct the rule-based modules, as well as data used for historical validation and alignment, are drawn from another six files plus miscellaneous sources. Graphical output is implemented via Excel using scriptable macros in Visual Basic. Scriptable tables are generated using Table Production Language. An extensive historical database is connected to the model.

## 3.1 Demographic processes

The demographic processes in the model are outlined below. Both individuals and families are generated and maintained. Persons are linked within families.

The person changes simulated comprise fertility, immigration and mortality. Person attributes simulated include age, race, sex, duration of residence, and citizenship. Family changes include first marriage, remarriage, assortative mating, divorce, custody of children, and leaving home. Finally, family attributes include headship assignment, relation to head, duration of marital status, number of divorces, and marriage/family links

## 3.2 Kinship

As noted above, CORSIM retains information linking all persons who have ever lived in the same family. Hence, attributes for one person can be related to attributes of other persons linked by blood or marriage (as long as these ties were formalised by living together in families for at least one year). For example, children who no longer live with their fathers nevertheless can be tabulated by the income of fathers, though children who never lived with their fathers cannot. Similarly, the achievements of adults can be compared to those of their siblings.

## 3.3 Schooling

Schooling is divided into ten discrete levels, and the overall process of individual schooling attainment is treated as an organised sequence of processes of two types: first, continuing to attend a particular level of schooling vs dropping out; and, second, completing a particular level of schooling vs not completing. In any year for any person, the model provides information on whether the person is currently attending school, the level of school the person is currently attending, and the person's level of already completed schooling. The two schooling processes simulated therefore are grade attendance and grade completion.

Separate equations are estimated for each level of school, and by sex and race groupings. The role of parental education links generations and creates intergenerational immobility. The other factors in the completion equations introduce life course contingencies into the overall educational attainment process.

## 3.4 Earnings

For each person over 17 in the sample, CORSIM generates an annual value for weeks worked and annual earnings, via a detailed set of probit/regression equations constructed by the economists Professors Alice and Masao Nakamura specifically for CORSIM. The equations were tested and estimated using data drawn from the Panel Study on Income Dynamics.

## 3.5 Assets

CORSIM simulates the ownership and accumulation of fourteen kinds of assets. A set of logistic equations generates probabilities of ownership of each type of asset, followed by regression equations which generate the market value of each type of asset. The equations were estimated from longitudinal data drawn from the 1983 and 1986 Surveys of Consumer Finance. Asset transfers at death and divorce are also generated.

## 3.6 Transfer income

Three generic types of transfer income are generated:

- employment-related;
- welfare-related; and
- pension-related.

Specific programs within each generic category are not separately represented, although it is intended that someday they will be added. One set of logistic equations generates the probability of receiving each type of transfer income; a second set generates amounts received for recipients.

## 3.7 Pre-tax family income

The sum across family members of earnings, asset income and transfers yields total pre-tax family income.

## 3.8 Taxes

Five taxes are represented in CORSIM:

- social insurance tax;
- state income;
- federal income
- property; and
- estate taxes.

The tax modules do not at this stage represent taxes with the detail required for policy-relevant tax simulations.

## 3.9 After-tax family income

Once total pre-tax family income is processed by the required tax routines, the taxes are subtracted to yield total family after-tax (disposable) family income. Disposable income then enters the next module - consumption.

### 3.10 Consumption

CORSIM allocates after-tax disposable family income between savings (which can be negative) and consumption. Once determined, total consumption is allocated among major consumption categories. The consumption module thus generates the following three outcomes:

- savings;
- consumption; and
- detailed consumption.

### 3.11 Risk factors

CORSIM includes four person-level risk factors. These are cigarette smoking, alcohol consumption, diabetes and sugar consumption.

## 4. Why Dynamic and Not Static Microsimulation?

Static and dynamic microsimulation modelling strategies are much less similar than the shared surname implies. In fact, they are as different as the proverbial apples and oranges. The major difference lies in the sharply different ageing strategies.

Static microsimulation treats an historical sample, on which it carries out policy analyses, as exogenously generated, by unknown generating processes. If static microsimulation requires a sample to be aged, the ageing occurs according to exogenously derived control totals. In contrast, dynamic microsimulation treats an historical sample as the outcome of a set of generating processes; the intent is to make the sample and its ageing processes endogenous, by modelling the generating processes as closely as possible. Static modellers are interested in data as an end in itself; dynamic modellers treat data as clues to underlying processes. Static modellers are interested in whether an historical sample accurately represents the population at one point in time, not in the processes which generated the sample. Dynamic modellers are interested in whether an historical sample accurately represents the population, but also in discovering the underlying processes generating a sample. Dynamic and static microsimulation both provide useful tools for policy analysis, though in different ways (Wolfson, 1991; Zedlewski et al., 1990). Arguably, only dynamic microsimulation is useful for basic research as well as policy analysis.

# 5. Debugging and Validation

*Debugging* is microsimulation's skeleton in the closet, seldom discussed but absolutely critical nonetheless. Bugs slither into the best code; unless detected, they wreck havoc with validation and experimentation and everything else. To reduce the damage done by bugs, an elaborate debugging apparatus is embedded within CORSIM. The debugger automatically calculates simulated means for all input and output variables for each equation in the model, so that bugs can be discovered and traced at the source. These massive input-output calculations enable the user to directly 'watch' the model carry out its detailed calculations, and are part of the routine output of each run. The calculations slow the model somewhat, but their value in damage control is more than worth this modest cost.

Moving to *validation*, four general classes of empirical validation can be applied to large-scale DM models, and especially to their component modules. Each class of validation can be regarded as posing a particular kind of test which the model - and its component modules - should undergo to earn credibility. Modelling strategies ought to be judged partly according to the degree of empirical validation to which they can be routinely exposed. Modelling strategies routinely exposed to all four types of empirical validation are likely to produce the most credible and useful models. One of the strengths of the DMM strategy is it can be exposed to all four types of empirical validation. In the case of DMM, the exceptional variety of outcomes generated by any realisation - ranging from aggregated time series to cross-sectional covariances to life histories - creates the potential for uniquely exacting empirical validations against a variety of historical data. The four validation tests are described briefly below.

## 5.1 In-sample validation

In conventional analytic research, the in-sample fit is typically the only systematic encounter between the model and relevant empirical data. The validation indicators are conventional measures of fit and of statistical significance (R-squared, t- and F-statistics).

## 5.2 Out-of-sample validation

A serious weakness of in-sample validation lies in the failure to confront a fitted module with 'fresh' data - data that has not been used during the original testing and estimation of the module. A confrontation with fresh data provides a more realistic measure of an equation's predictive power since, at the very least, the stochastic variation is new in the fresh data. When out-of-sample validation is performed, a fitted equation is confronted with fresh data - data not used originally in testing or estimating the equation - to assess its adequacy.

## 5.3 Out-of-type validation

Both in-sample and out-of-sample validation have the weakness that the validation plays to the data type-specific strengths of the fitted equation, rather than confronting its (potential) data type-specific weaknesses. For example, an equation fit to cross-section microdata might still perform reasonably well if confronted with fresh cross-section microdata; however the same equation might perform less well if confronted with data of a wholly different type, such as panel microdata or aggregate time series data. Similarly, an equation originally fit to time series data might perform reasonably well if confronted with fresh time series data, but might perform less well if confronted with cross-section or panel data.

For example, in principle - since a single health insurance coverage 'process' underlies the different forms of observed health insurance coverage data - a single module ought to account for all the different forms of coverage data, whether aggregate time series, cross-section microdata, or panel microdata. In practice, this versatility across data types seldom holds true. Partly this is because time series data, panel microdata, and cross-section microdata are usually analysed by different researchers, using different measures, different theories and different specifications. Consequently, to confront an equation with a new type of data, different from the type with which it was originally developed, is typically a very demanding test of performance. Perhaps for this reason, out-of-type validation is seldom performed in social science. Moreover, out-of-type validation is a complex procedure to implement, much more difficult than either in-sample validation or out-of-sample validation. In fact, out-of-type validation requires procedures like microsimulation.

## 5.4 Multiple-module validation

Each of the three types of validation described above assesses individual modules in isolation; none address the new questions raised when two or more interacting modules are linked together to generate outcomes jointly. Consider, for example, the interaction of two processes, marriage and health insurance coverage. Acting together, the two processes influence the size of the pool of covered married families (and uncovered married families, unmarried covered families, etc.). If either too few covered persons marry, or too few married persons are covered, the pool of covered married families is too small. Errors in the size of the pool therefore could signal errors in either - or both - processes. When considering errors which are attributable to more than one module, a researcher's first - and often most important - recourse is to return to the first three types of validation, attempting to make sure that each module - by itself - is performing as well as possible. However, insofar as improving separate modules does not eliminate errors from joint outcomes, the researcher next must consider interactions (ie. the joint performance of two or more modules). To achieve this, the researcher might follow several strategies - for example, forcing one module to track exact historical values - in an attempt to isolate the source of the discrepancies.

Multiple-module validation is largely uncharted territory, since large-scale DM models - which pose multiple module issues - are still relatively new. However, such validation ought to become more familiar in the future. Multiple-module validation should be seen as an unprecedented validation opportunity. When the interaction among modules produces additional information on sources of error in the overall model, the benefit for the analyst is additional leverage with which to improve the model as a whole, as well as its individual components.

To summarise, a major strength of DMM as compared to other modelling strategies is the potential for exposure to all four types of validation. While using the best in-sample and out-of-sample validation methods of standard single-equation analytic research in developing its component modules, DMM provides an opportunity to employ the even more exacting out-of-type and multiple-module validation methods as well. Having been exposed to all four types of validation, a particular DMM (like CORSIM) should provide a significantly more realistic, detailed and consistent representation of person and family behaviour than that provided by alternative large-scale modelling strategies or by unrelated pieces of analytic research.

## 6. Alignment

Evidence that multiple-module validation is uniquely exacting can be seen when a DMM is run over a historical period. The inevitable result is that simulated model outcomes do not track comparable historical values with perfect accuracy.

Consider as example one demographic group: non-white females ages 70-74. When CORSIM is run from 1960 to 1990, the simulated group-level mortality rates generated by summing across the simulated individual-level mortality probabilities for this group do not equal the historical mortality rates published by the US Office of Vital Statistics for non-white females 70-74. What factors give rise to these discrepancies? Why might the aggregated mortality rates from CORSIM not equal the officially collected, partly aggregated, mortality rates? Reasons why discrepancies are to be expected are given below:

1. The values of the independent variables in the micro-level mortality functions might be inaccurate. Potential sources of inaccuracy include the following:

- the initial representation of the population is (almost always) a sample, and the sample joint distribution of independent variables is subject to random differences from the population joint distribution;
- insofar as other equations in CORSIM are imperfect in form or parameters, they will generate inaccurate inputs to the mortality equation; and
- even if the other equations are perfect, outcomes are subject to Monte Carlo variation. Hence, any given run will produce variations from the joint distribution observed in the real historical world.

2. The micro-level mortality equations might be mis-specified. Violations of the following assumptions could cause mis-specification:

- the assumption that parameters are constant in time, whereas the factors influencing mortality could be changing in a magnitude in ways not captured by the equations;
- the assumption that period effects are absent or limited, whereas such historically specific impacts could be significant; and
- the assumption that there are no other specification flaws, whereas other specification flaws are inevitable.

3. The micro-level mortality equations might be estimated from imperfect microdata on mortality. Imperfections in the mortality data could arise from the following sources:

- the data are for a sample of the population of the United States and estimated parameters are subject to random variation around the population values;

- the data are almost certainly for a nonrandom sample of the United States population; consequently, estimated parameters are particularly vulnerable to bias in the presence of model mis-specification;

- the data are drawn from a population whose definition is not necessarily identical to that of the population to which the historical statistics are meant to apply (eg. in the treatment of the institutional population);

- the data sample the historical period in question unevenly, an especially serious problem if the underlying parameters of a process have been changing over the time period; and

- the data probably contain significant measurement error.

4. Even if the mortality equations are population equations they are stochastic processes and, hence, any particular simulation will be subject to random variation in the mortality experience of different groups.

5. Flaws might be present in the official historical statistics themselves. Hence, even if the predicted probabilities are perfectly accurate, which is highly unlikely, flawed historical aggregates might generate discrepancies.

6. A final possible source of discrepancies is program bugs.

Discrepancies between predicted aggregates and comparable historical aggregates are surely inevitable in any DMM. The important question is the size and pattern of these discrepancies and the issue of practical importance is to use these discrepancies as a guide to improving both the individual equations and the overall simulation. Assume that the task is to combine information from two different observational designs: longitudinal microdata, and partly aggregated rates. The general strategy followed in CORSIM for combining information from these two different data types proceeds in two stages.

The first stage relies solely upon the longitudinal microdata to test alternative specifications and subsequently to estimate a dynamic behavioural module. Typically, the final specification will include the lagged endogenous value as an important right-hand side variable, in order to incorporate the state-dependent nature of the dynamic process.

The second stage utilises partially aggregated, national historical rates as a basis for estimating a second set of parameters - termed group-level parameters - which are used to modify the original function.

The purpose of the second stage of the process - termed alignment - is to enhance the backcasting and forecasting accuracy of the behavioural modules, by drawing upon the strengths of additional data resources. In effect, the historical rates are used to estimate new parameters to strengthen the modules constructed originally from longitudinal microdata alone.

One implementation of this strategy is to use a simple multiplicative group parameter to scale the predicted pre-alignment probabilities directly, so that the scaled probabilities will track group-level historical aggregates as closely as possible. To implement this alignment strategy requires that the mean of the pre-alignment person-specific probabilities for group j in year t be known, in order to calculate each post-alignment person-specific probability. This cannot be accomplished with only one pass through the data, since the mean of a set of equation-derived probabilities cannot be known unless all the probabilities themselves are first calculated. A technique adopted in CORSIM to overcome this problem is to require the computer to pass twice through the sample data for each year. The first pass is used simply to calculate, for each group and each event, the relevant pre-alignment quantities.

# 7. Health Policy Modelling

Thinking broadly about health policy requires taking into account five distinct sets of processes from the consumer, or demand-side perspective. (Providers and group-level payers are key actors in the complete system, of course; there is no reason to preclude representing such actors in future DM models.) These five sets of factors are listed below.

- *Socioeconomic* variables are critical in structuring well-being, including health, in all societies. In addition, policy analysis has to take account of winners and losers along socioeconomic lines.

- *Risk factors* are behaviours over which an individual exerts some measure of control, and which are potentially important to preventive health policies.

- *Prices* of medical services are determined exogenously from an individual or family perspective, but the effective cost of a service to any family is also a function of *insurance* coverage.

- *Medical service use* is an outcome of the first three sets of factors, directly and/or indirectly, as well as of health conditions themselves. Any reform needs to know how service use will be affected.

- *Health conditions* are influenced by medical service use, risk factors and socioeconomic variables directly and indirectly, as well through other factors. Reforms should know how actual health is affected.

This strategy has been implemented in an application of CORSIM to dental conditions, service use and expenditures (Brown et al., 1992, 1995). Detailed information about conditions and service use at the family, person and individual tooth level are dynamically updated with functions based largely upon the RAND Health Insurance Experiment data. To carry out this modelling, CORSIM is linked to a dental application module consisting of approximately 160 stochastic equations and 1300 parameters, and requiring socioeconomic inputs from CORSIM. The aim of the full linked model is to represent dental conditions as the outcome of a complex sequence of interrelated social, economic, dentition and dental care use processes. Closely allied with this aim is the goal of modelling service use and expenditures for all services. In the full model the dental module is connected via one-way linkages to CORSIM - in the sense that outputs from CORSIM feed into the dental module, but outputs from the dental module do not feed into CORSIM. The model as a whole assumes that the social, economic and demographic attributes of families and persons significantly influence dental outcomes, but that dental attributes do not significantly influence the social, economic and demographic attributes of families and persons. The dental module dynamically updates the dental attributes every three years, even though CORSIM is updated every year.

For dental issues, the virtues of dynamic microsimulation are many. Multiple processes - dental, demographic, risk factor, economic, policy - are connected. The output of a demographic process is an input to a risk factor, whose output is an input to a dental condition, whose output conditions service use, etc. Diverse forms of knowledge are synthesised, eg. rule-based algorithms governing insurance benefits, binomial logistic equations predicting tooth transitions and multinomial probit equations generating PI scores co-exist peacefully. Complex representations appropriate to the underlying processes, including nonlinearities and discontinuities, are incorporated. Reciprocal feedbacks among behaviours are widespread; dental conditions affect dental care service use one period, which affect dental conditions the next period. Teeth are embedded within mouth, person, family and other contexts. Given the explicit representation of context, careful attention is paid to distinguishing selection into context from causation due to context. Highly heterogeneous responses (endemic to humans) to an identical policy context are captured.

Aggregation is implemented by adding across teeth, mouths, individuals and families without requiring simplifications imposed for the sake of analytic tractability. The entire US populations (within sampling limits) of teeth, mouths, individuals and families are represented. Policy parameters are linked to tooth, mouth, person and family outcomes. Longitudinal life paths are generated for each unit, which can incorporate simultaneous age-based, period-based and cohort-based changes. Distributional analyses (winners and losers) are easily carried out. From a methodological perspective, dynamic microsimulation exploits the testing and estimation power of important microdata files. Finally, from the standpoint of creating and organising a body of dental knowledge, dynamic microsimulation integrates and reconciles diverse bodies of data, pinpoints knowledge gaps and helps establish research priorities.

## 8. Generic Benefits of DMM

The generic benefits of DMM, which apply to basic research at least as much as they apply to applied research, are listed below (Caldwell, 1988a):

- connect multiple behaviours;
- synthesise diverse forms of knowledge;
- incorporate full complexity in behavioural relationships;
- represent reciprocal feedbacks among behaviours;
- embed actors within contexts;
- distinguish selection into context from causation due to context;
- capture heterogeneous responses to context;
- aggregate over time, actors and behaviours without imposing simplifications;
- model an entire population of actors of one type;

- model several populations of actors of different types;
- include structures linking actors (networks, markets);
- connect institutional forces to person/family outcomes;
- generate person and family life histories;
- include change over the life course (age-based), time (period-based) and cohorts (cohort-based);
- generate distributional analyses (winners/losers);
- produce detailed, flexible descriptions of actors;
- exploit the testing and estimation power of microdata;
- derive aggregates from individual outcomes;
- trace role of chance in bringing about person and family outcomes;
- integrate and reconcile diverse existing data;
- organise design and collection of new data; and
- pinpoint knowledge gaps and establish research priorities.

## 9. Next Steps

This chapter is a progress report from a project intended to last for some years to come. The project will advance both by incrementally improving the current version of CORSIM and also by developing qualitatively new versions. Improving CORSIM would involve the following types of activities:

- examining much more of CORSIM's full response surface, ie. the full range of model output as it is conditioned by Monte Carlo variation, changes in the initial sample, parameter and specification changes, and other changes in model inputs;
- validating CORSIM against ever more detailed and varied historical data;
- respecifying and re-estimating individual modules for CORSIM with stronger hypotheses and data; and

- aligning the full CORSIM model using all possible relevant group-level historical data.

## References

Brown, R. J., S.B. Caldwell and S.A. Eklund (1992), 'Microsimulation of dental conditions and dental service utilisation', in: J. G. Anderson (ed.), Proceedings of the Simulation in Health Care and Social Services Conference, Society for Computer Simulation, San Diego.

Brown, R. J., S.B. Caldwell and S.A. Eklund (1995), 'How fee and insurance changes could affect dentistry: Results from a microsimulation model', *Journal of the American Dental Association*, Vol. 126.

Caldwell, S. B. (1987), 'Microsimulation: The next thirty years', Paper presented for Conference on Income Transfer Policy, Carnegie-Mellon University and Urban Institute, Washington, DC.

Caldwell, S.B. (1988a), 'Small findings, large questions: synthesis in policy research', Paper presented to Thematic Session on Sociology and Social Policy, American Sociological Association Annual Meetings, Atlanta.

Caldwell, S.B. (1988b), 'A large-scale modelling strategy for regional science', Paper prepared for the Seventeenth Annual Meeting of the Northest Regional Science Association, Ithaca, New York.

Caldwell, S.B. (1988c), 'The Cornell model: Portable software for dynamic microsimulation', Paper presented to Conference on Public Data, International Association of Social Science Information Services and Technologies, Washington, DC.

Caldwell, S.B. (1991), 'CORSIM: The Cornell dynamic microsimulation model', Paper presented to the Symposium on Microsimulation, Stockholm, Sweden.

Merz, J. (1991), 'Microsimulation - a survey of principles, developments and applications', *International Journal of Forecasting*, Vol. 7, No. 1.

Orcutt, G.H., M. Greenberger, J Korbel and R. Wertheimer (1961), *Microanalysis of Socioeconomic Systems: A Simulation Study*, Harper and Row, New York.

Orcutt, G.H., S.B. Caldwell and R. Wertheimer, (1976), *Policy Exploration Through Microanalytic Simulation*, The Urban Institute, Washington, DC.

Wolfson, M.C. (1991), 'A system of health statistics: Toward a new conceptual framework for integrating health data', *Review of Income and Wealth*, Vol. 36, No. 1, March.

Zedlewski, S.R., R.O. Barnes, M.R. Burt, T.D. McBride and J.A. Meyer (1990), *The Needs of the Elderly in the 21st Century*, Urban Institute Report, Urban Institute Press, Washington DC.

# Index

## Notes on Contributors

*Leif Andreassen* is a Research Fellow at the Research Department of Statistics Norway, Oslo.

*Hans Baekgaard* was Head of Section at the Danish Ministry of Economic Affairs and is currently a Research Fellow at the National Centre for Social and Economic Modelling at the University of Canberra.

*Gérard Ballot* is Professor of Economics at Université Paris II and Director of ERMES, a research institute affiliated to this University and the Centre National de la Recherche Scientifique (CNRS).

*Keith Blackburn* is Senior Advisor, Economic Analysis, at theAustralian Bureau of Statistics.

*Steven Caldwell* is Professor of Sociology and Public Policy, Cornell University.

*Richard Eason* is Assistant Director with the Statistics and Economic Office at the UK Inland Revenue.

*Bengt Eklind* is Deputy Assistant Under-Secretary at the Economic Department, Swedish Ministry of Finance.

*Gunnar Eliasson* was President of the Industrial Institute for Economic and Social Research and is currently Professor of Industrial Economics at the Royal Technical Institute, Sweden.

*Ingemar Eriksson* is with the Economic Department of the Swedish Ministry of Finance.

*Jane Falkingham* is a lecturer in Population Studies in the Department of Social Policy at the London School of Economics and a Research Associate of the Welfare State Program at the London School of Economics.

*Dennis Fredriksen* is a Senior Executive Officer with the Research Department of Statistics Norway, Oslo

*Heinz Galler* is a Professor at the Faculty of Economics, Martin Luther-University, Halle Wittenburg, Germany.

*Anil Gupta* is a Senior Tax Policy officer with the Tax Policy Branch of the Canadian Department of Finance

*Ann Harding* is Professor of Applied Economics and Social Policy at the University of Canberra and inaugural Director of the University's National Centre for Social and Economic Modelling.

*Joakim Hussénius* is with the Economic Department of the Swedish Ministry of Finance.

*Vishnu Kapur* is a Senior Policy Analyst with the Federal-Provincial Relations and Social Policy Branch of the Canadian Department of Finance.

*Anders Klevmarken* is Professor of Econometrics at Uppsala University, Sweden.

*Lieve de Lathouwer* is a Research Assistant at the Centre for Social Policy, University of Antwerp, Belgium.

*Olav Ljones* is Assistant Director General, Economic Statistics Department at Statistics Norway, Oslo.

*Tony Meagher* is a Senior Research Fellow at the Centre of Policy Studies, Monash University, Australia.

*Joachim Merz* is a Professor at the Department of Economics and Social Sciences, University of Lüneburg, Germany.

*Mikael Müller* is with the Economic Department of the Swedish Ministry of Finance.

*Jan Nelissen* is Assistant Professor at the Department of Social Security Studies of Tilburg University, The Netherlands.

*Paul Olovsson* is a Research Assistant and Computer Specialist at the Department of Economics, University of Göteborg, Sweden.

*Stephen Pudney* is Tyler Professor of Economics at the Economics Department of the University of Leicester, England.

*Warren Richter* is Assistant Statistician, Data Management Branch at the Australian Bureau of Statistics.

*Aino Salomaki* is a researcher at the Government Institute for Economic Research (VATT), Finland.

*Holly Sutherland* is Director of the Microsimulation Unit at the Department of Applied Economics, University of Cambridge, England.

*Elizabeth Symons* is a Lecturer at the University of Keele in the UK.

*Erol Taymaz* is Professor of Economics at the Middle East Technical University, Ankara, and former Research Associate, Industrial Institute for Economic and Social Research (IUI), Stockholm.

*Klaas de Vos* was affiliated with the Department of Economic Sociology and Psychology at Erasmus University, Rotterdam and is currently a researcher at the Economics Institute in Tilburg.

*Neil Warren* is Senior Lecturer in Economics, Department of Economics, University of New South Wales, Australia.

*Michael Wolfson* is currently Director General of the Institutions and Social Statistics Branch at Statistics Canada and is a Fellow of the Canadian Institute for Advanced Research.

*Asghar Zaidi* was affiliated with the Department of Economic Sociology and Psychology at Erasmus University, Rotterdam and is currently a researcher at the Economics Institute, Tilburg.

Printed and bound by CPI Group (UK) Ltd, Croydon, CR0 4YY
18/09/2026
14964964-0003